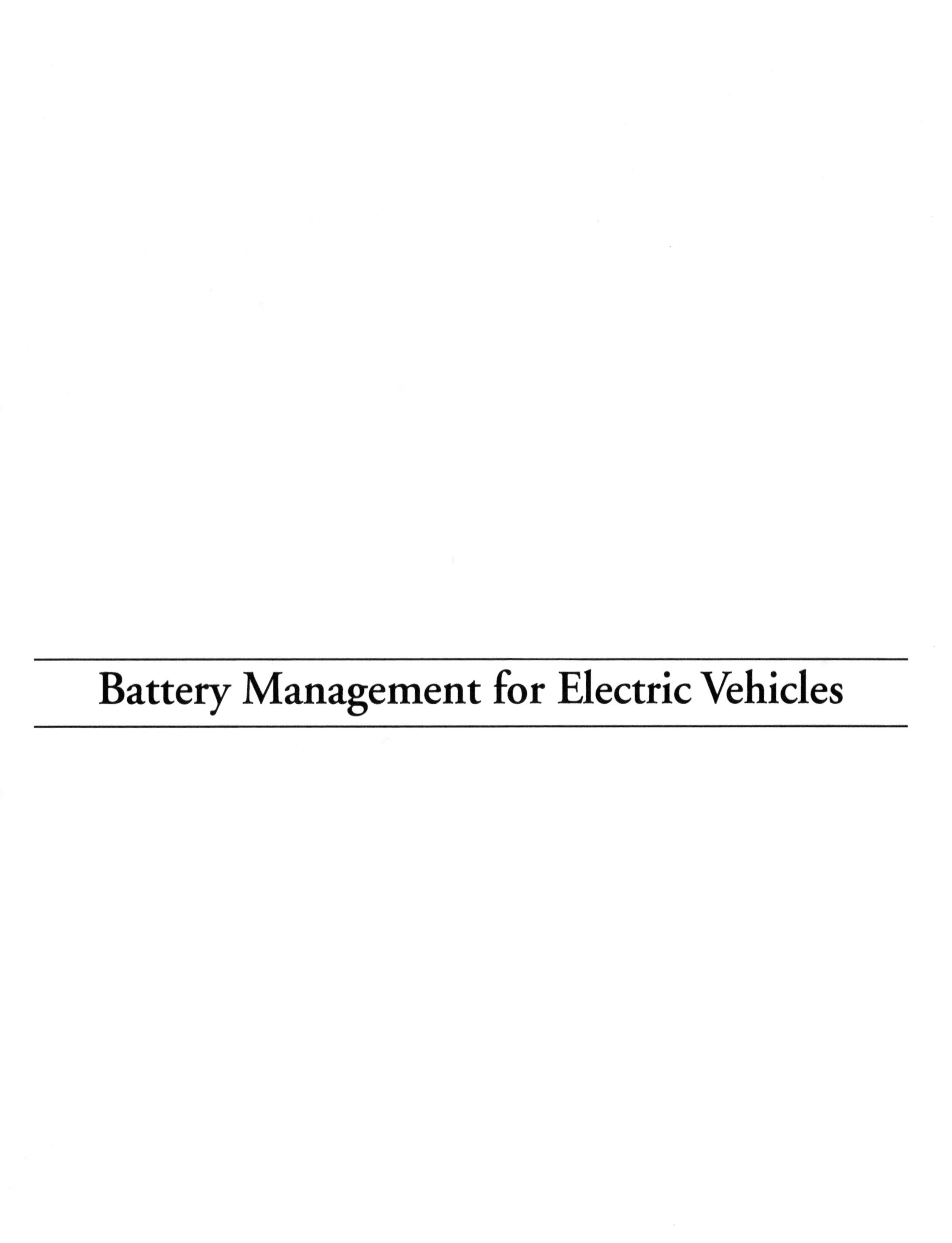

Battery Management for Electric Vehicles

Battery Management for Electric Vehicles

Editor: Nelson Cox

New York

Published by NY Research Press
118-35 Queens Blvd., Suite 400,
Forest Hills, NY 11375, USA
www.nyresearchpress.com

Battery Management for Electric Vehicles
Edited by Nelson Cox

International Standard Book Number: 978-1-64725-461-2 (Hardback)

Cataloging-in-Publication Data

Battery management for electric vehicles / edited by Nelson Cox.
p. cm.
Includes bibliographical references and index.
ISBN 978-1-64725-461-2
1. Electric vehicles--Batteries. 2. Electric vehicles. 3. Battery management systems.
4. Battery charging stations (Electric vehicles). I. Cox, Nelson.
TL220 .B38 2023
629.229 3--dc23

Contents

Preface

The main aim of this book is to educate learners and enhance their research focus by presenting diverse topics covering this vast field. This is an advanced book which compiles significant studies by distinguished experts in the area of analysis. This book addresses successive solutions to the challenges arising in the area of application, along with it; the book provides scope for future developments.

An electric vehicle (EV) refers to a type of vehicle that utilizes one or more electric motors for propulsion. It can use a battery to run on its own or a collector system providing it with electricity from extravehicular sources. Electric vehicles are classified into different types such as battery electric vehicles (BEV), hybrid electric vehicles (HEV), fuel cell electric vehicles (FCEVs), and plug-in hybrid electric vehicles (PHEV). The need for an energy storage system is one of the pressing issues related to the current EV technology. Some of the major types of batteries which are used within electric vehicles are lithium-ion batteries, nickel-metal hydride batteries and lead-acid batteries. This book contains some path-breaking studies related to battery management for electric vehicles. It presents researches and studies performed by experts across the globe. As the research on electric vehicles is emerging at a rapid pace, the contents of this book will help the readers understand the modern concepts and applications of this area of study.

It was a great honour to edit this book, though there were challenges, as it involved a lot of communication and networking between me and the editorial team. However, the end result was this all-inclusive book covering diverse themes in the field.

Finally, it is important to acknowledge the efforts of the contributors for their excellent chapters, through which a wide variety of issues have been addressed. I would also like to thank my colleagues for their valuable feedback during the making of this book.

Editor

1

Battery Energy Management of Autonomous Electric Vehicles using Computationally Inexpensive Model Predictive Control

Kyoungseok Han [1], Tam W. Nguyen [2] and Kanghyun Nam [3,*]

[1] School of Mechanical Engineering, Kyungpook National University, Daegu 41566, Korea; kyoungsh@knu.ac.kr
[2] Department of Aerospace Engineering, University of Michigan, Ann Arbor, MI 48109, USA; twnguyen@umich.edu
[3] Department of Mechanical Engineering, Yeungnam University, Gyeongsan 38541, Korea
* Correspondence: khnam@yu.ac.kr

Abstract: With the emergence of vehicle-communication technologies, many researchers have strongly focused their interest in vehicle energy-efficiency control using this connectivity. For instance, the exploitation of preview traffic enables the vehicle to plan its speed and position trajectories given a prediction horizon so that energy consumption is minimized. To handle the strong uncertainties in the traffic model in the future, a constrained controller is generally employed in the existing researches. However, its expensive computational feature largely prevents its commercialization. This paper addresses computational burden of the constrained controller by proposing a computationally tractable model prediction control (MPC) for real-time implementation in autonomous electric vehicles. We present several remedies to achieve a computationally manageable constrained control, and analyze its real-time computation feasibility and effectiveness in various driving conditions. In particular, both warmstarting and move-blocking methods could relax the computations significantly. Through the validations, we confirm the effectiveness of the proposed approach while maintaining good performance compared to other alternative schemes.

Keywords: self-driving car; model predictive control; dynamic programming; prediction horizon; move-blocking; warmstarting

---

1. Introduction

For the last few years, many experts in the field of automotive research have claimed that self-driving cars will become commonplace in the near future. Although there are still technical difficulties for commercializing self-driving cars, at least a certain degree of automation in the vehicles is expected to be available soon [1,2].

With the advent of new features in self-driving cars, novel approaches to optimize the energy efficiency of vehicles have become available. In general, human-driver behaviors are based on their own driving habits on the roads, which are not consistent with the traffic flow [3]. Self-driving cars, in contrast, can plan optimally their trajectories by exploiting real-time traffic information [4–6]. For example, many short and long-range traffic information using vehicle-to-vehicle (V2V), vehicle-to-infrastructure (V2I), and, more generally, vehicle-to-everything (V2X) communication (see Figure 1) provide useful traffic information, which the self-driving cars can effectively use in real time. For example, the geometry information of the route is available through the pre-loaded maps and the future traffic density is also available through the V2X technologies. Therefore, unlike human drivers, self-driving cars can make suitable decisions based on real-time traffic flow so as to optimize their energy efficiency.

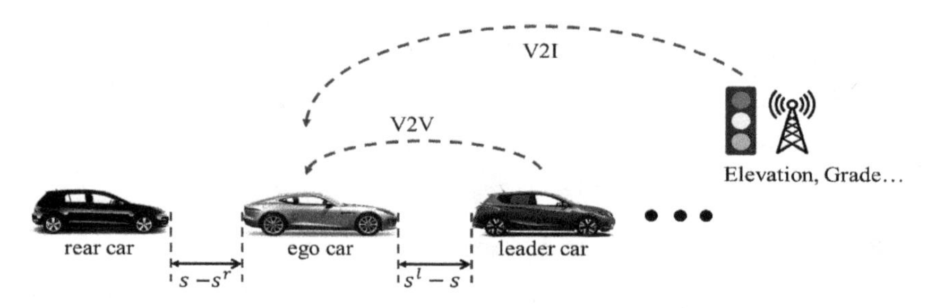

Figure 1. Increased levels of automation and connectivity through vehicle-to-everything (V2X) communication.

However, since the traffic dynamics are complex and nonlinear, there is a non-negligible computational burden associated with long-term real-time traffic prediction. Moreover, since the traffic predicted by the self-driving car in the long run become uncertain (e.g., traffic disturbances, unexpected changes in human-driver behavior), it is required to choose an appropriate length of the prediction horizon to effectively exploit the current traffic information.

Although there have been numerous researches that show the benefits of vehicle speed- and position-trajectory planning in terms of energy efficiency, to the best of our knowledge, practical aspects, such as the length of the prediction horizon and its relative computational burden, have not been discussed in the existing literature on self-driving cars. In order to achieve high performance while ensuring computational feasibility, this paper deals with the above-mentioned practical concerns for real production vehicles.

Also, in this paper, the battery-electric vehicle (BEV) is assumed to be ego vehicle that needs to be controlled, so, using connectivity technologies (i.e., V2X), the optimal battery energy management is our ultimate goal.

1.1. Literature Review

So far, there have been many research interests in the optimization of energy efficiency using constrained control theories. In general, both vehicle- and powertrain-level optimizations are used to achieve specific control objectives while ensuring constraint satisfaction. For example, the vehicle speed and position are planned in a way that minimizes the energy consumption, while at the same time the operating points in the powertrain such as engine, motor, and transmission are moved to the optimum area.

In particular, the BEV that is entirely powered by the electric motors is promising platform for the autonomous vehicles due to its environmental benefits and low operating cost. In [7], as compared to the internal-combustion engine vehicle, the benefit of BEV in terms of energy requirement per unit distance has been confirmed when both powertrain types were utilized as the fully autonomous vehicles. In addition, the synergies between shared autonomous vehicle fleets and electric vehicle has also been proposed in [8]. For this reason, this paper also employs the BEV as our vehicle platform and the minimization of battery energy consumption of BEV has been pursued.

To achieve this goal, one representative method is to adopt dynamic programming (DP), which ensures global optimality by exploring all possible combinations of the trajectories. DP requires, however, the entire traffic prediction before the trip. In [9], global optimum speed trajectory of the vehicle could be obtained using DP, but future traffic environments even including pre-known of the leading vehicle's speed profiles are assumed to be given initially. That is, DP is mainly suitable for *offline* computation and is thus not implementable in real time due to the heavy computational burden. However, the control performance using DP is meaningful as a benchmark to investigate the maximum achievable control performance.

To manage the computational complexity of DP, another well-known trajectory optimization approach is Pontryagin Minimum Principle (PMP), which has been utilized in [10], for example. Note that only the necessary condition for optimality is guaranteed, and the method still requires a long-term prediction. Moreover, two-point boundary value problem associated with PMP conditions cannot be easily solved numerically. Unfortunately, most literature that employ the PMP [10,11] do not handle the tractability of their optimal controller rigorously.

Although these two approaches, i.e., DP and PMP, have shown the effectiveness when planning the optimal trajectory, these methods are usually effective for the deterministic system. However, the interactions between the road participants in the traffic cannot be modelled in deterministic way. To handle the such uncertainties in the traffic prediction, stochastic DP [12], Q-learning-based PMP [13], and reinforcement learning [14,15] have been considered, but the analysis on the practical aspects like computational tractability is omitted. Additionally, since these methods are based on Markov decision process, where the relationship between the control action and probability are pre-determined, these methods cannot ensure robustness when the vehicle is in an unexpected situation [16].

The above-mentioned two approaches, that is, DP and PMP, are the most popular methods and provide an optimal-state-trajectory framework. However, as discussed, the heavy computational burden prevents the introduction of such techniques in production vehicles. Also, the requirement of long-term preview information in DP and PMP is not usually available in the real-world traffic since the traffic is significantly influenced by the unexpected uncertainties. To address this, the method that re-calculate the optimal trajectory once the new traffic information become available can be considered. More appropriately, the popular receding-horizon control, model predictive control (MPC) [17,18], can be used to update a short-term traffic prediction at each time instant, which effectively approximates the long-term traffic prediction of DP and PMP by means of real-time trajectory corrections using the current state estimate.

Unlike the internal-combustion engine vehicle where the pulse-and-glide driving strategy is assumed to the optimal way to increase the fuel economy, it is well-known that speed smoothing of the BEV can optimize the energy efficiency of the electric motor [19]. Therefore, the speed trajectory of the BEV should be flatten as much as possible for the prediction horizon to prevent battery spikes, while satisfying the safety constraints. Also, the battery temperature that significantly influences the battery efficiency should be maintained in the appropriate ranges. Taking these all into consideration, we can conclude that the MPC that can handle the state and control constraints is appropriate method for the battery energy management of the autonomous BEV.

In [20–23], MPC is designed using short-term prediction and it has been confirmed that, to a certain degree, this method is effective to improve energy efficiency. However, the computational burdens depending on the several important factors in MPC such as sampling-time and length of the prediction time were not analyzed thoroughly. To the best of our knowledge, computational load reduction in MPC for real-time computational feasibility has not been discussed in the literature of self-driving cars, and this paper will reveal the critical factors determining the computational burden in MPC. Finally, several remedies to increase the feasibility of the MPC will be suggested.

1.2. Research Contribution

The primary contributions of this paper can be summarized as follows.

First, to overcome the technical shortcomings of a nominal MPC, several methods that enable the real-time implementation of the optimizer are proposed, and its effectiveness is verified through various case studies. Depending on the driving conditions, the computational-time constraint violation rate is reduced around 35% compared to the one of a nominal MPC, but the control performance is similar to each other, which is our central argument.

Secondly, an appropriate prediction-horizon selection for MPC for BEVs control is studied by comparing the performances with the one of DP. In most works employing MPC for trajectory optimization and control, the procedure determining the horizon length is omitted, but our approach starts with this important aspect considering the implementation in real hardware. As compared to the control results of nominal MPC and DP, the proposed method can reduce the computational burden significantly while minimizing the performance degradation in terms of the battery state-of-charge (SOC) reduction.

1.3. Paper Organization

This paper is organized as follows. Section 2 introduces the vehicle- and battery-dynamics models, and the optimization control problem is formulated in Section 3. Next, the central part of this paper is described in Section 4, which proposes a method to reduce the computational complexity of MPC. Section 6 demonstrates the effectiveness of our approach through simulations, and Section 7 concludes the paper.

2. Vehicle and Battery Dynamics

Since the interest of this study is to optimize the energy efficiency of the BEVs, relevant vehicle and powertrain dynamics are modeled first. Note that, compared to other types of vehicles, BEVs have a simple structure, as shown in Figure 2. Since the power is transmitted through a single path, the energy loss is only determined by the efficiency of each actuator depending on the operating points. Before designing the controller, the central parts of BEVs are simplified and modeled in this section.

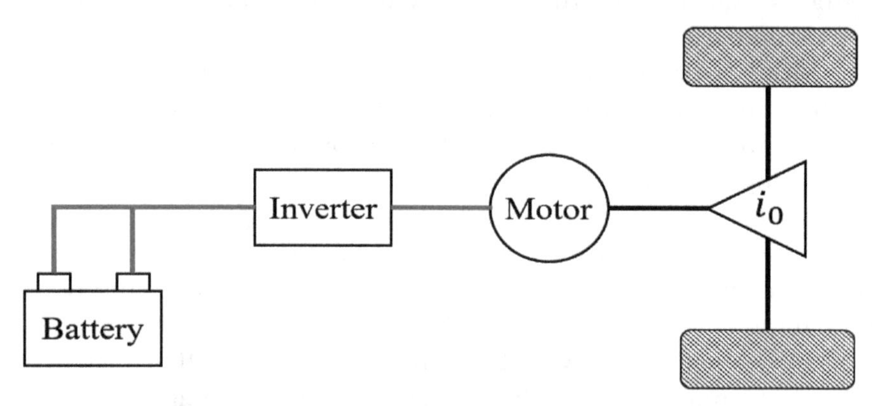

Figure 2. Schematic of a battery-electric vehicle.

2.1. Vehicle Longitudinal Dynamics

Consider the point-mass longitudinal dynamics of the vehicle. We assume that the effective mass of the vehicle, which normally accounts for the vehicle static and rotational effect together, is approximately equivalent to the vehicle mass [24]. Accordingly, the continuous-time longitudinal dynamics are given by

$$\dot{s} = v, \tag{1}$$

$$\dot{v} = \left(\frac{T_\mathrm{m} + T_\mathrm{b}}{mr}\right) i_0 - \frac{\rho A_\mathrm{f} C_\mathrm{d} v^2}{2m} - g\sin\theta - \gamma g \cos\theta, \tag{2}$$

where $s \in \mathbb{R}$ is the vehicle position, $v \in \mathbb{R}$ is the vehicle speed, $m > 0$ is the vehicle mass, $T_\mathrm{m} \in \mathbb{R}$ and $T_\mathrm{b} \in \mathbb{R}$ are the motor and friction brake torques, respectively, $r > 0$ is the wheel radius, $i_0 > 0$ is the final gear ratio, $\rho > 0$ is the air density, $A_\mathrm{f} > 0$ is the frontal area of the vehicle, $C_\mathrm{d} \geq 0$ is the aerodynamic drag coefficient, $\theta \in (-\pi/2, \pi/2)$ is the road inclination that influences the energy consumption significantly but assumed to be zero in this paper, $g > 0$ is the gravitational constant, and $\gamma \geq 0$ is the coefficient of rolling resistance. Note that θ is a function of the position, that is, $\theta(s)$, and that the rolling friction is kinetic, that is, $\gamma = 0$ for $v = 0$. In the remainder of this paper, we consider $T_\mathrm{b} = 0$ without a significant loss of generality.

In actual fact, several parameters that are assumed to be constant such as m and θ influence the vehicle's energy efficiency significantly. Also, the other constant parameters vary depending on driving conditions, so the parameter estimations are needed for the accurate energy-efficiency calculation. However, for simplicity, this paper does not consider the varying-parameters, but it should be treated in the future.

To apply digital control methods, the continuous-time mode (Equations (1) and (2)) is discretized using the Euler-forward method

$$x_{k+1}^{\mathrm{v}} = x_k^{\mathrm{v}} + g^{\mathrm{v}}(x_k^{\mathrm{v}}, u_k^{\mathrm{v}})\, T_{\mathrm{s}}, \tag{3}$$

where $x_k^{\mathrm{v}} \triangleq [s_k\ v_k]^{\mathrm{T}}$, $u_k^{\mathrm{v}} \triangleq T_{m,k}$, $g^{\mathrm{v}} : \mathbb{R}^2 \times \mathbb{R} \to \mathbb{R}^2$ is the vehicle dynamics function vector that aggregates in a column the right-hand side of Equations (1) and (2), and $T_{\mathrm{s}} > 0$ is the sampling time.

2.2. Battery Dynamics

2.2.1. Continuous-Time SOC Dynamics

The continuous-time battery SOC dynamics is given by

$$\dot{\mathrm{SOC}} = -\frac{I_{\mathrm{b}}}{C_{\mathrm{b}}}, \tag{4}$$

where $\mathrm{SOC} \in [0,1]$ is the state of charge of the battery, $I_{\mathrm{b}} \in \mathbb{R}$ is the battery current, and $C_{\mathrm{b}} > 0$ is the battery capacity and is assumed to be constant.

The battery current I_{b} is a function of the open-circuit voltage $V_{\mathrm{oc}} \in \mathbb{R}$, the battery resistance $R_{\mathrm{b}} > 0$, and the battery-power consumption $P_{\mathrm{b}} \in \mathbb{R}$ as,

$$I_{\mathrm{b}} = \frac{V_{\mathrm{oc}} - \sqrt{V_{\mathrm{oc}}^2 - 4R_{\mathrm{b}}P_{\mathrm{b}}}}{2\,R_{\mathrm{b}}}, \tag{5}$$

where

$$P_{\mathrm{b}} = \begin{cases} \frac{P_{\mathrm{m}}}{\eta_{\mathrm{b}}^{+}} & \text{for} \quad P_{\mathrm{m}} \geq 0, \\ \frac{P_{\mathrm{m}}}{\eta_{\mathrm{b}}^{-}} & \text{for} \quad P_{\mathrm{m}} < 0, \end{cases} \tag{6}$$

$\eta_{\mathrm{b}}^{+} \in (0,1)$ is the battery-depletion efficiency, $\eta_{\mathrm{b}}^{-} > 1$ is the battery-recharge efficiency, and P_{m} is the motor-output power.

The SOC dynamics is captured more accurately by accounting for the variations of V_{oc} and R_{b}, which are functions of SOC, as illustrated in Figure 3.

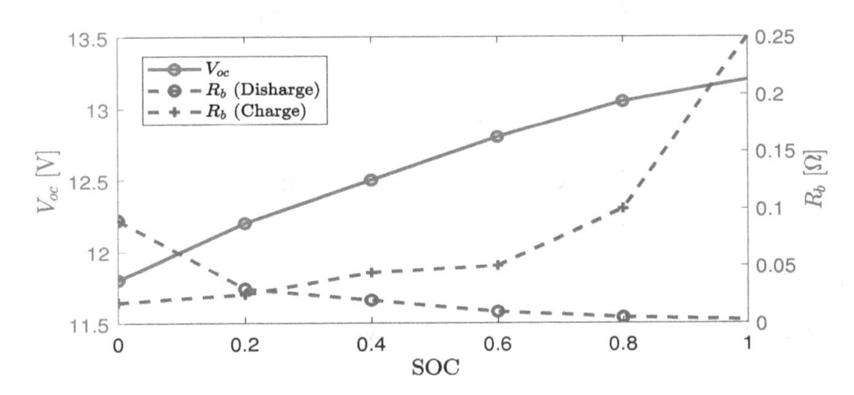

Figure 3. Variation of the open-circuit voltage V_{oc} and internal resistance R_{b} as functions of state-of-charge (SOC).

Note that, in practice, the variations of V_{oc} and R_{b} are relatively small compared to P_{b} during driving time. Therefore, the battery-output power P_{b} mainly contributes to the battery SOC changes in Equation (4). In general, P_b is used as a control input in the battery dynamics. Note that P_b depends also on the motor-output power P_m, whose dynamics are described hereafter.

2.2.2. Motor Dynamics

The motor-output power is determined by the motor torque T_m, motor speed $\omega_m \triangleq \frac{v}{r} i_0$, and motor efficiency η_m as

$$P_m = \frac{T_m \omega_m}{\eta_m(T_m, \omega_m)}. \quad (7)$$

Note that the motor efficiency η_m is a function of the motor operating points, which depend on T_m and w_m (see Figure 4). Furthermore, we assume that the motor efficiency is mirrored in terms of propulsion and regenerative braking, that is, $\eta_m(T_m, \omega_m) = \eta_m(-T_m, \omega_m)$.

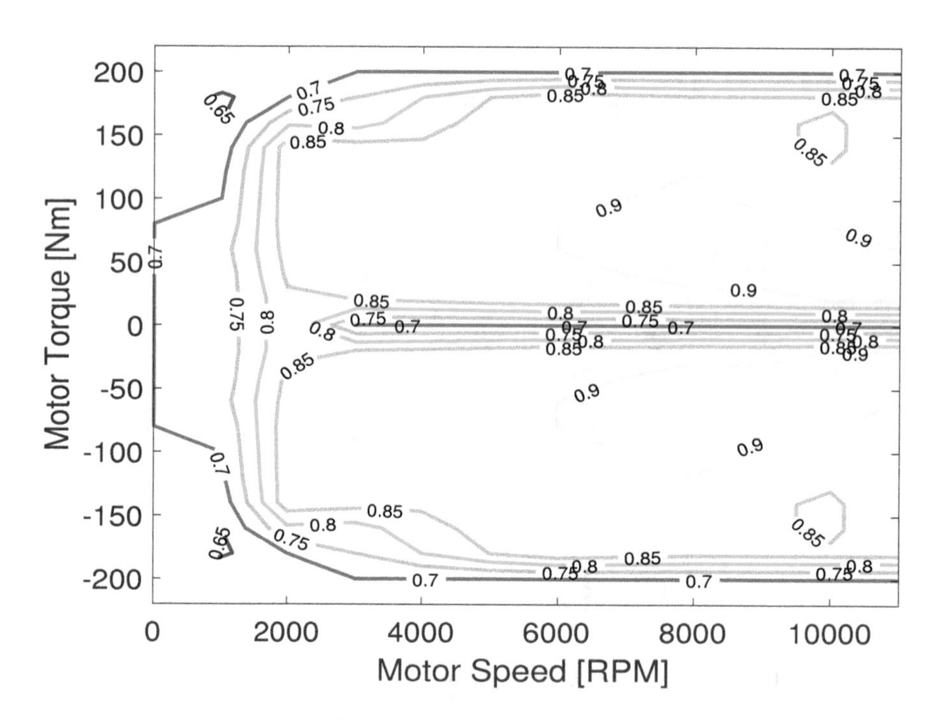

Figure 4. Motor efficiency map.

2.2.3. Discrete-Time Battery Dynamics

By combining Equations (4)–(7) and using the Euler-forward method, the continuous-time SOC dynamics Equation (4) can be discretized as

$$\mathrm{SOC}_{k+1} = \mathrm{SOC}_k + g^s(\mathrm{SOC}_k, T_{m,k})T_s, \quad (8)$$

where $g^s : [0,1] \times \mathbb{R} \to [0,1]$ is the SOC dynamics function vector, which aggregates as a column the right-hand side of Equations (4) and (5).

2.3. Discrete-Time Model for Implementation

The discrete-time model of the vehicle and battery dynamics for digital implementation are given by

$$x_{k+1} = x_k + f(x_k, u_k)T_s, \quad (9)$$

where $x_k \triangleq [s_k \; v_k \; \mathrm{SOC}_k]^T$, $u_k \triangleq T_m$, and $f \triangleq [(g^v)^T \; (g^s)^T]^T$. In this paper, we assume that, for all $k \geq 0$, the state x_k is available, and thus, an observer needs not to be built. Based on Equation (9), the optimization control problem and several optimization methods will be described in Section 3.

3. Optimal Control Problem Formulation

In this paper, we assume that the controlled car (ego car) is driving between a leader and rear car. Additionally, the ego car is able to send and receive information through V2X communication (see Figure 1). In this case scenario, the ego car can plan its future trip by predicting the traffic dynamics while ensuring safety constraints between the leader and rear cars. Assuming that the rear car behaves rationally, it is possible to ignore its behavior in the optimization formulation [25]. In addition, this paper assumes only a single lane, which means that no lane changing is considered. However, in reality, such interactions between the ego vehicle and adjacent vehicles might occur frequently and influence on the ego vehicle's trajectory planning. In this paper, however, we exclude the above-mentioned situations, which are left to our future work.

The objective of the optimization formulation is to minimize the battery-output power over a given route by optimizing the speed profile.

Let $N > 0$ be the discrete-time prediction horizon, define $\hat{x}_{i|k} \triangleq [\hat{s}_{i|k}\ \hat{v}_{i|k}\ \hat{\mathrm{SOC}}_{i|k}]^{\mathrm{T}}$ as the i-step-ahead state predicted at step k, let $\hat{\omega}_{\mathrm{m},i|k}$ be the i-step-ahead motor speed predicted at step k, and define $\hat{U}_k \triangleq [\hat{u}_{0|k}\ \cdots\ \hat{u}_{N-1|k}]^{\mathrm{T}} \in \mathbb{R}^N$ as the sequence of control inputs computed at step k. Using Equation (9), the optimization problem at step k is given by

$$\min_{\hat{U}_k} J_k(\hat{U}_k) = \sum_{i=0}^{N-1} P_{b,i}(\hat{u}_{i|k}, \hat{x}_{i|k}), \tag{10a}$$

s.t.

$$\hat{x}_{i+1|k} = \hat{x}_{i|k} + f(\hat{x}_{i|k}, \hat{u}_{i|k})T_{\mathrm{s}}, \tag{10b}$$

$$\hat{x}_{0|k} = [s_k\ v_k\ \mathrm{SOC}_k]^{\mathrm{T}}, \tag{10c}$$

$$v_{\min} \leq \hat{v}_{i+1|k} \leq v_{\max}, \tag{10d}$$

$$\tau_{\min}(\hat{v}_{i+1|k} + \delta) \leq s^l_{i+1} - \hat{s}_{i+1|k} \leq \tau_{\max}(\hat{v}_{i+1|k} + \delta) \tag{10e}$$

$$\mathrm{SOC}_{\min} \leq \hat{\mathrm{SOC}}_{i+1|k} \leq \mathrm{SOC}_{\max} \tag{10f}$$

$$T^{\min}_{\mathrm{m},i}(\hat{\omega}_{\mathrm{m},i|k}) \leq \hat{u}_{i|k} \leq T^{\max}_{\mathrm{m},i}(\hat{\omega}_{\mathrm{m},i|k}) \tag{10g}$$

$$i = 0, \cdots, N-1, \tag{10h}$$

where s^l is the deterministic position of leader car provided by driving cycle, $0 \leq v_{\min} < v_{\max}$, $0 < \tau_{\min} < \tau_{\max}$ are the minimum/maximum time headways, $\delta > 0$ is the velocity bound, $\mathrm{SOC}_{\min} < \mathrm{SOC}_{\max} \in [0,1]$, and $T^{\min}_{\mathrm{m},i} : \mathbb{R} \to \mathbb{R}$ and $T^{\max}_{\mathrm{m},i} : \mathbb{R} \to \mathbb{R}$ are given constraint functions at step j that depend on $\hat{\omega}_{\mathrm{m},i|k}$.

Note that the problem formulation (10a)–(10h) is a receding-horizon problem formulation if the problem must be solved for all $k \geq 0$. The terms (10d)–(10f) are the state constraints, and (10g) is the control constraint. The constraint bounds are reasonable values according to the driving conditions on US roads [26]. In particular, depending on the ego car's speed, (10e) indicates the reasonable varying position constraint between the ego car and leader car.

4. Practical Considerations for Real-Time Implementation of MPC

The receding-horizon control problem (10a)–(10h) can be solved using standard nonlinear solvers. We expect superior control performance in MPC for long prediction horizons in the case where the model matches the plant and the solver gives the global optimal solution at each step. However, for implementation reasons, an appropriate prediction-horizon length should be chosen.

In this section, we first give the performance benchmark of DP when the prediction horizon is the entire trip of the ego car. Next, for real-time implementation in MPC, we propose a simplification of the cost function in order to reduce computational complexity. A numerical analysis is carried out to highlight the impact of an increase in N on performance and computational feasibility of MPC.

All simulations presented in this paper are carried out using MATLAB R2020a, Windows 10, Intel Core i7-9700 CPU @ 3.00 Ghz, and 16 GB RAM. Furthermore, the continuous-time dynamics are integrated by *ode45* and the control is constant between samples.

4.1. Dynamic Programming

In this subsection, the deterministic DP algorithm is employed to solve (10a)–(10h) for the entire trip. In general, DP is not applicable in practice because DP requires all future reference inputs and disturbances in advance, indicating that DP is a noncausal controller [27,28]. For example, the entire speed trajectory of the leader car should be provided, which is not reasonable in reality. However, in many applications, the control performance using DP is usually assumed to be the best solution that guarantees global optimality. Therefore, employing DP is meaningful as a benchmark, with which the control performance of other causal controllers can be compared. In this paper, we will compare MPC with DP benchmark performance.

DP solves the optimization problem (10a)–(10h) by setting $k = 0$ and $N = N_{\mathrm{dp}}$, where $N_{\mathrm{dp}} > 0$ is the discrete-time prediction horizon of the entire trip. Note that, as DP is used once and thus does not depend on the actual time step k, we omit the index k in (10a)–(10h) in this subsection.

For all $j \geq 0$, let $\mathcal{X}_j$ be the feasible state-constraint set (10d)–(10f) at predicted step j, let $\mathcal{X}_N$ be the terminal state-constraint set, and let $\mathcal{U}_j$ be the feasible control-constraint set (10g) at predicted step j. Next, define the discretized-state set $\tilde{X} \triangleq \{\tilde{x}_1, \ldots, \tilde{x}_i, \ldots, \tilde{x}_\ell\} \subset \mathcal{X}_j$, for all $0 \leq j \leq N$, where $\tilde{x}_i$ is the i-th node of $\tilde{X}$, $i \in \{1, \ldots, \ell\}$, and $\ell > 0$ is the total number of discretized states in $\tilde{X}$.

Using the principle of optimality [29], the optimal control policy for the node $\tilde{x}_i$ is obtained by backward induction, where $\hat{x}_0 = \tilde{x}_i$ in (10c). In particular, starting from the optimal final stage cost

$$\mathcal{J}_N(\tilde{x}_i) = \min_{\hat{u}_N \in \mathcal{U}_N} P_{\mathrm{b},N}(\hat{u}_N, \tilde{x}_i), \tag{11}$$

where $N = N_{\mathrm{dp}}$, the DP backward recursion is given by

$$\begin{aligned} \mathcal{J}_j(\tilde{x}_i) = & \min_{\hat{u}_j \in \mathcal{U}_j} [P_{b,j}(\hat{u}_j, \tilde{x}_i) \\ & + \mathcal{J}_{j+1}(\tilde{x}_i + f(\tilde{x}_i, \hat{u}_j) T_{\mathrm{s}})], \end{aligned} \tag{12}$$

where $j \in \{N-1, \ldots, 0\}$. Accordingly, the optimal control policy π for each node $\tilde{x}_i$ is mapped by

$$\pi = \{\mu_0(x), \mu_1(x), \cdots, \mu_{N_{\mathrm{dp}}-1}(x)\}, \tag{13}$$

where $\mu_i : \mathbb{R}^3 \to \mathbb{R}$ is the mapping function and $i = 0, \ldots, N_{\mathrm{dp}} - 1$. Accordingly, Equation (13) can be used to find the optimal control input in a forward simulation.

In this paper, we use the generic DP function in Matlab, that is, *dpm* [30]. An appropriate grid number for the state and control variables must be specified to avoid the curse of dimensionality. Doing so, the computation is completed approximately in five minutes while satisfying all constraints, as illustrated in Figure 5. The entire vehicle speed trajectories for two drive cycles, i.e., WLTC, US06, are optimized so as to minimize the cost and enforce the constraints. One distinguished feature is that the speed profile when applying DP is flattened as much as possible to prevent spikes in the battery current.

The battery SOC consumption for each drive cycle is compared in Table 1, which shows that DP can significantly improve the energy efficiency of BEV around 14 to 20%. These control performances are considered ideal, so we will use these as our benchmark. However, as mentioned, DP is not implementable in real time due to the computational complexity and strong assumptions on command and disturbance preview.

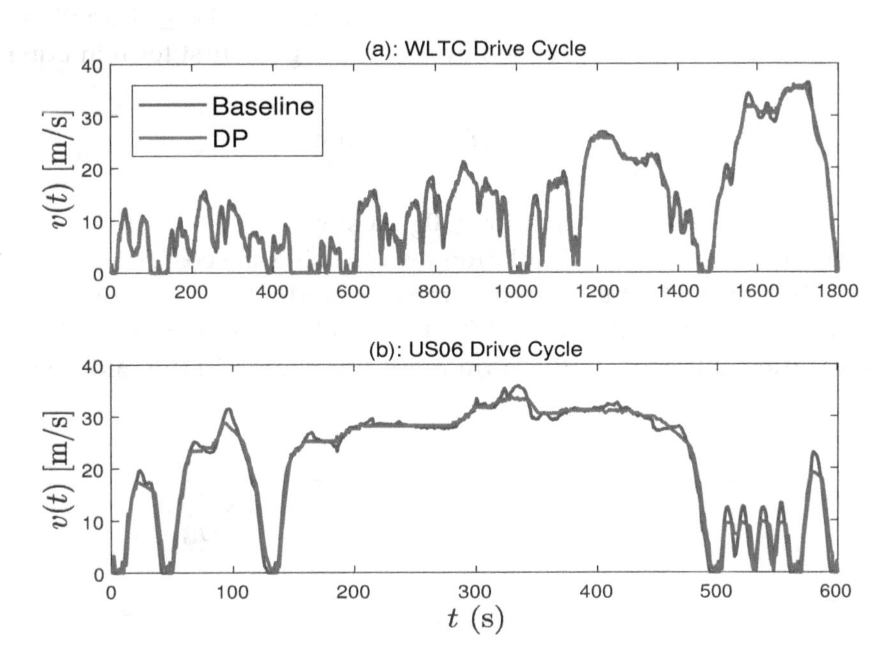

Figure 5. Comparison of the speed trajectories obtained by dynamic programming (DP) and human-driver maneuvers with (**a**) WLTC and (**b**) US06.

Table 1. Comparison of the battery state-of-charge (SOC) consumption (%) with the speed profiles obtained by dynamic programming (DP) and nominal speed.

Drive Cycle	Baseline	DP	Improvement (%)
WLTC	17.55	14.96	14.76
US06	13.41	10.74	19.90

Since the cost function is nonconvex and the dynamics are nonlinear, several local minima may be presented. However, as DP computes all costs for every node by backward induction, it can be said that the obtained performance ensures global optimality.

4.2. Model Predictive Control

4.2.1. Quadratic Cost Simplification

In the previous section, the nonlinear function P_b is used as a cost when applying DP, which gives a globally optimal result by backward induction. However, for real-time computation, a computationally tractable method is required. In particular, we show in this section that the nonlinear cost P_b can be approximated by a quadratic cost.

In [19], by assuming constant actuator efficiencies, the motor power P_b can be simplified as

$$P_\text{b} \approx P_\text{m} \approx \frac{T_\text{m} v}{r} i_0 + \alpha T_\text{m}^2 \tag{14}$$

where $\alpha > 0$ is a tunable motor parameter. Note that, even though the cost is greatly simplified compared to the original P_b, Equation (14) is still nonlinear and depends on the value of α, which modifies the overall control performance significantly.

In this paper, the cost function is further simplified as follows. Since the objective of this study is to minimize the battery SOC reduction, the fluctuation of I_b should be minimized according to Equation (4). It also can be interpreted that the sudden changes in the control input u, which cause I_b

fluctuations, should be avoided as much as possible. Therefore, for all $k \geq 0$, another minimization of a quadratic cost in place of Equation (10a) is used by eliminating the first term in Equation (14), that is,

$$\min_{\hat{u}_{i|k} \in \mathcal{U}_{i|k}} \sum_{i=0}^{N-1} P_{b,i}(\hat{x}_{i|k}, \hat{u}_{i|k}) \approx \min_{\hat{u}_{i|k} \in \mathcal{U}_{i|k}} \sum_{i=0}^{N-1} \hat{u}_{i|k}^2, \tag{15}$$

where $\mathcal{U}_{i|k}$ is the i-step-ahead constraint Equation (10g) at step k.

To verify the above assumption, the simulation results using the cost functions (10a) and (15) are compared in Figure 6. It is important to note that the SOC reductions for both cases are almost the same. Due to the nonconvexity in P_b, the control fluctuates a lot to search the optimum point at every step, and is sometimes stuck at the local minimum points, as shown in the middle plot of Figure 6.

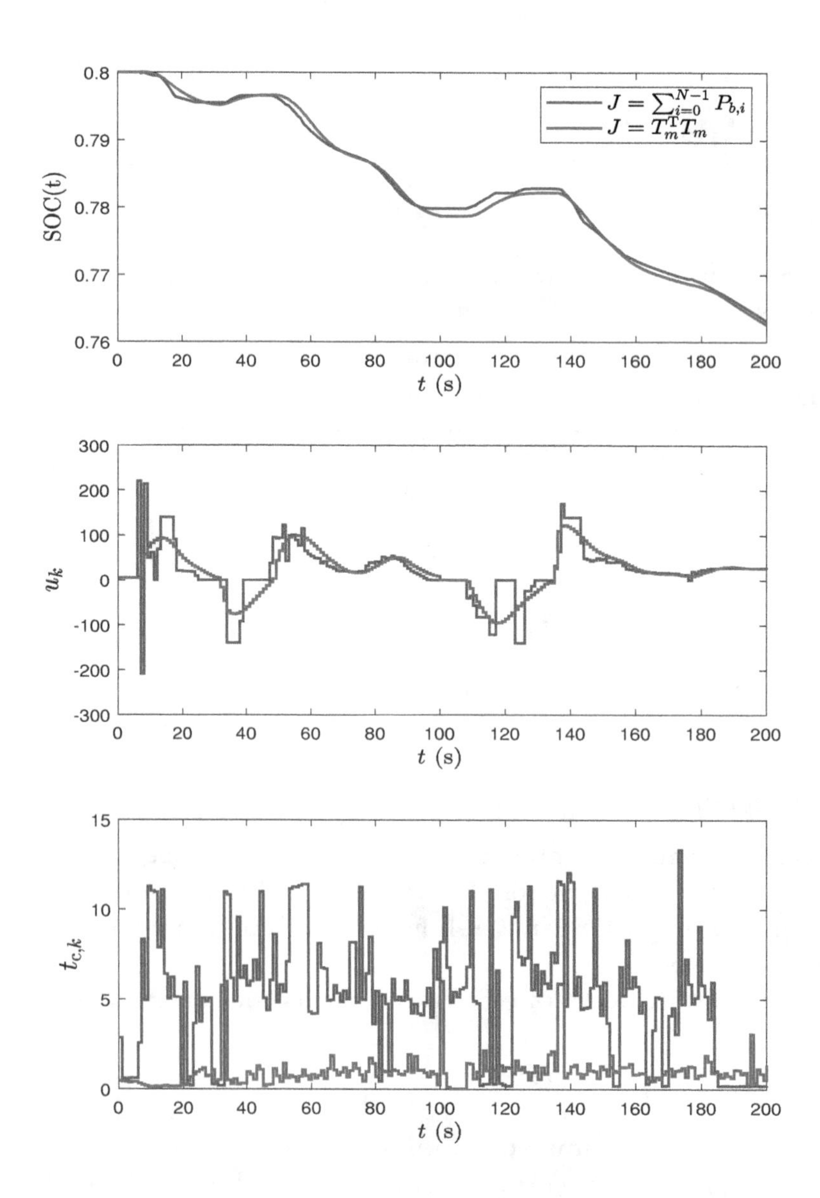

Figure 6. Comparison of the simulation results using the different cost functions with $N = 10$ and $T_s = 1$ s. The top plot shows the battery SOC trajectories, the middle plot shows the control input evolutions, and the bottom plot shows the computation times $t_{c,k}$ at each step k.

In contrast, by exploiting the quadratic form (15), the computation time at each step is significantly reduced compared to the original form, as shown in the bottom plot of Figure 6.

Therefore, it can be concluded that the nonlinear cost (10a) can be replaced with the quadratic cost (15) without sacrificing too much performance in terms of the battery SOC reduction.

4.2.2. Real-Time Computational Feasibility

One of the major issues in MPC is the computational burden when we handle large or fast dynamics-related systems. Moreover, a long prediction horizon increases the computational complexity significantly, even though a longer horizon provides better performance in the ideal case.

In this subsection, an appropriate prediction horizon for our problem that takes into account the relationship between performance and real-time implementation feasibility is analyzed. The nonlinear solver *fmincon* in MATLAB is used to solve the optimization control problem (10a)–(10h), where the sampling time is specified as $T_s = 1$ s in all cases. That is, the computation should be completed in 1 s at each step.

By solving the optimization problem, we obtain the optimal sequence $\hat{U}_k^*$ at each step k and only the first element $\hat{u}_{0|k}^*$ is applied, which is the principle of receding horizon control. To see the relationship between the control performance and the prediction horizon, we gradually increase the prediction horizon while observing the required computations for each horizon. Note that it is possible to achieve superior performance with faster computation by using other advanced solvers, e.g., tailored NLP solvers using inexact and real-time iteration solution strategies [31]. However, for the purpose of comparison, we utilize the same solver for all simulations.

As shown in Figure 7, better control performance in terms of the battery SOC reduction can be achieved by increasing N. However, the required average computation time for each step exceeds the sample time T_s, meaning that *fmincon* is not implementable in real time.

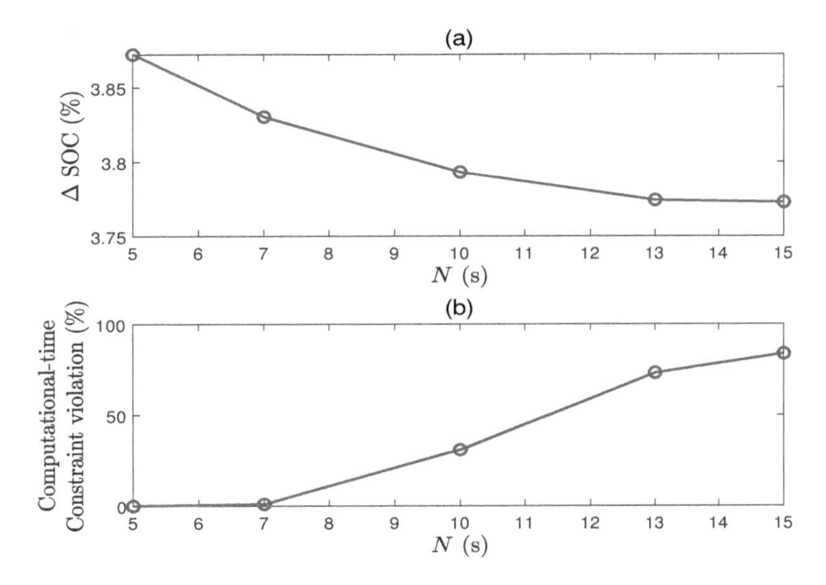

Figure 7. (**a**) Battery SOC reductions with respect to N, and (**b**) computational burdens with respect to N. The computational-time constraint violation represents the ratio between the number of occurrences where $t_c > T_s$ and the number of samples of the entire experiment.

It can be concluded that $N = 10$ is sufficient to optimize the energy efficiency of our BEV, but the computational burden should be reduced for implementation. A collection of methods to deal with the computational complexity of MPC is proposed in the next section.

5. Strategies for Computational Load Reduction in MPC

In the previous section, an appropriate prediction horizon for MPC is determined heuristically. The computational load is, however, unmanageable with the specified horizon. In this section, we suggest a collection of strategies, which potentially reduce the computational load of MPC.

In particular, we describe sampling-time adjustment, warmstarting, and move-blocking strategies, and explain their possible impact upon performance.

5.1. Sampling-Time Adjustment

A sampling-time increase is expected to reduce the computational load of MPC as, for a similar preview of the trajectories, the prediction horizon N decreases. However, note that, as T_s increases, crucial intersampled information might be ignored during prediction causing potential performance degradation of MPC, especially regarding constraint satisfaction. In the following, we numerically investigate the impact of larger sampling times in MPC.

As shown in Figure 8, the position constraints between the ego car and leader car, which are hard constraints, are sometimes violated as T_s increases. As expected, we observe that the average computation time proportionally decreases as T_s increases, and the result is omitted since it is obvious. In fact, due to the inaccuracy of intersampling model information, MPC cannot handle the constraints when the vehicle is close to the bounds (see e.g., at 40 s in Figure 8b,c).

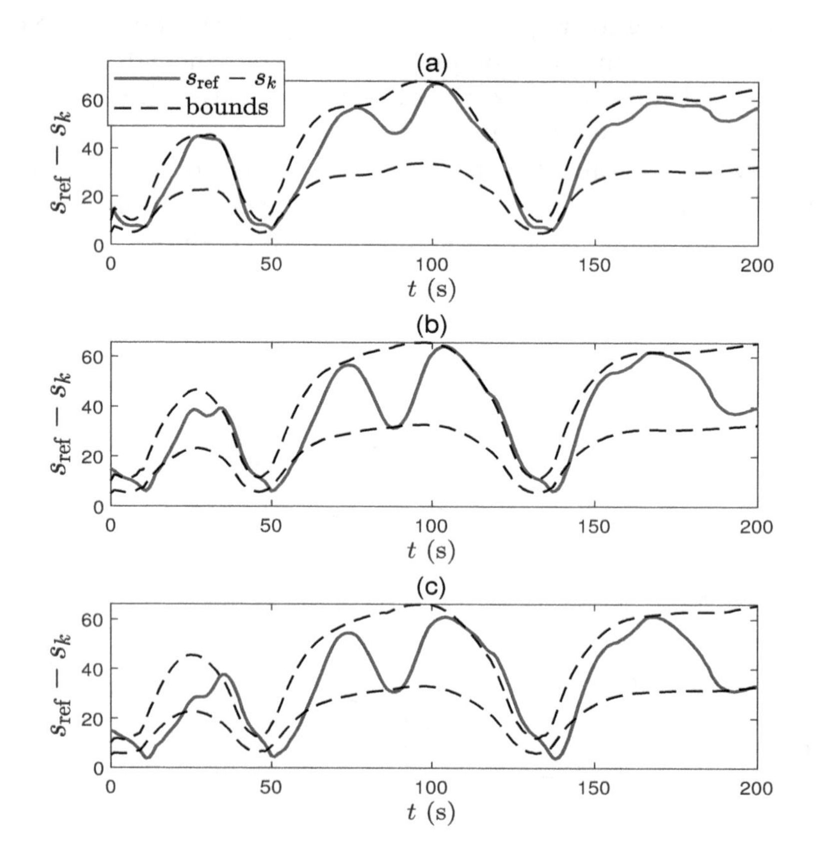

Figure 8. Distance difference between ego car and leader car with (**a**) $T_s = 1$ s, (**b**) $T_s = 2$ s, and (**c**) $T_s = 3$ s.

5.2. Warmstarting

Another remedy for computational burden reduction is warmstarting [32]. Warmstarting is a method that supplies an initial guess of the sequence of controls to the solver rather than using a constant initial control sequence, typically, a vector of zeros. In general, supplying an initial guess is useful when dealing with nonconvex problems and nonlinear solvers because the solvers can easily get stuck to a local minimum point. In addition, this scheme also prevents cases with singularities by avoiding regions close to the initial condition.

Specifically, instead of starting the optimizer with the initial control sequence $\hat{U}_{k,0} = [0 \ \cdots \ 0]^{\mathrm{T}}$, we use the previous optimal solution, which is computed at $k-1$, at the current step k as

$$\hat{U}_{k,0} = \begin{cases} [0 \ \cdots \ 0]^{\mathrm{T}}, & k = 0, \\ [\hat{U}^*_{k-1}(2:\text{end}) \ \hat{U}^*_{k-1}(\text{end})]^{\mathrm{T}}, & k > 0, \end{cases} \tag{16}$$

where $\hat{U}^*_{k-1}(2:\text{end}) \triangleq [u^*_{1|k-1} \ \cdots \ u^*_{N-1|k-1}]^{\mathrm{T}} \in \mathbb{R}^{N-1}$ is the optimal control sequence computed at step $k-1$, which is truncated from the second component to the N-th component. Note that the last element $\hat{U}^*_{k-1}(\text{end})$ is copied twice in $\hat{U}_{k,0}$ so as to match the number of components in the array to N components.

The effectiveness of adopting warmstarting strategy is summarized in Table 2. Although the improvement is not very significant, it can be said that adopting warmstarting is usually helpful.

Table 2. Comparison of the average computation time with and without warmstarting.

	Without Warmstarting	With Warmstarting
$N = 15$	2.5336 s	2.5210 s
$N = 20$	2.5646 s	2.5512 s

5.3. Move Blocking

To reduce the computational load of MPC, a move-blocking strategy is considered [33]. Move-blocking strategy is similar to control-horizon strategy, where the control is free to move for the first initial prediction steps and then blocked. The difference is that, after the free initial moves of the control, move-blocking strategy blocks the control for several consecutive intervals, whereas control-horizon strategy blocks the last control for the remainder of the prediction horizon.

Control-horizon strategy is described as follows. At step k, instead of solving the optimization problem with the optimal control sequence $\hat{U}_k \in \mathbb{R}^N$, we use the alternative sequence $\hat{U}'_k \in \mathbb{R}^{N_c+1}$, where N_c is the control horizon and $0 < N_c < N$. The relationship between $\hat{U}_k$ and $\hat{U}'_k$ is given by

$$\hat{U}_k = T_c \hat{U}'_k, \tag{17}$$

where $T_c \in R^{N\times(N_c+1)}$ is the control-horizon matrix and is defined by

$$T_c \triangleq \text{blkdiag}(I_{N_c}, 1_{(N-N_c)\times 1}) \in \mathbb{R}^{N\times(N_c+1)}, \tag{18}$$

where I_n is the n-by-n identity matrix, and $1_{m\times n}$ is the m-by-n matrix of ones.

A variation of the control-horizon strategy is adopted in this paper, which is the move-blocking strategy. The motivation of this variation is that, since the reference position of the leader car can vary significantly in the future, blocking the control over a large interval yields constraint violation and, in some cases, suboptimality or infeasibility. Therefore, we divide the prediction horizon into several blocking intervals. Accordingly, the control-horizon matrix is modified to the move-blocking matrix

$$T_b = \text{blkdiag}(I_{k_b}, 1_{k_b\times 1}, \ldots, 1_{k_b\times 1}, 1_{k_r\times 1}) \in \mathbb{R}^{N\times N_b}, \tag{19}$$

where $0 < k_b < N$ is the number of control moves blocked at each interval (except the first interval), k_r is the remainder of the division N/k_b (if zero, then this term disappears), and $N_b \triangleq \text{ceil}(N/k_b) - 1 + k_b$ is the length of $\hat{U}'_k$ using move-blocking strategy. With this move-blocking matrix, the control is free to move for the first k_b steps, and, afterwards, blocked for k_b steps consecutively within $N_b - k_b$ intervals. An example of move-blocking sequence using the proposed method is shown in Figure 9 with $k_b = 3$ and $N = 10$.

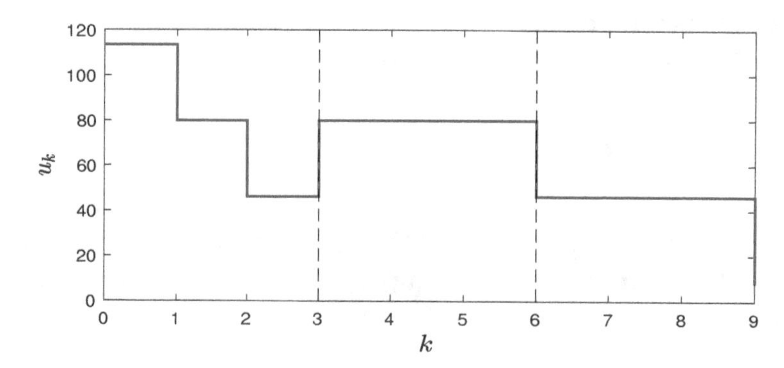

Figure 9. Control sequence using move-blocking strategy, where $k_b = 3$ and $N = 10$. The first three moves are free, then blocked in consecutive blocks of three controls. Note that $N_b = 6$, $k_r = 1$, and the last control is within a separate block from $k = 9$ to $k = 10$.

6. Simulation Results

So far, several methods to reduce the computational burden of MPC have been discussed. In this section, the effectiveness of the proposed strategies is verified for two driving cycles, namely, WLTC and US06. The parameters of the simulations are listed in Table 3. The model parameters are extracted from the Ford Focus BEV model in ADVISOR [34], which is a high-fidelity simulator with the aim of analyzing the vehicle-energy efficiency.

Table 3. Simulation Parameters.

Symbol	Description	Value (Unit)
m	Vehicle total mass	1445 (kg)
r	Wheel radius	0.3166 (m)
A_f	Vehicle frontal area	2.06 (m^2)
C_d	Aerodynamic drag coefficient	0.312
ρ	Air density	1.2 (kg/m^3)
θ	road inclination	0 (°)
γ	Rolling resistance coefficient	0.0086
i_0	Final drive ratio	4.2
v^{min}, v^{max}	Acceptable range of speed	(0, 150} (km/h)
C_b	Battery capacity	55 (Ah)
η_b^+	Battery-depletion efficiency	0.9
η_b^-	Battery-recharge efficiency	1.11
τ_{min}, τ_{max}	max/min time headway	{1,2} (s)
N	Prediction horizon	10
T_s	Sampling time	1 (s)
k_b	Control moves blocked	3

The following simulations compare control performance between the proposed method and other strategies, i.e., nominal MPC and DP. As mentioned earlier, the performance using DP can be considered as ideal. Therefore, if the performance using our approach closely matches the one of DP with a computationally manageable complexity, it can be said that the main objective of this study is achieved. Also, compared to the nominal MPC, the computational complexity should be reduced with the almost same control performance.

Simulation 1: WLTC driving cycle. The results in urban-driving scenario obtained from the proposed method and DP are compared in Figure 10. Since the state and control trajectories using the nominal MPC are very similar to those of our approach, we omit the plots with the nominal MPC.

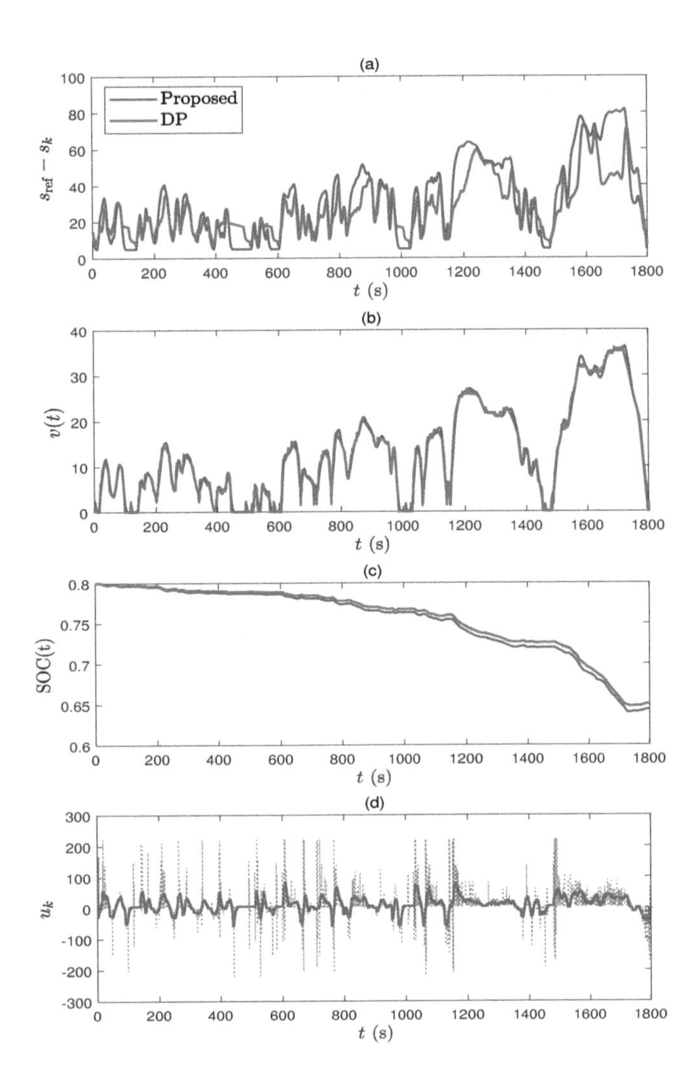

Figure 10. Comparison between model prediction control (MPC) and DP for the WLTC driving cycle, (**a**) distance difference between ego car and leader car, (**b**) speed profiles, (**c**) battery SOC trajectories, and (**d**) control trajectories.

Note that the appropriate distances between the self-driving car and leader car are maintained for both methods, as shown in Figure 10a. Depending on the self-driving car speed, the time-varying upper- and lower- bounds (10e) are specified in the predictions. No constraint violation occurs for MPC and DP. The entire speed trajectories are very similar to each other, but slightly different in Figure 10b. The battery SOC trajectories are depicted in Figure 10c and the SOC consumption using the proposed method is very similar to that of DP, which verifies the effectiveness of our approach. Since DP explores all possible combinations of control and state pairs, the control input in Figure 10d sometimes changes rapidly because DP is just designed to minimize the cost function without considering the physical relationship between the actuator and vehicle. Therefore, the control performance using DP is exaggerated to some extent, but it can serve as a benchmark that only considers the minimization of the cost.

The trajectories of s, v, and SOC are shown in Figure 11a–c. From the battery SOC trajectory in Figure 11c, the proposed MPC, which is computationally tractable, can give a solution that is very close to the globally optimum solution obtained by DP. Moreover, the trajectories obtained by DP sometimes physically do not make sense and should be corrected for hardware implementation. Since the nonlinear model used in this paper only captures the primary dynamics of the vehicle and powertrain, the control input using DP in Figure 11d cannot ensure a globally optimum solution in real production vehicle. Therefore, in reality, we expect that the performance gap between DP and our approach is much closer to each other, but the computational burden is significantly reduced with our approach, which is the central argument of this paper.

Simulation 2: US06 drive cycle. The US06 drive cycle represents the case of a highway-driving scenario. The results are shown in Figure 11.

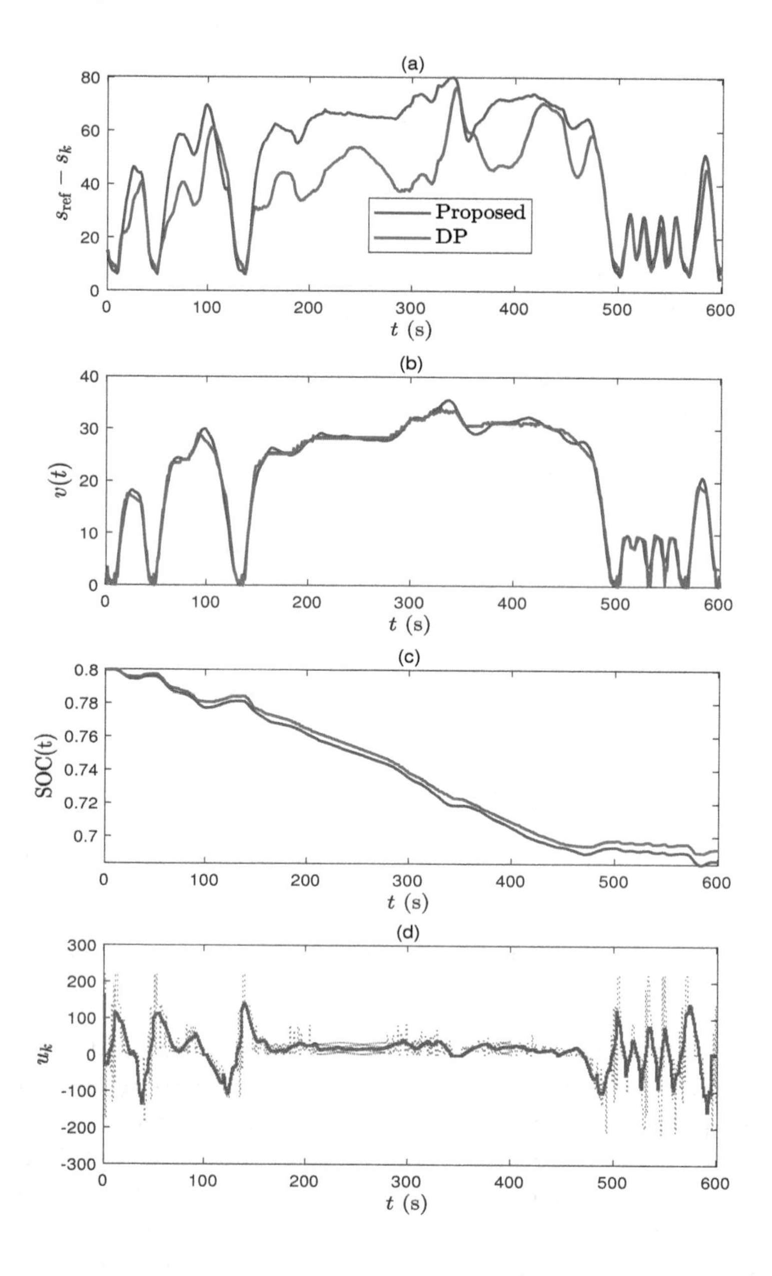

Figure 11. Comparison between MPC and DP in US06 driving cycle, (**a**) distance difference between ego-car and leader-car, (**b**) speed profiles, (**c**) battery SOC trajectories, and (**d**) control trajectories.

Discussion. Table 4 and Figure 12 support the above claims. As expected, the speed modification that minimizes T_{m}^2 is effective in terms of battery SOC reduction. As expected, the proposed MPC with move-blocking method does not outperform the nominal MPC but is very close to that of nominal MPC in terms of battery SOC reduction performance (see Table 4). Unfortunately, the control performances of our approach and nominal MPC shows the significant difference by comparing with DP result, but DP is not implementable in the real-time and trajectories from the DP do not consider the physical limitation of the actuator.

Table 4. ΔSOC obtained by different optimized speed trajectories (the values in parentheses describe the improvements compared to the baseline).

	WLTC	US06
Baseline	17.55	13.41
Proposed	15.64 (10.88%)	11.42 (14.83%)
Nominal MPC	15.42 (12.14%)	11.30 (15.73%)
DP	14.96 (14.76%)	10.74 (19.90%)

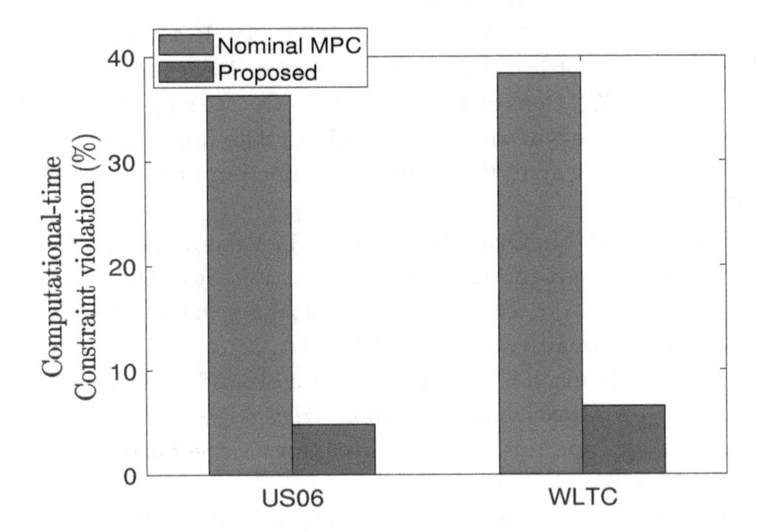

Figure 12. Comparison of the computational-time constraint violation rates obtained by nominal MPC and proposed method.

Note that the computational-time constraint violation rate is significantly reduced with the proposed method (see Figure 12). Even though our approach sometimes exceeds the sampling period T_s, some remedies to handle this issue are possible. For example, we can further reduce the computation time by replacing the language from MATLAB to C, which is essentially needed for hardware implementation. Also, constraint relaxation and the introduction of faster solvers instead of *fmincon* can be alternative solutions, which are left to our future works.

7. Conclusions

There have been extensive approaches in the literature to apply MPC in automotive control applications, but its computational feasibility was not clearly reviewed. This paper investigated several methods to reduce the computational complexity of a nominal MPC implemented in a battery-electric vehicle (BEV). First, an appropriate prediction horizon for our optimization control problem was determined with a nominal MPC, and we concluded that the control performance of nominal MPC closely matches the one of DP, which is assumed to be globally optimal. Next, to ensure real-time feasibility, adopting move-blocking strategy, with warmstarting and large sampling times, significantly reduces the computational burden compared to a nominal MPC without sacrificing too much of the control performance in terms of battery state-of-charge (SOC) reduction. Methods that provide further reduction of the computation load for real-time implementation of MPC in hardware will be our future work. In addition, a lot of constraint violations are expected to occur with the proposed approach when the reference value varies significantly. To overcome such a shortcoming, the relaxation of the constraints with the proposed approach is also left to our future work.

Author Contributions: Conceptualization, K.H., T.W.N., and K.N.; Data curation, K.H. and T.W.N.; Formal analysis, K.H.; Funding acquisition, K.N; Investigation, K.H. and T.W.N.; Methodology, K.H., and T.W.N.; Software, K.H. and T.W.N.; Supervision, K.N.; Visualization, T.W.N.; Writing—original draft, K.H. and T.W.N.; Writing—review & editing, K.N. All authors have read and agreed to the published version of the manuscript.

References

1. Wadud, Z.; MacKenzie, D.; Leiby, P. Help or hindrance? The travel, energy and carbon impacts of highly automated vehicles. *Transp. Res. Part A Policy Pract.* **2016**, *86*, 1–18. [CrossRef]
2. Masood, K.; Molfino, R.; Zoppi, M. Simulated Sensor Based Strategies for Obstacle Avoidance Using Velocity Profiling for Autonomous Vehicle FURBOT. *Electronics* **2020**, *9*, 883. [CrossRef]
3. Li, N.; Oyler, D.W.; Zhang, M.; Yildiz, Y.; Kolmanovsky, I.; Girard, A.R. Game theoretic modeling of driver and vehicle interactions for verification and validation of autonomous vehicle control systems. *IEEE Trans. Control Syst. Technol.* **2017**, *26*, 1782–1797. [CrossRef]
4. Han, K.; Li, N.; Kolmanovsky, I.; Girard, A.; Wang, Y.; Filev, D.; Dai, E. Hierarchical Optimization of Speed and Gearshift Control for Battery Electric Vehicles Using Preview Information. In Proceedings of the IEEE American Control Conference (ACC), Denver, CO, USA, 1–3 July 2020; pp. 4913–4919.
5. Han, J.; Sciarretta, A.; Ojeda, L.L.; De Nunzio, G.; Thibault, L. Safe-and eco-driving control for connected and automated electric vehicles using analytical state-constrained optimal solution. *IEEE Trans. Intell. Veh.* **2018**, *3*, 163–172. [CrossRef]
6. Ersal, T.; Kolmanovsky, I.; Masoud, N.; Ozay, N.; Scruggs, J.; Vasudevan, R.; Orosz, G. Connected and automated road vehicles: State of the art and future challenges. *Veh. Syst. Dyn.* **2020**, *58*, 672–704. [CrossRef]
7. Tate, L.; Hochgreb, S.; Hall, J.; Bassett, M. *Energy Efficiency of Autonomous Car Powertrain*; Technical Report; SAE Technical Paper; SAE International: Warrendale, PA, USA, 2018.
8. Chen, T.D.; Kockelman, K.M.; Hanna, J.P. Operations of a shared, autonomous, electric vehicle fleet: Implications of vehicle & charging infrastructure decisions. *Transp. Res. Part A Policy Pract.* **2016**, *94*, 243–254.
9. Zeng, X.; Wang, J. Globally energy-optimal speed planning for road vehicles on a given route. *Transp. Res. Part C Emerg. Technol.* **2018**, *93*, 148–160. [CrossRef]
10. Chen, D.; Kim, Y.; Stefanopoulou, A.G. State of charge node planning with segmented traffic information. In Proceedings of the IEEE Annual American Control Conference (ACC), Milwaukee, WI, USA, 27–29 June 2018; pp. 4969–4974.
11. Wan, N.; Vahidi, A.; Luckow, A. Optimal speed advisory for connected vehicles in arterial roads and the impact on mixed traffic. *Transp. Res. Part C Emerg. Technol.* **2016**, *69*, 548–563. [CrossRef]
12. McDonough, K.; Kolmanovsky, I.; Filev, D.; Szwabowski, S.; Yanakiev, D.; Michelini, J. Stochastic fuel efficient optimal control of vehicle speed. In *Optimization and Optimal Control in Automotive Systems*; Springer: Berlin, Germany, 2014; pp. 147–162.
13. Mehta, P.; Meyn, S. Q-learning and Pontryagin's minimum principle. In Proceedings of the 48h IEEE Conference on Decision and Control (CDC) held jointly with 2009 28th Chinese Control Conference, Shanghai, China, 15–19 December 2009; pp. 3598–3605.
14. Lee, H.; Song, C.; Kim, N.; Cha, S.W. Comparative Analysis of Energy Management Strategies for HEV: Dynamic Programming and Reinforcement Learning. *IEEE Access* **2020**, *8*, 67112–67123. [CrossRef]
15. Lee, H.; Kang, C.; Park, Y.I.; Kim, N.; Cha, S.W. Online Data-Driven Energy Management of a Hybrid Electric Vehicle Using Model-Based Q-Learning. *IEEE Access* **2020**, *8*, 84444–84454. [CrossRef]
16. Li, S.; Li, N.; Girard, A.; Kolmanovsky, I. Decision making in dynamic and interactive environments based on cognitive hierarchy theory: Formulation, solution, and application to autonomous driving. *arXiv* **2019**, arXiv:1908.04005.
17. Kouvaritakis, B.; Cannon, M. *Model Predictive Control*; Springer: Charm, Switzerland, 2016.
18. Nguyen, T.W.; Islam, S.A.U.; Bruce, A.L.; Goel, A.; Bernstein, D.S.; Kolmanovsky, I.V. Output-Feedback RLS-Based Model Predictive Control. In Proceedings of the American Control Conference (ACC), Denver, CO, USA, 1–3 July 2020.
19. Han, J.; Vahidi, A.; Sciarretta, A. Fundamentals of energy efficient driving for combustion engine and electric vehicles: An optimal control perspective. *Automatica* **2019**, *103*, 558–572. [CrossRef]
20. Prakash, N.; Cimini, G.; Stefanopoulou, A.G.; Brusstar, M.J. Assessing fuel economy from automated driving: Influence of preview and velocity constraints. In Proceedings of the ASME 2016 Dynamic Systems and Control Conference, Boston, MA, USA, 6–8 July 2016.
21. Seok, J.; Wang, Y.; Filev, D.; Kolmanovsky, I.; Girard, A. Energy-Efficient Control Approach for Automated HEV and BEV With Short-Horizon Preview Information. In Proceedings of the ASME 2018 Dynamic Systems and Control Conference, Atlanta, GA, USA, 30 September–3 October 2018.

22. HomChaudhuri, B.; Vahidi, A.; Pisu, P. Fast model predictive control-based fuel efficient control strategy for a group of connected vehicles in urban road conditions. *IEEE Trans. Control Syst. Technol.* **2016**, *25*, 760–767. [CrossRef]
23. Jia, Y.; Jibrin, R.; Itoh, Y.; Görges, D. Energy-optimal adaptive cruise control for electric vehicles in both time and space domain based on model predictive control. *IFAC-PapersOnLine* **2019**, *52*, 13–20. [CrossRef]
24. Kim, Y.; Figueroa-Santos, M.; Prakash, N.; Baek, S.; Siegel, J.B.; Rizzo, D.M. Co-optimization of speed trajectory and power management for a fuel-cell/battery electric vehicle. *Appl. Energy* **2020**, *260*, 114254. [CrossRef]
25. Brackstone, M.; McDonald, M. Car-following: A historical review. *Transp. Res. Part F Traffic Psychol. Behav.* **1999**, *2*, 181–196. [CrossRef]
26. National Highway Traffic Safety Administration. *Summary of State Speed Laws Ninth Edition: Current as of January 1, 2006*; National Committee on Uniform Traffic Laws and Ordinances: Washington, DC, USA, 2006.
27. Angel, E. Dynamic programming for noncausal problems. *IEEE Trans. Autom. Control* **1981**, *26*, 1041–1047. [CrossRef]
28. Kalia, A.V.; Fabien, B.C. On Implementing Optimal Energy Management for EREV using Distance Constrained Adaptive Real-Time Dynamic Programming. *Electronics* **2020**, *9*, 228. [CrossRef]
29. Bellman, R. Dynamic programming. *Science* **1966**, *153*, 34–37. [CrossRef]
30. Sundstrom, O.; Guzzella, L. A generic dynamic programming Matlab function. In Proceedings of the IEEE Control Applications, (CCA) & Intelligent Control, (ISIC), St. Petersburg, Russia, 8–10 July 2009; pp. 1625–1630.
31. Walker, K.; Samadi, B.; Huang, M.; Gerhard, J.; Butts, K.; Kolmanovsky, I. *Design Environment for Nonlinear Model Predictive Control*; Technical Report; SAE Technical Paper; SAE International: Warrendale, PA, USA, 2016.
32. Liao-McPherson, D.; Huang, M.; Kolmanovsky, I. A regularized and smoothed fischer–burmeister method for quadratic programming with applications to model predictive control. *IEEE Trans. Autom. Control* **2018**, *64*, 2937–2944. [CrossRef]
33. Cagienard, R.; Grieder, P.; Kerrigan, E.C.; Morari, M. Move blocking strategies in receding horizon control. *J. Process Control* **2007**, *17*, 563–570. [CrossRef]
34. Wipke, K.B.; Cuddy, M.R.; Burch, S.D. ADVISOR 2.1: A user-friendly advanced powertrain simulation using a combined backward/forward approach. *IEEE Trans. Veh. Technol.* **1999**, *48*, 1751–1761. [CrossRef]

2

State of Charge Estimation in Lithium-Ion Batteries: A Neural Network Optimization Approach

M. S. Hossain Lipu [1,*], M. A. Hannan [2,*], Aini Hussain [1], Afida Ayob [1], Mohamad H. M. Saad [3] and Kashem M. Muttaqi [4]

[1] Department of Electrical, Electronic and Systems Engineering, Universiti Kebangsaan Malaysia, Bangi 43600, Malaysia; draini@ukm.edu.my (A.H.); afida.ayob@ukm.edu.my (A.A.)
[2] Department of Electrical Power Engineering, College of Engineering, Universiti Tenaga Nasional, Kajang 43000, Malaysia
[3] Department of Mechanical and Manufacturing Engineering, Universiti Kebangsaan Malaysia, Bangi 43600, Malaysia; hanifsaad@ukm.edu.my
[4] School of Electrical, Computer and Telecommunications Engineering, University of Wollongong, Wollongong, NSW 2522, Australia; kashem@uow.edu.au
* Correspondence: lipu@ukm.edu.my (M.S.H.L.); hannan@uniten.edu.my (M.A.H.)

Abstract: The development of an accurate and robust state-of-charge (SOC) estimation is crucial for the battery lifetime, efficiency, charge control, and safe driving of electric vehicles (EV). This paper proposes an enhanced data-driven method based on a time-delay neural network (TDNN) algorithm for state of charge (SOC) estimation in lithium-ion batteries. Nevertheless, SOC accuracy is subject to the suitable value of the hyperparameters selection of the TDNN algorithm. Hence, the TDNN algorithm is optimized by the improved firefly algorithm (iFA) to determine the optimal number of input time delay (UTD) and hidden neurons (HNs). This work investigates the performance of lithium nickel manganese cobalt oxide ($LiNiMnCoO_2$) and lithium nickel cobalt aluminum oxide ($LiNiCoAlO_2$) toward SOC estimation under two experimental test conditions: the static discharge test (SDT) and hybrid pulse power characterization (HPPC) test. Also, the accuracy of the proposed method is evaluated under different EV drive cycles and temperature settings. The results show that iFA-based TDNN achieves precise SOC estimation results with a root mean square error (RMSE) below 1%. Besides, the effectiveness and robustness of the proposed approach are validated against uncertainties including noise impacts and aging influences.

Keywords: time-delay neural network; improved firefly algorithm; lithium-ion battery; state of charge; electric vehicle

---

1. Introduction

Environmental issues such as global warming, climate change, and carbon emissions drive the necessity to deploy battery storage technologies [1]. Lithium-ion batteries are extensively employed in the automotive industry due to their attractive characteristics such as low self-discharge, long life cycle, high voltage, and high energy density [2]. However, the lithium-ion battery has some issues such as performance degradation with aging cycles, temperature rise, accurate charge estimation, over-charging, and over-discharging [3]. Thus, further investigation is required on the lithium-ion battery charge estimation under a safe temperature region in electric vehicle (EV) applications.

EV has a battery management system (BMS) that executes operations such as state of charge (SOC) monitoring, battery health estimation, remaining life prediction, temperature management, battery equalization, and fault diagnosis [4,5]. SOC is a crucial parameter of BMS which defines the residual charge existing inside a battery cell [6,7]. SOC in EV applications has become an increasingly popular research topic and is of great importance for improving battery lifecycles. An accurate SOC calculation technique confirms the safe driving operation and protects the battery from many abnormalities such as over-charged, over-discharged, and over-heating problems. Nonetheless, SO is influenced by various factors such as the cathode material, material degradation, aging cycles, and temperatures [8]. Hence, advanced research is concerned greatly with developing an accurate and robust SOC estimation algorithm.

1.1. Related Works

A lot of studies have reported different approaches for SOC estimation of lithium-ion batteries under different operating conditions. The conventional methods like coulomb counting (CC) [9] and open-circuit voltage (OCV) [10] are the straightforward approaches for estimating SOC. Nevertheless, the accuracy of the CC method is affected by sensor precision which accumulates during each current integration. The OCV method obtains SOC by looking up the OCV vs. SOC curve; nevertheless, it needs a long rest time and cannot operate in online conditions. To overcome these concerns, model-based SOC estimation methods have been proposed. The model-based methods utilize a battery model incorporated with adaptive filter algorithms such as the extended Kalman filter (EKF) [11], unscented Kalman filter (UKF) [12], particle filter (PF) [13], H-infinity filter [14] and sliding mode observer [15] to estimate SOC. The KF is a popular approach for SOC estimation; however, the accuracy of KF may diverge badly in a highly nonlinear system. PF has fast execution and can deliver accurate SOC results. Nevertheless, PF has a complex mathematical computation. The H-infinity filter is suitable for moderate accuracy and a low computational cost. However, the performance would abruptly decrease owing to temperature and aging effects. In recent years, data-driven SOC estimation approaches have gained a lot of attention around the world because of their robust computation capabilities for handling highly non-linear lithium-ion battery characteristics. Besides, data-driven approaches examine SOC without exploring battery material structure, features, and associated chemical reactions [16]. The back-propagation neural network (BPNN) algorithm offers simple and easy execution but suffers from slow training operation [17]. The radial basis function neural network (RBFNN) achieves reasonable SOC accuracy with incomplete information; however, it has the shortcoming of lengthy training duration [18]. The extreme learning machine (ELM) is excellent with regard to fast learning speed and improved generalization performance, but the accuracy is influenced by the number of hidden layer neurons [19]. Although the wavelet neural network (WNN) has less complexity in training operation, it needs many hidden units. The support vector machine (SVM) can estimate SOC in the highly non-linear system, but has a complex execution process [20]. Gaussian process regression (GPR) can estimate SOC with model uncertainty; nonetheless, it has a drawback of poor efficiency in high dimensional spaces [21]. The nonlinear autoregressive with exogenous input neural network (NARXNN) is effective for mapping the lithium-ion battery non-linear characteristics, but the performance highly depends on the suitable value of the hyperparameters [22]. The long short-term memory (LSTM) [23] and gated recurrent unit (GRU) [24] can examine SOC under long-term dependencies; however, they require a large pool of data and an appropriate training operation to achieve accurate SOC estimation results. The adaptive neuro-fuzzy inference system (ANFIS) is an intelligent data-driven method that can obtain satisfactory SOC estimation solutions against changing environmental conditions, but it has a complex structure and lengthy computation process [25].

1.2. Major Contributions

The main contribution of this research is to design an optimized data-driven algorithm-based SOC estimation technique for lithium-ion batteries. In particular, a time-delay neural network (TDNN) algorithm optimized by the improved firefly algorithm (iFA) is proposed to elevate the accuracy and robustness in SOC evaluation. The significant contributions of this research work are explained below:

- The data pre-processing of the proposed iFA-based TDNN algorithm is simple and has easy execution which only requires sensors to monitor the battery variables including voltage, current, and temperature, thereby avoiding the need for an added filter.
- The TDNN has a self-learning algorithm that updates the learning parameters and employs input layer information in the previous time steps to estimate SOC in the future stage. In contrast, the model-based SOC estimation techniques require depth information and knowledge about battery internal characteristics as well as experience and time to develop a battery model and estimate related parameters accurately.
- The traditional TDNN algorithm examines SOC with a trial and error approach to determine the suitable values of input time delay (UTD) and hidden neurons (HNs) [26]. However, the trial and error method has some drawbacks such as inefficiency, data under-fitted, and over-fitted issues. Therefore, the TDNN algorithm is integrated with iFA to avoid the trial and error method and achieve accurate SOC estimation solutions.
- The generalization capability of the iFA-based TDNN algorithm is tested with two dissimilar types of lithium-ion batteries. Moreover, two suitable experimental tests are carried out to validate the proposed algorithm.
- Apart from the experiments, the accuracy of the proposed method is examined using three EV drive cycles such as the dynamic stress test (DST), federal urban drive schedule (FUDS), and US06. Accordingly, the variation of SOC estimation is monitored at three different temperature conditions.
- The influence of electromagnetic interference and low sensor precision might lead to inaccuracy in measured current and voltage values. Thus, this paper considers uncertainty issues such as noise impacts and aging profiles while estimating SOC. The robustness and effectiveness of the iFA-based TDNN method are verified against both bias noise and random noise. The performance of lithium-ion batteries deteriorates after the battery is repeatedly charged and discharged a hundred times. Therefore, the adaptability of the proposed method is assessed under 50, 100, 150, and 200 aging cycles.

The remainder of the paper is organized into six sections. Section two covers the explanation of the proposed algorithm framework. The lithium-ion battery test bench and related experimental tests are outlined in section three. Section four illustrates the design, methodological structure and the execution of the proposed algorithm. The SOC estimation results under various conditions are delivered in section five. The conclusion is presented in section six.

2. Theoretical Framework of SOC Algorithm

This section explains the theoretical strategy and framework of the proposed optimized data-driven algorithm for SOC estimation in lithium-ion batteries.

2.1. SOC Modeling with Time Delay Network Algorithm

TDNN is a modified version of the feedforward neural network (FNN) which is simple, fast, and efficient. TDNN is a supervised machine learning algorithm that is well-suited to model a dynamic system with large time delays. Generally, FNN has no internal memory for storing past information, which is ineffective for solving time-series problems. TDNN overcomes this issue by having a past memory with tapped delay lines. TDNN exhibits dynamic memory and employs multiple layers under a necessary interconnection between units which confirms the capability to address non-linear and complex decisions [27]. TDNN has intelligent self-learning skills and robust computation abilities, and hence it is suitable for SOC estimation.

The structure of TDNN is designed using one input layer, one or more hidden layers, and one output layer, as shown in Figure 1. This research uses one hidden layer to estimate SOC. TDNN employs a series of data such as an input vector with UTD ($x = x_{k,} x_{k-1}, \ldots x_{k-i+1}$), HNs ($h = h_{1,} \ldots h_L$), and output vector y_k to examine SOC. The output of the network can be expressed mathematically using the following equations:

$$y_k = \varphi_0 \left(\sum_{j=1}^{L} w_{oj} \cdot \varphi_j \left(\sum_{i=1}^{M} x_{k-i+1} w_{ji} + b_j \right) + b_o \right) \tag{1}$$

where φ_j and φ_0 denote the output of the hidden layer and output layer, respectively; w_{ji} is the weight between the input layer and hidden layers, w_{oj} is the weight between the hidden layer and output layer; b_j and b_o denote the hidden layer bias and output layer bias, respectively; M and L represent the number of inputs and hidden neurons, respectively.

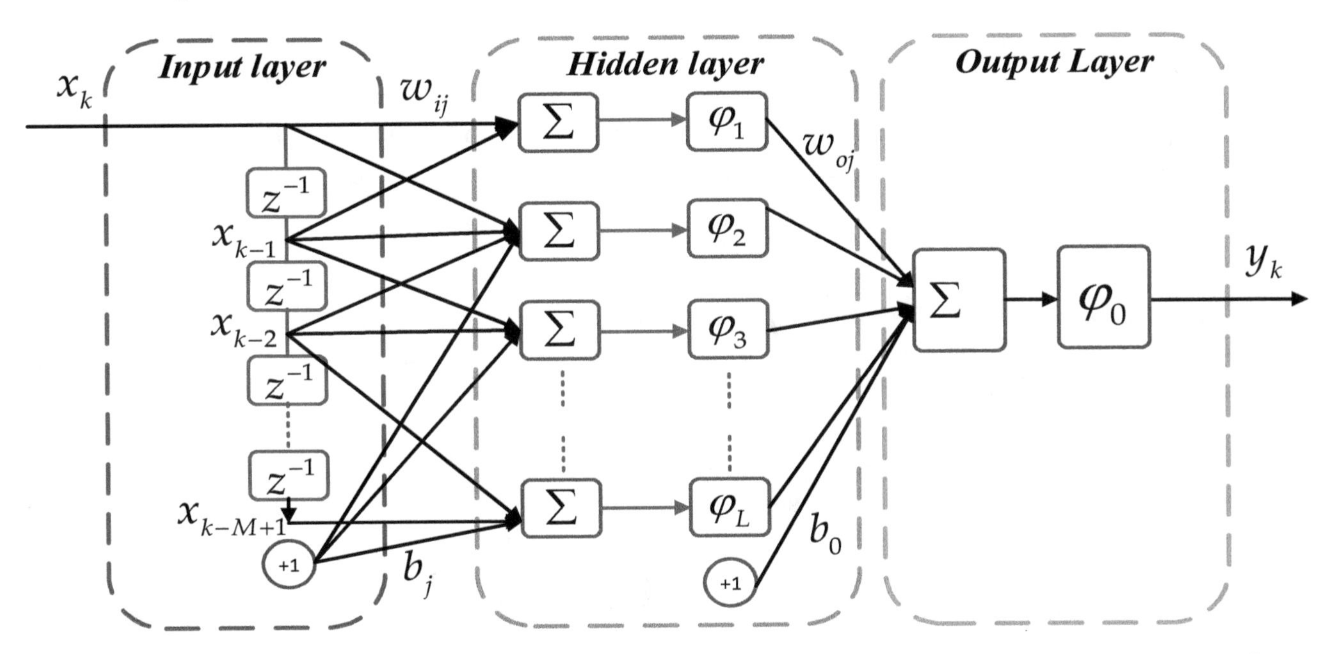

Figure 1. The time-delay neural network (TDNN) configuration for state-of-charge (SOC) estimation of the lithium-ion battery.

The sigmoid activation function is represented by $\varphi(t)$ and it is expressed as,

$$\varphi(t) = \frac{1}{1 + e^{-1}} \tag{2}$$

The back-propagation learning algorithm is applied to update network parameters of the TDNN algorithm including weights and bias by propagating the output error from the output layer to the hidden layer [28]. The Levenberg–Marquardt (LM) optimization technique is employed to train the TDNN due to its high accuracy and fast response toward the training operation [29]. LM updates the parameters adaptively to minimize the sum of the squared error, as shown in the following equations:

$$E(w) = \bar{e}(w)^T \bar{e}(w) = \sum_{k=0}^{\infty} \left\| e_k(w) \right\|^2 \tag{3}$$

where $\bar{e}(w)$ is the vector with element $e_k(w)$; $e_k(w)$ denotes the error in the k-th epoch.

A first-order Taylor series is employed to expand the error vector if the difference between the past weight vector and newly estimated weight vector becomes small [30]. The Newton method is used to reduce the function $E(w)$ with respect to w, as presented in the following equation.

$$\Delta w = -\left[\nabla^2 E(w) \right]^{-1} \nabla E(w) \tag{4}$$

where $\nabla E(w)$ stands for gradient and $\nabla^2 E(w)$ denotes the Hessian matrix which can be obtained using the following equation:

$$\nabla E(w) = J^T(w)\bar{e}(w) \tag{5}$$

$$\nabla^2 E(w) = J^T(w)J(w) + \sum_{k=1}^{N} e_k(w)\frac{\partial^2 e_k(w)}{\partial w_i \partial w_j} \tag{6}$$

where $J(w)$ denotes the Jacobian matrix which includes the first derivative of the network error. Since the second term of Equation (6) is trivial in comparison to the product of the Jacobian matrix, then Equation (4) can re-written as,

$$\Delta w = -\left[J^T(w)J(w)\right]^{-1} J^T(w)\bar{e}(w) \tag{7}$$

If the simplified Hessian matrix is found not invertible, then LM algorithm is adjusted and accordingly, Gauss-Newton method becomes,

$$\Delta w = -\left[J^T(w)J(w) + \gamma I\right]^{-1} J^T(w)\bar{e}(w) \tag{8}$$

where γ is the parameter to confirm the term $\left[J^T(w)J(w) + \gamma I\right]$ is positive and invertible, I is the identity matrix. The selection of the appropriate value of γ is essential for the LM function to ensure the steadiness and convergence speed.

2.2. *Improved Firefly Algorithm*

The concept of the FA is derived based upon the flashing characteristics of fireflies. The fireflies use different flashes to commutate among themselves as well as attract potential prey and mating partners [31]. Three statements are applied to develop FA. The first statement mentions that the attraction between two fireflies is independent since all the fireflies are unisex. The second statement declares that the brighter fireflies attract the less-bright fireflies. The fireflies with the same brightness travel randomly inside the boundary. The third statement says that the brightness of the fireflies will define the fitness function [32].

The standard FA has slow convergence issues when the fireflies are located far away in the early phases of generation. Moreover, the technique used to find the potential prey or communicate with mating partners in standard FA is mimicked. Also, the influence of environmental factors on the visibility of the flashing light is ignored. Generally, the attractiveness/ brightness of FA relies on many factors, such as the type and shape of the landscape, the distance between two fireflies, and a few environmental factors. For instance, the brightness visibility is reduced in the presence of fog, while the attractiveness/brightness rises as darkness increases. Likewise, the brightness of fireflies cannot be seen in the presence of high-intensity light. To address these concerns, Ball et al. [33] proposed a new improved FA (iFA) algorithm (Algorithm 1) to increase the convergence speed by updating the brightness of the fireflies. The exploration and exploitation capacities are enhanced in iFA algorithm by adding two new terms: brightness visibility (L_{mod}) and environmental factors (ξ).

Generally, the probability of brightness visibility increases in the lower-dimension landscape and easy optimization problems. In that case, L_{mod} is assigned to a high value. In contrast, L_{mod} is set to lower value when the brightness visibility decreases in the high dimensional and complex optimization problems. The impacts of environmental factors also affect the brightness visibility of iFA. Thus, two environmental variables are introduced, including ambient darkness (σ_1) and ambient fogginess (σ_2). The connection between these two variables can be expressed as follows:

$$\zeta = \frac{1}{1 + e^{-\left(\frac{\sigma_2}{\sigma_1 \ln\left(i_{gen}\right)}\right)}} \tag{9}$$

where i_{gen} denotes the generation number varying between 0 and i_{MaxGen}. The brightness of fireflies can be written as follows:

$$\beta(r) = \beta_0{}^{\text{Proposed}} e^{-\gamma r^m}, (m \geq 1) \tag{10}$$

where the attractiveness for $r = 0$ is represented by β_0; γ is the absorption coefficient of light, r is the distance between fireflies. The distance between two fireflies, r_{ij} is based on the cartesian coordinate system that can be written as follows:

$$r_{ij} = \|x_i - x_j\| = \sqrt{\sum_{k=1}^{d}\left(x_{i,k} - x_{j,k}\right)^2} \tag{11}$$

where $x_{i,k}$ and $x_{j,k}$ are the spatial coordinate of the i-th and j-th fireflies towards the k-th component.

The proposed attractiveness of fireflies ($\beta_0{}^{\text{Proposed}}$) is defined in the following equations:

$$\beta_0{}^{\text{Proposed}} = (\beta_0 \times L_{\text{mod}}) + \zeta \text{ when } rand \leq \zeta_{seed} \tag{12}$$

$$\beta_0{}^{\text{Proposed}} = \beta_0 \text{ when } rand > \zeta_{seed} \tag{13}$$

If the ambient darkness increases, the contributions from the second term of Equation (12) will be reduced, resulting in a lower environmental effect on $\beta_0{}^{\text{Proposed}}$. In opposition, if the intensity of ambient darkness decreases, then the impact of the environment becomes stronger, yielding lower brightness visibility of fireflies. The rise of ambient fogginess in the environment leads to a stronger impact of the environmental factor (ζ) on the proposed brightness of fireflies ($\beta_0{}^{\text{Proposed}}$).

During the early iteration of iFA, the fireflies are located far away from one another, and accordingly, the effect on environmental factors becomes stronger. As the iteration starts to increase, the fireflies are situated close to each other, and subsequently their brightness visibility increases. In this case, the environmental factor has less impact on the brightness visibility of iFA. Therefore, a probability index (ζ_{seed}) is formulated using i_{gen} and normalized between 0 and 1.

$$\zeta_{seed} = 1 - \frac{i_{gen}}{i_{MaxGen}}; 1 \leq i_{gen} \leq i_{MaxGen} \tag{14}$$

If the random number $rand$ is equal to or lower than ζ_{seed}, then Equation (12) will be executed; otherwise, Equation (13) will be selected. The brighter firefly j is attracted by the lesser bright firefly i, as expressed by,

$$x_{i\_new} = x_{i\_old} + \beta_0^{\text{Proposed}} e^{-\gamma r_{ij}^m}\left(x_j - x_{i\_old}\right) + \alpha \in_i \tag{15}$$

where α is the randomization parameter and $\in_i$ defines the random numbers located between '0' and '1'.

The pseudocode of iFA is illustrated below:

Algorithm 1 Improved Firefly Algorithm (iFA)

Start
Define the fitness function $f(x)$, $x = (x_1, \ldots, x_d)^T$
Create initial population of fireflies $i = 1, 2, \ldots, Size_{population}$
Assign γ α β_0 L_{mod} σ_1 *and* σ_2
Assess fitness function of individual fireflies $f(x), i = 1, 2, \ldots, Size_{population}$
While $t < Max_{genertion}$
Assess ξ_{seed} *with Equation (14)*
if $\xi_{seed} \geq rand$
Assess $\beta_0{}^{Proposed}$ *with Equation (12)*
else
Assess $\beta_0{}^{Proposed}$ *with Equation (13)*
end if
for $i = 1 : Size_{population}$
for $j = 1 : Size_{population}$
if $f(x_j) > f(x_i)$
Move firefly i toward j
End if
Update the attractiveness of fireflies (β) *with Equation (10)*
Assess new solutions and update light intensity with Equation (15)
end for *j*
end for i
Rank the fireflies and find the current best population
$t = t + 1$
end while

3. Lithium-Ion Battery Experiments and Data Preparation

This section covers the type of lithium-ion cells used in this research, collected dataset, experimental arrangement, experimental procedures, data partition, and evaluation criteria.

3.1. Lithium-Ion Battery Cell

In this research, two popular EV lithium-ion battery cells are used; one is a 3200 mAh rated NCR18650B manufactured by Panasonic and the other is a 2600 mAh rated ICR18650-26F developed by Samsung. The positive electrodes of NCR18650B and ICR18650-26F are made using $LiNiCoAlO_2$ (LiNCA) and $LiNiMnCoO_2$ (LiNMC), respectively, while the graphite is used as a negative electrode in both batteries. LiNMC is popular due to its extended lifespan, while LiNCA offers a high level of specific energy. Table 1 provides the information and specifications of the two lithium-ion batteries [6].

Table 1. Specifications of the lithium-ion battery.

Parameters	$LiNiCoAlO_2$	$LiNiMnCoO_2$
Nominal capacity (Ah)	3.2 Ah	2.6 Ah
Nominal Voltage (V)	3.6	3.7
Min/Max voltage (V)	2.5/4.2	2.75/4.2
Charging method	CC-CV	CC-CV
Charging time (hours)	4	3
Charging current (mA)	1625	1300
Specific Energy (Wh/kg)	200–260	150–220
Weight (g)	48.5	47.0
Lifespan (cycles)	500	1000–2000
Thermal runaway (temperature)	150 °C	210 °C

3.2. Battery Experimental Setup

A lithium-ion battery test bench model is established to monitor the key parameters including capacity, current and voltage, power, and cycle number. The developed test bench model is separated into two sections: one is the hardware section and the other is the software section, as depicted in Figure 2. The hardware section includes two segments; the measurement unit and the control unit. The measurement unit is designed using NEWARE BTS-4000 and LiNCA, LiNMC battery cells. The maximum capacity of voltage and current for NEWARE BTS-4000 are 5 V and 6 A, respectively. The NEWARE BTS-4000 has 8 channels that are independent of each other and it is capable of sensing and recording the battery parameters each second with an accuracy of ±0.05% full scale (FS) [34]. The control unit can keep the charging and discharging voltage under the defined cut-off values, thus protecting the battery from overcharging and over-discharging issues. In contrast, the software section includes the battery testing system (BTS) software version 7.6 related to NEWARE BTS-4000 hardware and MATLAB 2015a. Both pieces of software are installed on the host computer. Each step of the SDT and HPPC experiment is executed though the BTS software. After, the recorded data are transferred to the host computer. Besides, the charging and discharging control of LiNCA and LiNMC batteries are implemented through the appropriate software function of the BTS software. Finally, MATLAB 2015a software is used to execute the iFA-based TDNN algorithm code to examine SOC.

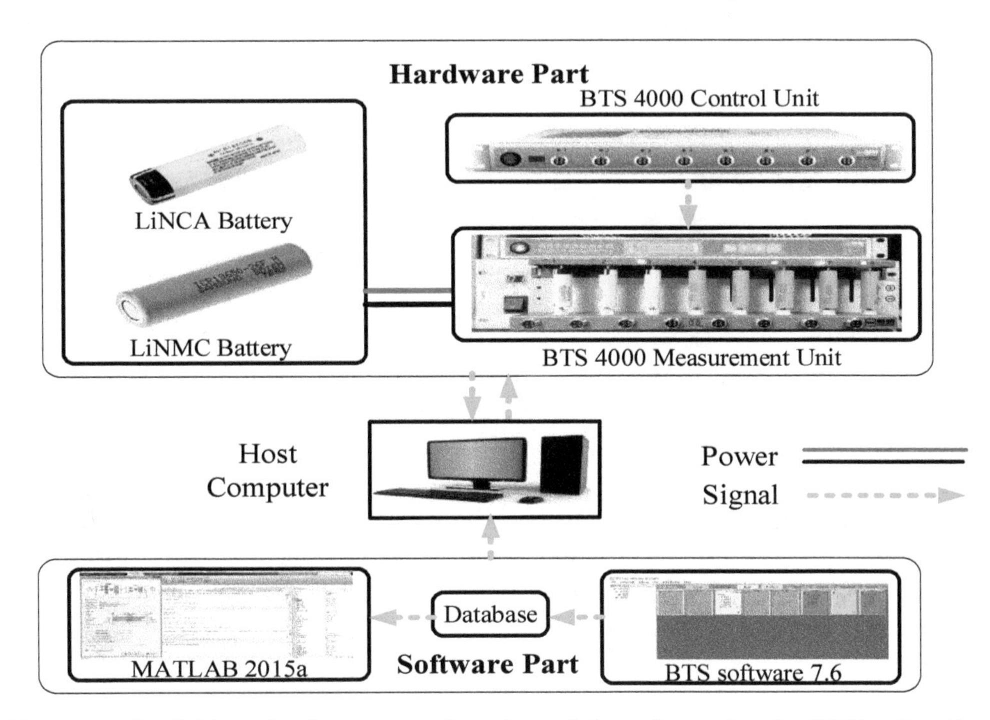

Figure 2. The lithium-ion battery test bench model configuration for SOC estimation.

3.3. Battery Experimental Tests

The validation of the proposed algorithm is performed using a static discharge test (SDT) and a hybrid pulse power characterization (HPPC) test. Besides, the noise test and aging cycle tests are carried out to check the SOC accuracy against uncertainties. Before the experiments begin, the battery is completely charged, and accordingly, the initial SOC is assigned to 100%. The Samsung ICR18650-26F battery is used to explain the steps of each experimental test. These steps can also be applied to a Panasonic NCR18650B while maintaining the manufacturer's requirements in terms of current and voltage.

(1) SDT SDT uses the constant discharge current of the lithium-ion battery to evaluate SOC. The operation of SDT is explained using the steps mentioned below.

i. Firstly, a constant current (CC) of 1.3 A (0.5 C) is applied to charge the battery fully until the charge voltage increases to the maximum threshold of 4.2 V.
ii. Then, a constant voltage (CV) of 4.2 V is applied until a drop in the charge current to 0.13 A (0.05 C) is achieved.
iii. The battery being tested is kept idle for 1 h.
iv. A discharge current of 2.6 A (1 C) is applied until the voltage is reduced to 2.75 V.
v. The test ends if the battery voltage reaches the minimum threshold of 2.75 V; otherwise, step ii will continue.

(2) HPPC test The HPPC test consists of the array of charge and discharge current pulses arranged in an orderly manner. The following steps are used to describe the operation of HPPC.

i. The CC-CV method is employed to charge the battery completely until the battery current decreases to 0.13 A (0.05 C).
ii. The battery being tested is kept idle for 1 h.
iii. A discharge current of 1.3 A (0.5 C) is applied for 10 s.
iv. The battery being tested is kept idle for 3 min.
v. A charge current of 1.3 A (0.5 C) is applied for 10 s.
vi. The battery is kept idle for 3 min.
vii. A discharge current of 0.65 A (0.25 C) is applied for 24 min to decrease the SOC capacity of the battery by 10%.
viii. The experiment ends if the battery voltage reaches the lower cut off voltage; otherwise, step iii will start again.

(3) Noise test

The accuracy of SOC could deviate due to electromagnetic interference (EMI) noises and sensor precision. EMI noises take place when the switching of the power converter is operated at a high frequency, and hence they may combine with the measured current and voltage signals. On the other hand, current and voltage sensors have a common issue, equipment error, which leads to errors in measurements. Thus, this research work considers both EMI impacts and sensor precision by adding random noises and bias noises, respectively, to evaluate the proposed model's suitability in the real-world environment. Accordingly, the positive bias noises of 0.1 A and 0.01 V are injected into the current and voltage measurements, respectively. At the same time, the random noises having amplitude values of 0.1 A and 0.01 V are included in current and voltage measurements, respectively.

(4) Aging cycle test

Battery aging is a significant index term used to evaluate performance after continuous charge-discharge cycles. Usually, the capacity of the battery declines with the rise of aging cycles. However, capacity degradation does not occur in the same way in different types of lithium-ion batteries. In this research, LiNCA and LiNMC battery cells are employed to assess the aging effect on performance and predictable capacity, as well as to evaluate SOC. Four milestones of aging cycles are selected to validate the accuracy and robustness of the proposed method including 50, 100, 150, and 200 cycles. The steps of the aging cycle are highlighted as follows:

i. Firstly, the complete charge operation is executed based on the CC-CV method with a constant charge current of 1.3 A (0.5 C) until the battery voltage reaches 4.2 V. After, the charge voltage of 4.2 V is kept constant until the charging current declines to 0.13 A (0.05 C).
ii. The idle operation of the battery is performed for 15 min.
iii. A constant discharge current of 2.6 A (1 C) is applied until the battery voltage decreases to 2.75 V.

iv. The lowest point of the discharge voltage (2.75 V) of the battery is checked. The one aging schedule is completed when the battery reaches a cut-off voltage of 2.75 V; otherwise, step iii will begin again.

v. After the completion of one aging cycle, the battery is kept in an idle operation stage for one hour.

vi. Step i starts again to perform the second aging cycle test. The process continues until the defined cycles are achieved.

3.4. Dataset Training and Testing

The dataset achieved from the battery test bench is split into two groups; training and testing. The data partition is performed based on the cross-validation method using a 70-30 ratio [35]. The training operation is implemented in the offline phase while testing operation and SOC estimation is performed in the online phase. To improve training accuracy and speed, the dataset is normalized within the boundary of [−1, 1], as represented by the equation below,

$$x' = \frac{2(x - x_{\min})}{x_{\max} - x_{\min}} - 1 \tag{16}$$

where $x_{\max}$ and $x_{\min}$ are the maximum and the minimum values of variable x, respectively. The maximum iterations and performance goal are set to be 1000 and 0.000001, respectively. The learning rate is considered as 0.5. The host computer has sufficient computational power with 12 GB random access memory (RAM). The current and voltage load profiles of SDT and HPPC for the LiNCA and LiNMC batteries are shown in Figures 3 and 4, respectively. In this research, the positive current and negative current correspond to the charging process and the discharging process, respectively. The relationship between SOC and voltage is established where voltage increases with the rise of SOC, as illustrated in Figure 5.

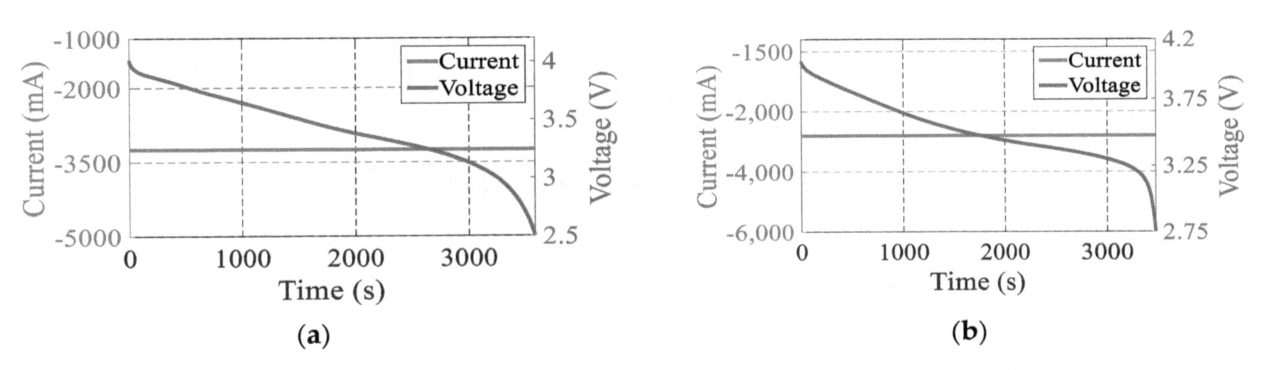

Figure 3. SDT load profile of (**a**) the LiNCA battery and (**b**) the LiNMC battery.

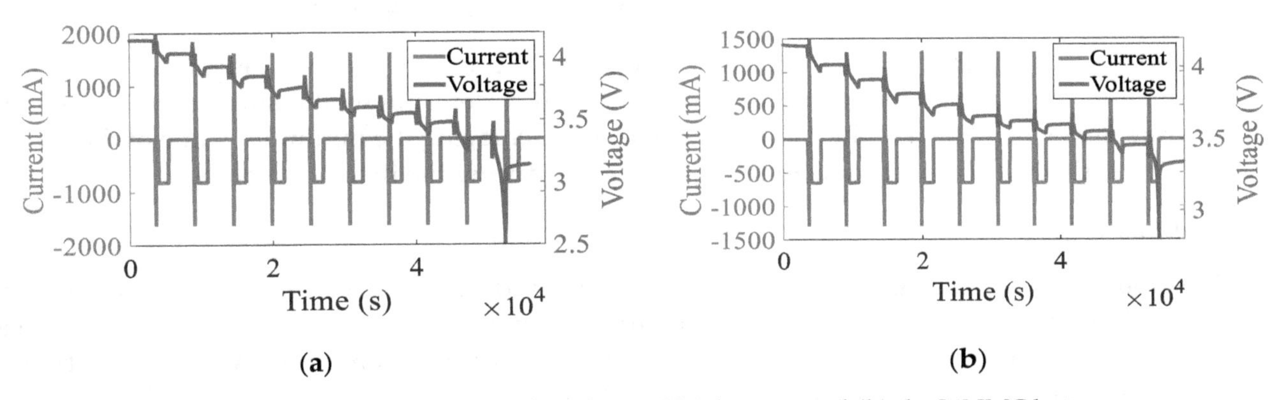

Figure 4. HPPC load profiles of (**a**) the LiNCA battery and (**b**) the LiNMC battery.

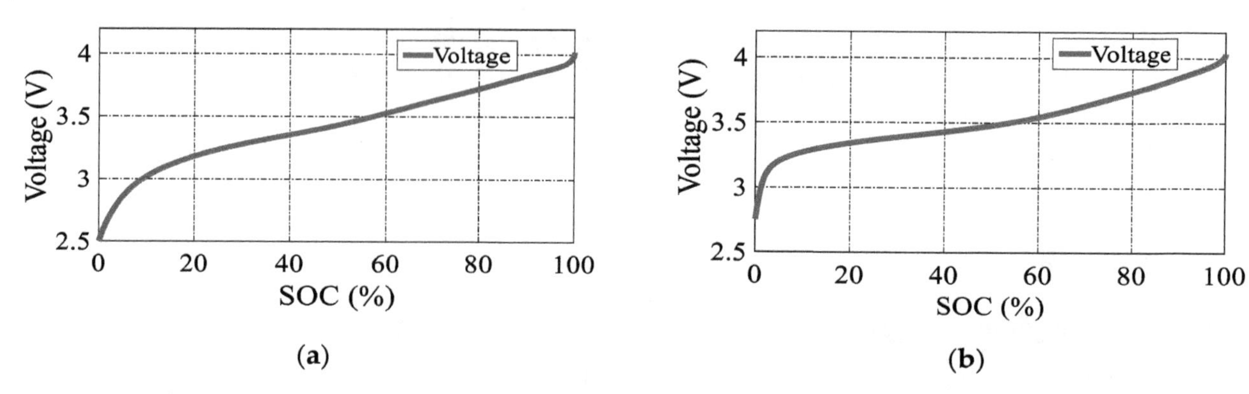

Figure 5. The relationship between SOC and voltage for (**a**) the LiNCA battery and (**b**) the LiNMC battery.

3.5. Measurement of SOC Effectiveness

The performance of the iFA-based TDNN algorithm is assessed by comparison with reference SOC values achieved through the CC method with an adjustable current sensor. The different statistical error rate terms shown in (17)–(22) are used to check the accuracy of the proposed algorithm [35].

$$SOC\ error = SOC_{a_i} - SOC_{es_i} \tag{17}$$

$$RMSE = \sqrt{\frac{1}{N}\sum_{i=1}^{N}\left(SOC_{a_i} - SOC_{es_i}\right)^2} \tag{18}$$

$$MSE = \frac{1}{N}\sum_{i=1}^{N}\left(SOC_{a_i} - SOC_{es_i}\right)^2 \tag{19}$$

$$MAE = \frac{1}{N}\sum_{i=1}^{N}\left(SOC_{a_i} - SOC_{es_i}\right) \tag{20}$$

$$MAPE = \frac{1}{N}\sum_{i=1}^{N}\left|\frac{SOC_{a_i} - SOC_{es_i}}{SOC_{a_i}}\right| \tag{21}$$

$$SD = \sqrt{\frac{1}{N-1}\sum_{i=1}^{N}\left(SOC\ error - \overline{SOC\ error}\right)^2} \tag{22}$$

where SOC_a is the actual/reference SOC and SOC_{es} is the estimated SOC by the proposed algorithm, N is the number of the data sample, MSE is mean square error, MAE is mean absolute error, MAPE is mean absolute percentage error, SD is the standard deviation, and $\overline{SOC_{error}}$ is the average error of SOC estimated values.

4. Design and Implementation of iFA Based TDNN Algorithm for SOC Estimation

The proposed algorithm is designed based on three vital components, including input dimension, fitness function, and constraints of optimization. The iFA determines the suitable values of the hyper-parameters which include the UTD and HNs while achieving the minimum value of the fitness function and satisfying all constraints in the optimization during the iterative method.

4.1. Input Information

The input information consists of a matrix that is designed using the number of rows and columns and defines the dimension and boundary of the hyperparameters. The number of problem dimension is characterized by the number of rows, while the population of hyperparameters is outlined by the number of columns in the matrix, as represented by the equation below:

$$D_{ij} = \begin{bmatrix} X_{11} & X_{12} & X_{13} & \cdots & X_{1j} \\ X_{21} & X_{22} & X_{23} & \cdots & X_{2j} \\ X_{31} & X_{32} & X_{33} & \cdots & X_{3j} \\ X_{41} & X_{42} & X_{43} & \cdots & X_{4j} \\ \vdots & \vdots & \vdots & \cdots & \vdots \\ X_{i1} & X_{i2} & X_{i3} & \cdots & X_{ij} \end{bmatrix} \quad (23)$$

where D_{ij} represents the matrix of the input data which can be defined by i and j, where $i = 1, 2, \ldots\ldots, P$, with P representing the population number; $j = 1, 2, \ldots\ldots, N$, with N denoting the problem dimension.

4.2. Fitness Function

The fitness function helps to achieve precise SOC estimation outcomes by finding the appropriate values of hyperparameters. The iFA aims to obtain the lowest error rate of fitness function through iterations that correspond to the optimum values of the UTD and HNs. Lithium-ion battery SOC estimation has a high volume of datasets where SOC error is distributed randomly. Thus, root means square error (RMSE) is selected as the fitness function which is shown in the following equation:

$$Objective\ function = \min\ (RMSE) \quad (24)$$

4.3. Optimization Constraints

The optimization constraints are determined by assigning the upper and lower number of the UTD and HNs of the

TDNN algorithm. If the population of hyperparameters is located outside the boundary, then the accuracy of iFA could deviate which may result in unsatisfactory performance during SOC estimation. Therefore, the values of hypermeters are checked in each iteration. If any values are located outside the boundary, then values will be reproduced and updated accordingly. The constraints of the hyperparameters must satisfy the following equation:

$$X_{i,j}^{k-1} < X_{i,j}^{k} < X_{i,j}^{k+1} \quad (25)$$

The above equation indicates that the value $X_{i,j}^{k}$ of the hyperparameter should be placed between $X_{i,j}^{k-1}$ and $X_{i,j}^{k+1}$.

4.4. Execution Process of iFA Based TDNN Algorithm

The implementation of the TDNN-iFA begins with the collection of battery datasets, including current and voltage, through experimental tests. After that, iFA is employed to find the appropriate value of UTD and HNs based on the lowest value of the fitness function. Finally, the accuracy and robustness of SOC estimation are assessed using different performance indicator terms. The proposed method is then advanced into different tests and uncertainties. The flow diagram of the proposed algorithm is shown in Figure 6. The methodology of the TDNN-iFA algorithm is divided into three stages.

In stage I, a battery test bench model is built to carry out the experiments as well as collect battery data. Then, data is pre-processed through normalization. After, the data is separated into training and testing groups.

In stage II, the implementation of TDNN starts with assigning the epoch number, performance goal, and learning rate. Subsequently, the process of iFA starts with the selection of suitable iteration number, population size, input dimension, fitness function, and optimization constraints. At first, the initial population of iFA is generated and the fitness function is assessed. After, the light intensity of the firefly is evaluated based on the fitness function, and accordingly, the best population is determined. Later, the probability index and proposed attractiveness are assessed using Equations (12) and (13). Next, the movement of the firefly from the brighter one toward the less-bright one is estimated. Later, attractiveness among fireflies is evaluated and light intensity is updated using Equations (10) and (15), respectively. The aforementioned processes continue until the highest number of iterations is completed. Finally, the optimal values of UTD and HNs are found and consequently proceed to the TDNN algorithm. Accordingly, the TDNN training process is executed using the LM method using Equations (3)–(8). Afterward, the network parameters of TDNN are updated using the backpropagation learning rule. Finally, the output of TDNN is computed using Equation (1).

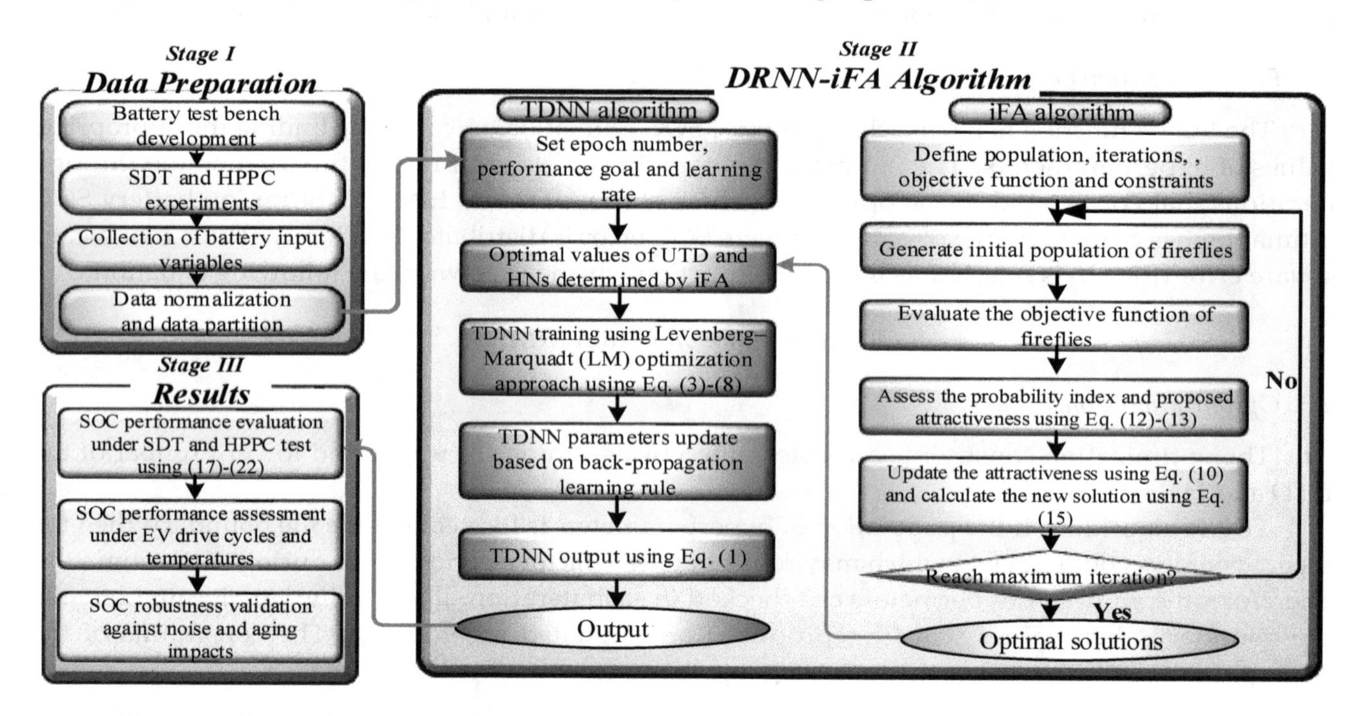

Figure 6. The methodological framework of the iFA-based TDNN algorithm for SOC estimation.

In stage III, the TDNN-iFA algorithm is verified through SDT and HPCC experiments based on statistical error rate terms, as indicated in Equations (17)–(22). In line with that, the effectiveness of the TDNN-iFA algorithm is assessed under EV drive cycles and varying temperature situations. Besides, the robustness of the proposed method is evaluated against the uncertainties, including noise tests and the aging cycle tests.

5. SOC Experimental Results and Validation

This section describes the experimental results of SOC estimation under different chemistries of lithium-ion battery cells. Also, the validation of SOC is carried out under EV drive cycles, noise effects, and aging impacts.

5.1. *Assessment of Fitness Function and Optimal Parameter*

The fitness function is evaluated using the optimization response curve in SDT and HPPC experiments for LiNCA and LiNMC batteries, as presented in Figures 7 and 8, respectively. The fitness function performance of iFA is compared with FA and particle swarm optimization (PSO). The optimization curve is generated using a population of 50 and 500 iterations.

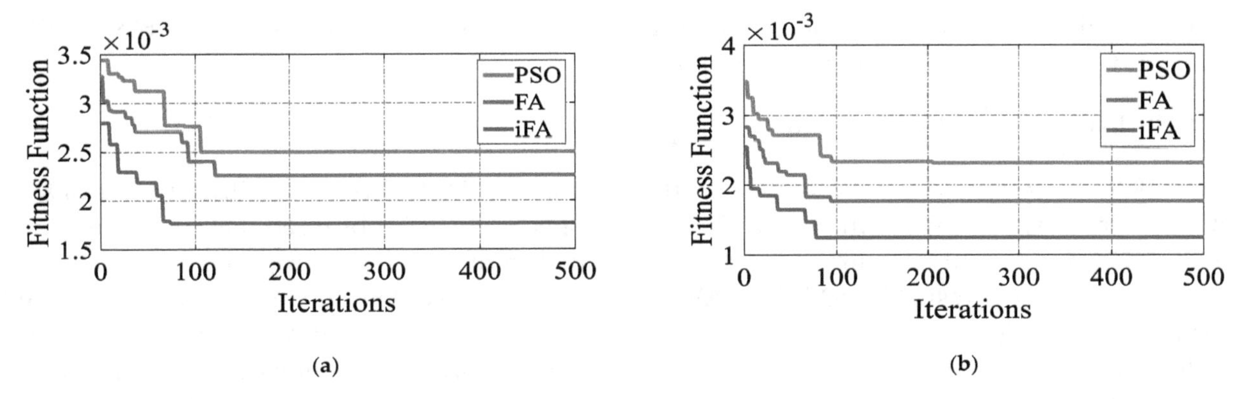

Figure 7. The response curve for optimization in the SDT load profile of (**a**) the LiNCA battery and (**b**) the LiNMC battery.

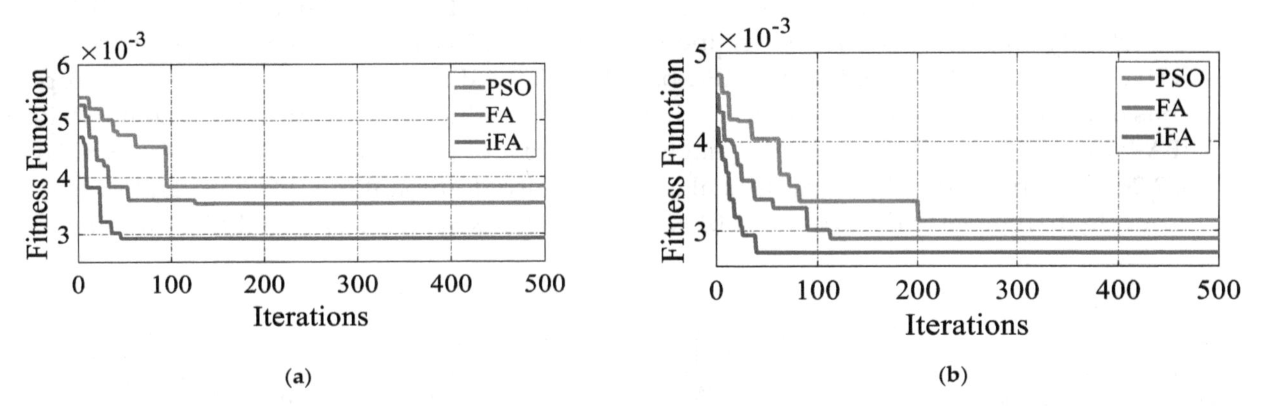

Figure 8. The response curve for optimization in the HPPC load profile of (**a**) the LiNCA battery and (**b**) the LiNMC battery.

The optimal values of UTD and HNs are determined via the response curve for optimization by identifying the lowest point of the fitness function. For example, in SDT, iFA achieves the lowest fitness function after 74 and 78 iterations for the LiNCA and LiNMC battery, respectively, which are smaller than the FA and PSO algorithms. The corresponding iteration number delivers the fitness functions of 0.177% and 0.125% and provides the best values of the UTD and HNs of 2, 3, and 12, 15, respectively. Likewise, in the HPPC test, iFA demonstrates excellent solutions in obtaining the minimum value of the fitness function compared to FA and PSO algorithms achieving 0.292% and 0.276% after 45 and 39 iterations, respectively, for the LiNCA battery and LiNMC battery. Accordingly, the optimal values of UTD and HNs of 4, 5, and 10, 18 are obtained after 45 and 39 iterations, respectively. The appropriate values of the UTD and HNs in SDT and HPPC tests are summarized in Table 2.

Table 2. Optimum hyperparameters of the TDNN-iFA algorithm in SDT and HPPC tests.

Battery Test	Optimal Hyperparameters	LiNCA Battery	LiNMC Battery
SDT	UTD	2	3
	HNs	12	15
HPPC	UTD	4	5
	HNs	10	18

5.2. Experimental Verification Results

The performance of the iFA-based TDNN algorithm is verified using SDT and HPPC experiments and results are compared and analyzed with three commonly-reported data-driven algorithms; namely, the backpropagation neural network (BPNN) and radial basis function neural network (RBFNN) and Elman neural network (ENN). The comparative analysis is performed using a similar length of input datasets for training and testing. In addition, iFA is employed to update the number of HNs of BPNN, RBFNN, and ENN to achieve unbiased comparison.

(1) SOC Estimation in LiNCA Battery

The SOC is examined for the LiNCA battery using the TDNN-iFA algorithm under SDT and HPPC experiments, as depicted in Figures 9 and 10, respectively. It is observed that the reference SOC values are placed adjacent to estimated SOC values which proves that the proposed approach obtains accurate solutions. The results of RMSE, MAE, MAPE, and SD also demonstrate the superiority of the iFA-based TDNN algorithm over iFA-based BPNN, iFA-based RBFNN, iFA-based ENN as denoted in Table 3. In SDT, RMSE is estimated to be 0.5844% in the proposed algorithm, which is a 32.2%, 54.9%, and 19% decrease when compared to the iFA-based BPNN, iFA-based RBFNN, iFA-based ENN, respectively. Moreover, the proposed algorithm computes MAE of 0.2374% which is reduced by 60.8%. 78.7% and 56.7% from those obtained from the iFA-based BPNN, iFA-based RBFNN, iFA-based ENN, respectively. Similarly, MAPE is declined by 29.9%, 54.1% and 38.7% in comparison to those values derived using the iFA-based BPNN, iFA-based RBFNN, iFA-based ENN. Besides, the proposed method has a maximum SOC error under 3%; however, high fluctuations are observed in iFA-based BPNN, iFA-based RBFNN, iFA-based ENN, having SOC error of [−5.19%, 6.45%], [−4.36%, 8.89%] and [−3.32%, 2.26%], respectively. The proposed algorithm is also dominant in HPPC test and possesses a lower error rate than those from iFA-based BPNN, iFA-based RBFNN, iFA-based ENN methods.

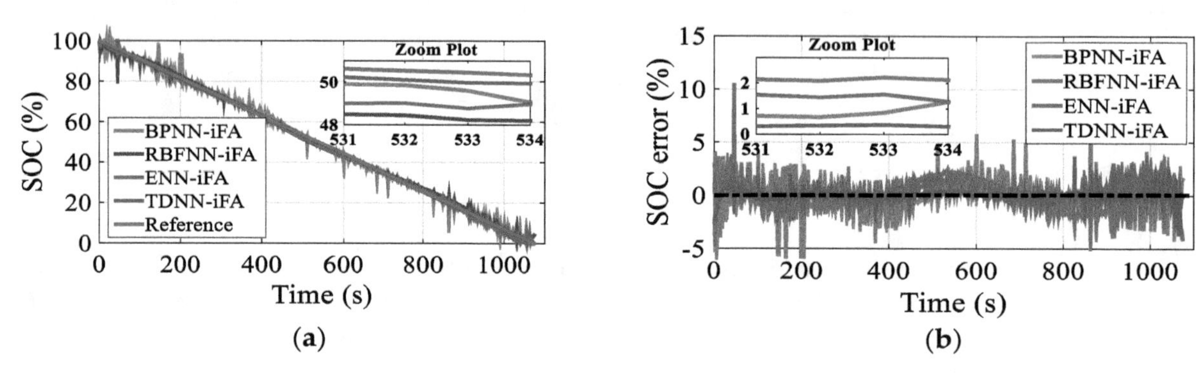

Figure 9. The experimental outcomes of SDT for LiNCA battery (**a**) SOC (**b**) SOC error.

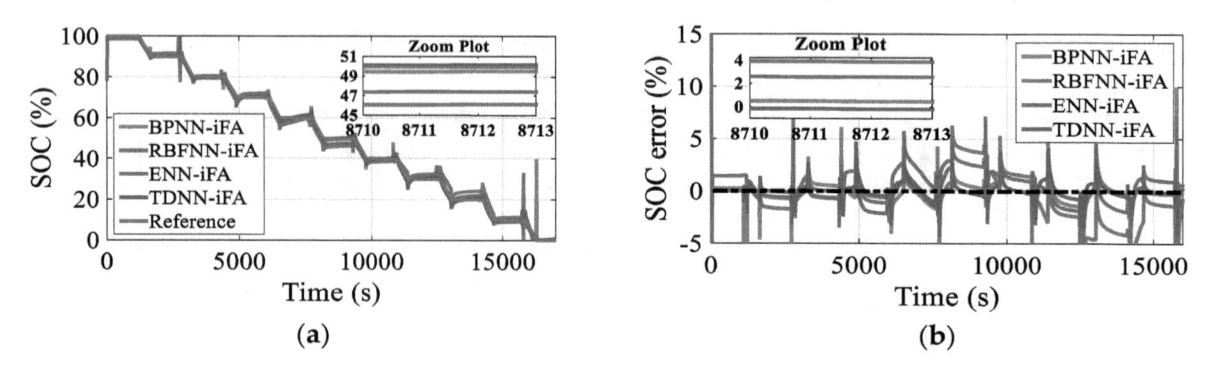

Figure 10. The experimental outcomes of HPPC for the LiNCA battery: (**a**) SOC, (**b**) SOC error.

Table 3. SOC performance evaluation in the LiNCA battery.

SOC Method	BPNN-iFA		RBFNN-iFA		ENN-iFA		TDNN-iFA	
Load Profile	SDT	HPPC	SDT	HPPC	SDT	HPPC	SDT	HPPC
RMSE (%)	0.8620	1.4124	1.2961	2.5155	0.7215	1.6524	0.5844	0.8512
MSE (%)	0.0074	0.0199	0.0168	0.0633	0.0052	0.0273	0.0034	0.0072
MAE (%)	0.6059	0.6659	1.1145	1.997	0.5479	1.2294	0.2374	0.4652
MAPE (%)	3.6939	6.2650	5.6405	10.4826	4.2235	7.5826	2.5864	3.5624
SD (%)	0.8610	1.1685	1.2815	2.4878	0.6876	1.4869	0.5841	0.8505
SOC error bound (%)	[−5.19, 6.45]	[−5.45, 9.98]	[−4.36, 8.89]	[−15.28, 12.32]	[−3.32, 2.26]	[−5.09, 7.17]	[−2.58, 2.05]	[−4.31, 4.73]

(2) SOC Estimation in LiNMC Battery

The performance of SOC estimation results is also evaluated for the LiNMC battery under SDT and HPPC experiments, as illustrated in Figures 11 and 12, respectively. From Table 4, it can be observed that RMSE is decreased by 60.3%, 70.8%, and 49.7% in the proposed algorithm under SDT compared to the iFA-based BPNN, iFA-based RBFNN, iFA-based ENN, respectively. Similarly, 76.2%, 84.3%, and 68.6% reductions are reported in the proposed algorithm in comparison to iFA-based BPNN, iFA-based RBFNN, and iFA-based ENN, respectively, while calculating MAE. Besides, the iFA-based TDNN method demonstrates excellent results in the aspects of SOC error, MSE and MAPE. For instance, the maximum SOC error is noted to be 1.38% in the proposed approach, while it is 3.31%, 5.44%, and 3.02% in iFA-based BPNN, iFA-based RBFNN, and iFA-based ENN, respectively. The proposed algorithm also obtains a narrow SOC error in the HPPC test, obtaining a maximum SOC error of 4.23%. Since TDNN has few mathematical complications in the testing stage, the execution time is small, indicating less than 30 milliseconds and 0.5 s in SDT and HPPC tests, respectively. In conclusion, the aforesaid accurate outcomes and fast computation time in the testing phase demonstrate the suitability of TDNN-iFA algorithms in real-time BMS.

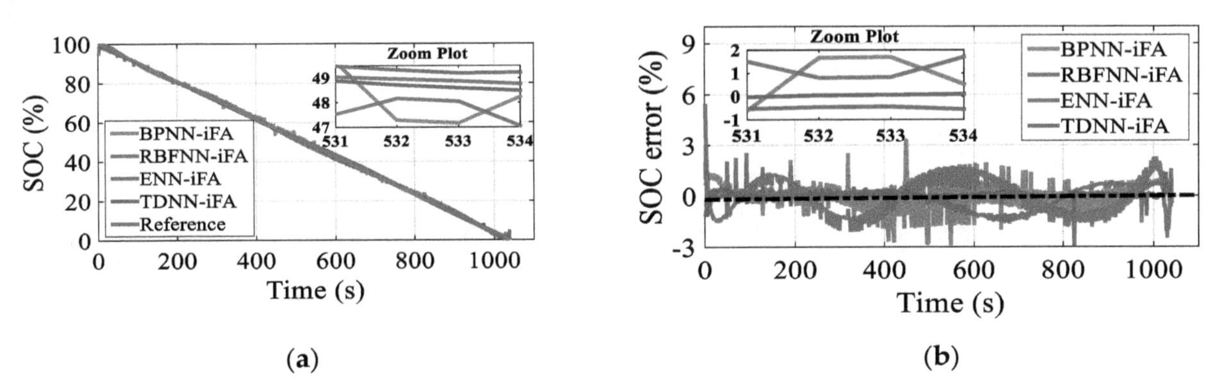

Figure 11. The experimental outcomes of SDT for LiNMC battery (**a**) SOC (**b**) SOC error.

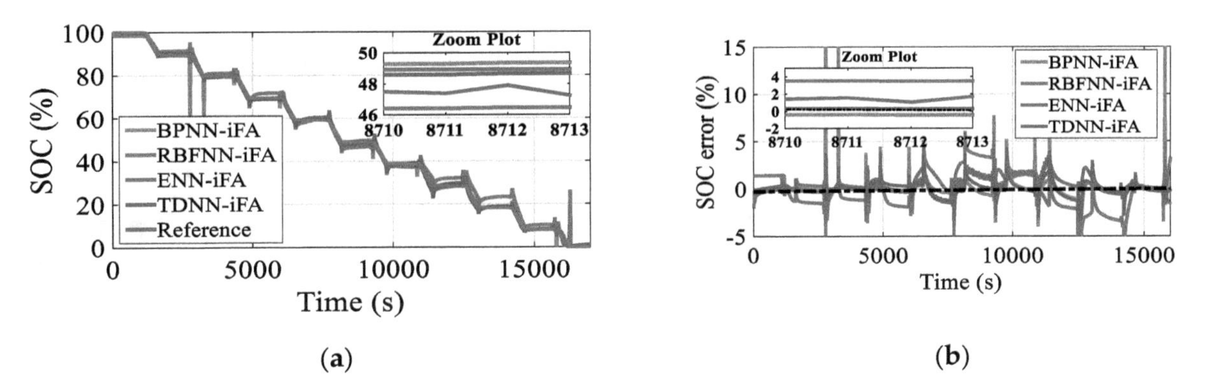

Figure 12. The experimental outcomes of HPPC for LiNMC battery (**a**) SOC (**b**) SOC error.

5.3. SOC Estimation under EV Drive Cycles and Temperatures

The accuracy and effectiveness of the iFA-based TDNN approach for SOC estimation are further evaluated under different EV drive cycles. EV drive cycle data are obtained on LiNMC battery with a rated capacity of 2.0 Ah [36]. The DST, FUDS, and US06 drive schedules are employed to assess the accuracy and robustness of the proposed method, as depicted in Figure 13. The aforementioned drive cycles are diverse in terms of current and voltage values. DST relates to the battery charging and discharging under the dynamic phase, FUDS relates to EV driving in urban areas and US06 corresponds to high acceleration driving with quick speed fluctuation. It is noticed that SOC estimation results under DST, FUDS, and US06 drive cycles are found to be located very near to the reference SOC values which confirms high robustness and low estimation error. In all drive cycles, SOC error is restricted under ±5%. Moreover, the temperature effects; 0 °C, 25 °C and 45 °C are taken into account under DST, FUDS, and US06 drive cycles, as shown in Figure 14. It is observed from Figure 13 that, iFA-based TDNN achieves RMSE and MAE below 0.8% and above 6%, respectively in DST drive cycle at 0 °C. The RMSE and MAE in the FUDS cycle are slightly higher than the DST cycle, indicating below 0.9% and 0.8%, respectively at 0 °C. US06 cycle obtains the highest error rates among all drive cycles due to high fluctuation of current values with RMSE and MAE below 1% and 0.8% respectively. It is observed that error rates in all drive cycles decrease with the rise of temperature from 0 °C to 45 °C. For instance, RMSE in DST cycles is achieved to be over 0.6% at 25 °C while it is below 0.6% at 45 °C. In the US06 cycle, iFA-based TDNN has MAE below 0.8% at 25 °C; however, MAE drops at 45 °C, indicating under 0.6%.

Table 4. SOC Performance Evaluation in LiNMC Battery.

SOC Method	BPNN-iFA		RBFNN-iFA		ENN-iFA		TDNN-iFA	
Load Profile	**SDT**	**HPPC**	**SDT**	**HPPC**	**SDT**	**HPPC**	**SDT**	**HPPC**
RMSE (%)	0.7775	1.2989	1.0576	2.1121	0.6137	1.0272	0.3084	0.7937
MSE (%)	0.0065	0.0169	0.0112	0.0446	0.0038	0.0106	0.0009	0.0063
MAE (%)	0.6091	0.4222	0.9242	1.6669	0.4620	0.7265	0.1452	0.3283
MAPE (%)	3.7937	7.7595	7.1818	14.3527	4.2617	7.2337	2.1826	5.5247
SD (%)	0.7770	1.2982	1.0556	2.1115	0.6123	0.9818	0.3041	0.7940
SOC error bound (%)	[−2.94, 3.31]	[−5.47, 15.87]	[−2.97, 5.44]	[−10.87, 6.04]	[−1.62, 3.02]	[−5.24, 8.04]	[−1.18, 1.38]	[−3.32, 4.23]

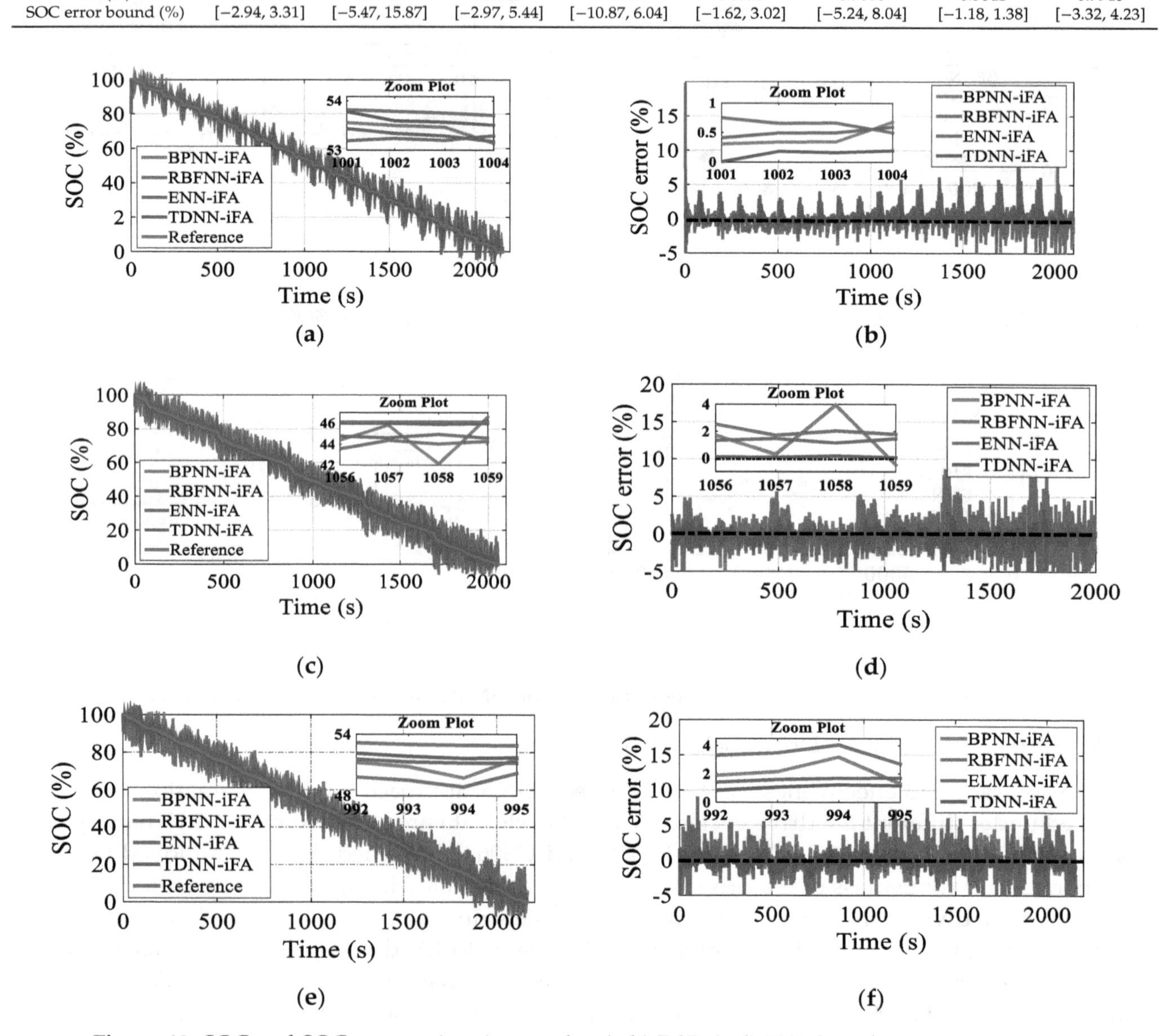

Figure 13. SOC and SOC error estimation results: (**a**,**b**) DST, (**c**,**d**) FUDS, and (**e**,**f**) US06 at 25 °C.

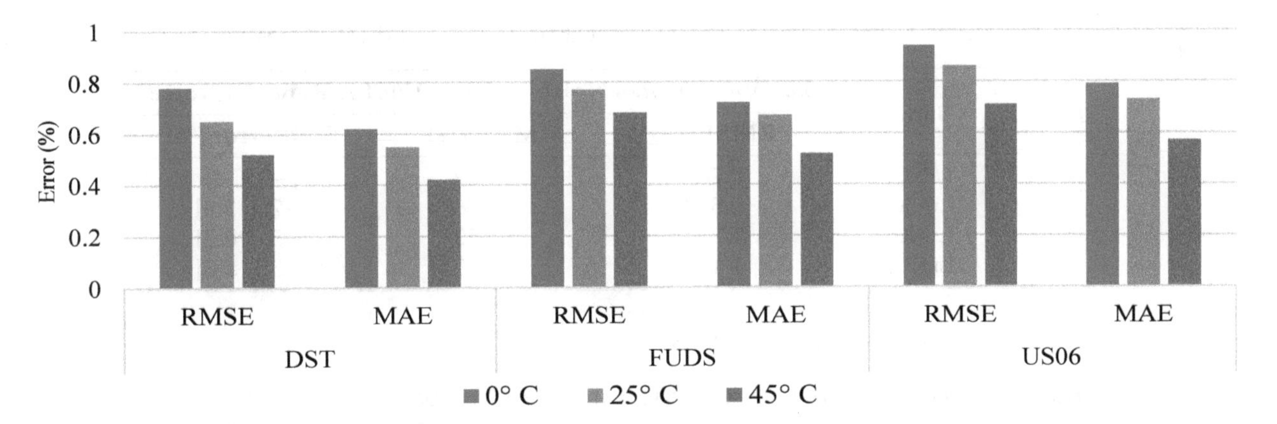

Figure 14. SOC performance comparison under different temperatures and EV drive cycles.

5.4. SOC Robustness Validation against Noise Effects

The robustness of the iFA-based TDNN algorithm for SOC estimation is validated against both bias noise and random noise, as presented in Figure 15 and Table 5. The results reveal that the combination of bias and random noise has a small influence on SOC estimation with regard to RMSE, MAE, and SOC error. It is noticed that the addition of bias and random noises elevates the SOC error bound a bit; however, SOC error rates have stayed inside the reasonable range. For example, iFA-based TDNN achieves the maximum SOC error of 3.5% and 5.8% in SDT and HPPC load profiles, respectively, for the LiNMC battery. In line with that, the proposed method obtains RMSE of 0.558% and 1.112%, respectively. The results are also satisfactory in the LiNCA battery, where the maximum SOC error is reported to be 4% and 6.3% in SDT and HPPC tests, respectively. The aforementioned results prove that the iFA-based TDNN algorithm demonstrates strong robustness against biased noises and random noises toward accurate SOC estimation.

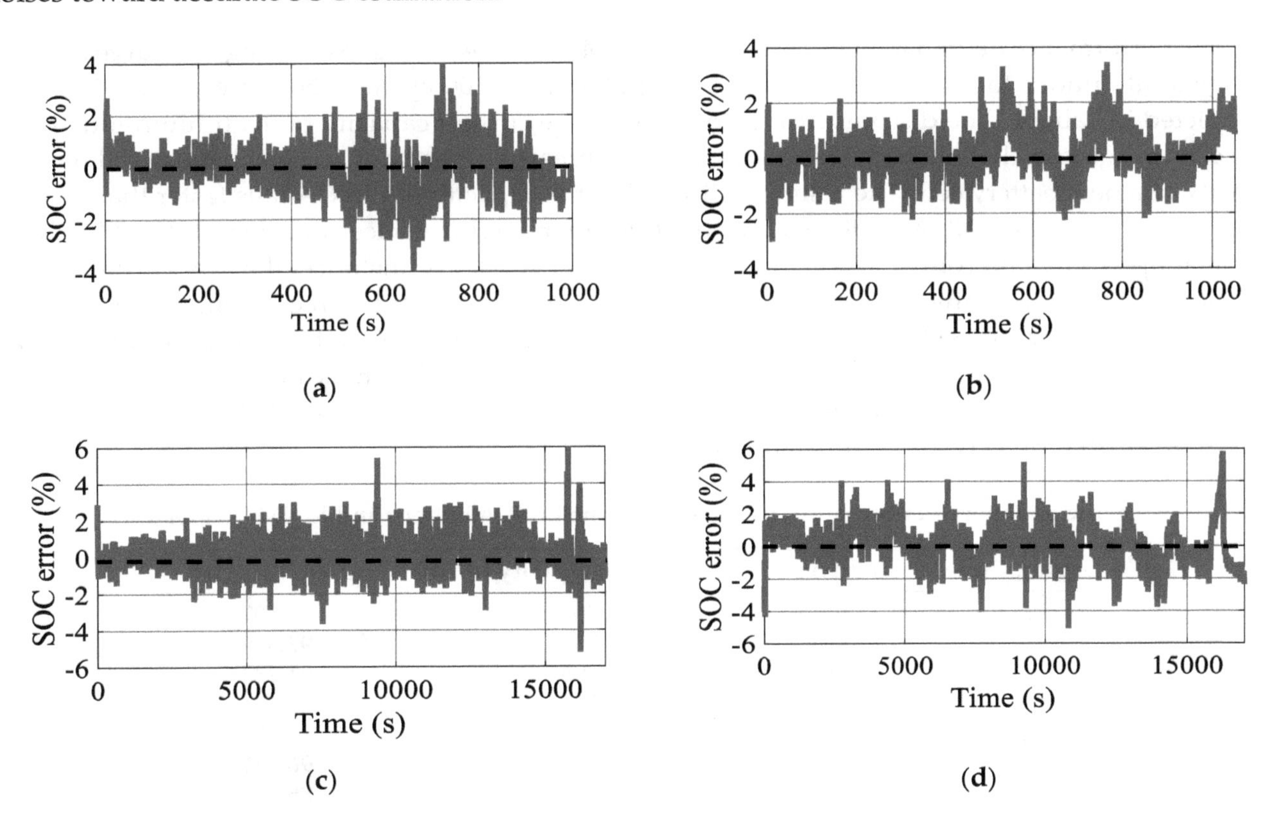

Figure 15. SOC error analysis at 0.01 V/0.1 A bias noise and 0.01 V/ 0.1 A random noise: (**a**) SDT for the LiNCA battery; (**b**) SDT for the LiNMC battery; (**c**) HPPC for the LiNCA battery; and (**d**) HPPC for the LiNMC battery.

Table 5. SOC estimation with noise effect under SDT and HPPC tests.

Test	Battery	0.01 V/0.1 A Bias Noise and 0.01 V/0.1 A Random Noise		
		RMSE (%)	MAE (%)	SOC Error (%)
SDT	LiNCA	0.765	0.482	[−3.9, 4]
	LiNMC	0.558	0.386	[−2.9, 3.5]
HPPC	LiNCA	1.287	0.852	[−5.2, 6.3]
	LiNMC	1.112	0.728	[−5.1, 5.8]

5.5. SOC Robustness Validation against Aging Impacts

The comparative performance between LiNCA and LiNMC batteries is further examined using the aging cycle test. The current and voltage characteristics of LiNCA and LiNMC batteries for one aging cycle are shown in Figure 16.

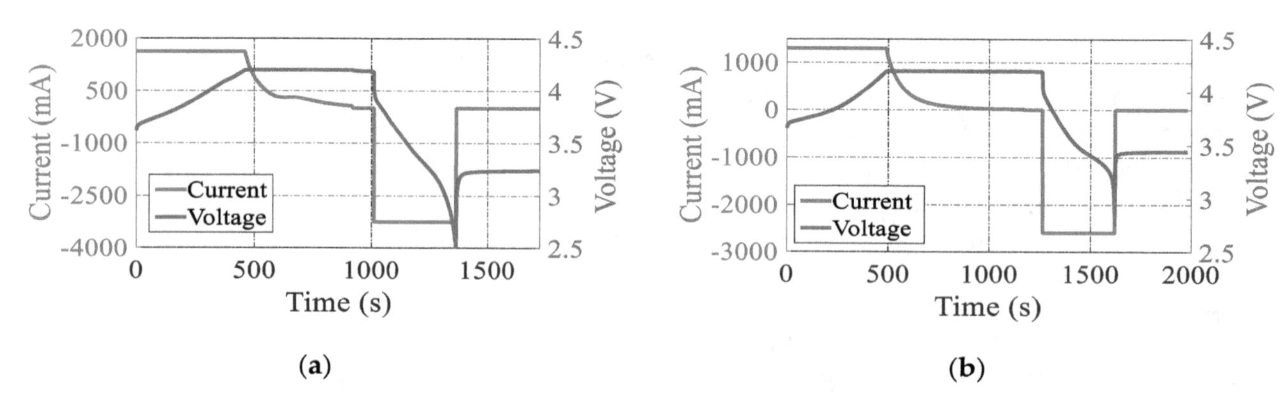

Figure 16. One aging cycle test: (**a**) the LiNCA battery and (**b**) the LiNMC battery.

The performance degradation of LiNCA and LiNMC batteries is assessed using discharge capacity and cycle life under four milestone aging cycles; 50 cycles, 100 cycles, 150 cycles, and 200 cycles, as depicted in Table 6. The discharge capacity denotes the current capacity of the lithium-ion battery after certain aging cycles while cycle life compares the present capacity of an aged battery cell with the capacity of a new battery cell. The results indicate that the capacity of LiNCA falls faster than LiNMC battery, which means LiNCA performs poorly under many aging cycles. In contrast, LiNMC shows strong adaptability and robustness against many aging cycles. For instance, the LiNCA battery has a capacity of 2763 mAh after 200 aging cycles, which is reduced by 9.5% in comparison to the value obtained after 50 aging cycles. However, only a 3.6% decrease in capacity is noted in the LiNMC battery. The results of the cycle life are also satisfactory in the LiNMC battery under many aging cycles. For example, LiNMC has a cycle life of 95.756% after 200 aging cycles; nevertheless, the cycle life of LiNCA is reduced quickly and estimated to be 85.923%.

Table 6. Battery performance degradation in the LiNCA battery and LiNMC battery.

Aging Cycles	Battery	Discharge Capacity (mAh)	Cycle Life (%)
50	LiNCA	3052	95.107
	LiNMC	2515	97.889
100	LiNCA	2951	91.282
	LiNMC	2477	97.231
150	LiNCA	2850	88.629
	LiNMC	2460	96.931
200	LiNCA	2763	85.923
	LiNMC	2425	95.756

At first, the LiNCA and LiNMC batteries are cycled for a definite number of cycles. Afte the completion of the aging cycle test, the lithium-ion battery is loaded with the HPPC test. The performance of the TDNN-iFA model is tested with the HPPC experimental dataset of aged LiNCA and LiNMC batteries for 50, 100, 150, and 200 cycles. The results of SOC and SOC errors under 50, 100, 150, and 200 aging cycles for LiNCA and LiNMC battery are presented in Figures 17 and 18, respectively. Moreover, the results of the RMSE, MAE, and SOC error bound are depicted in Table 7.

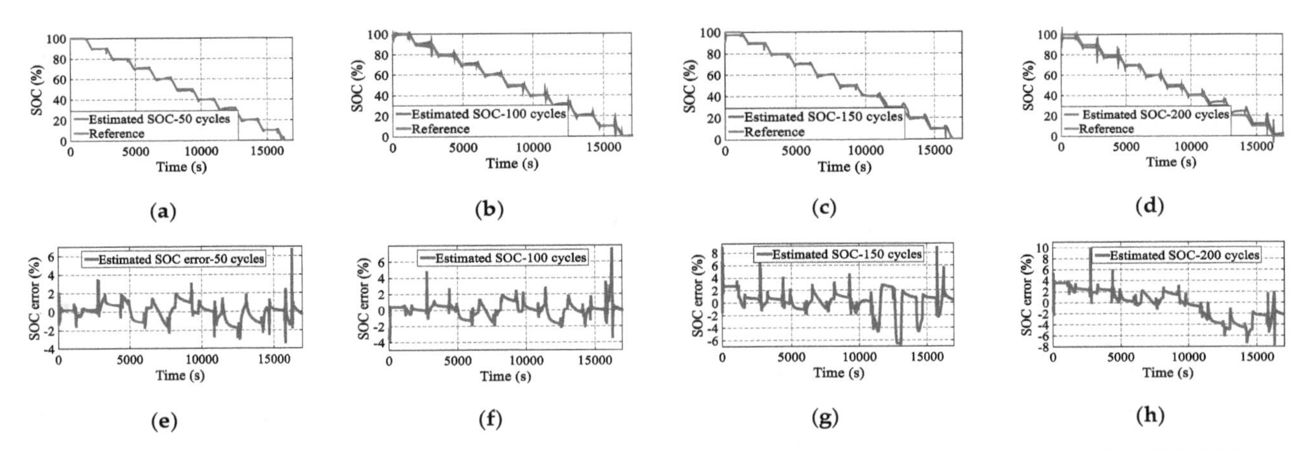

Figure 17. Aging performance for the LiNCA battery: (**a**) SOC estimation under 50 cycles, (**b**) SOC estimation under 100 cycles, (**c**) SOC estimation under 150 cycles, and (**d**) SOC estimation under 200 cycles; (**e**) SOC error under 50 cycles, (**f**) SOC error under 100 cycles, (**g**) SOC error under 150 cycles, and (**h**) SOC error under 200 cycles.

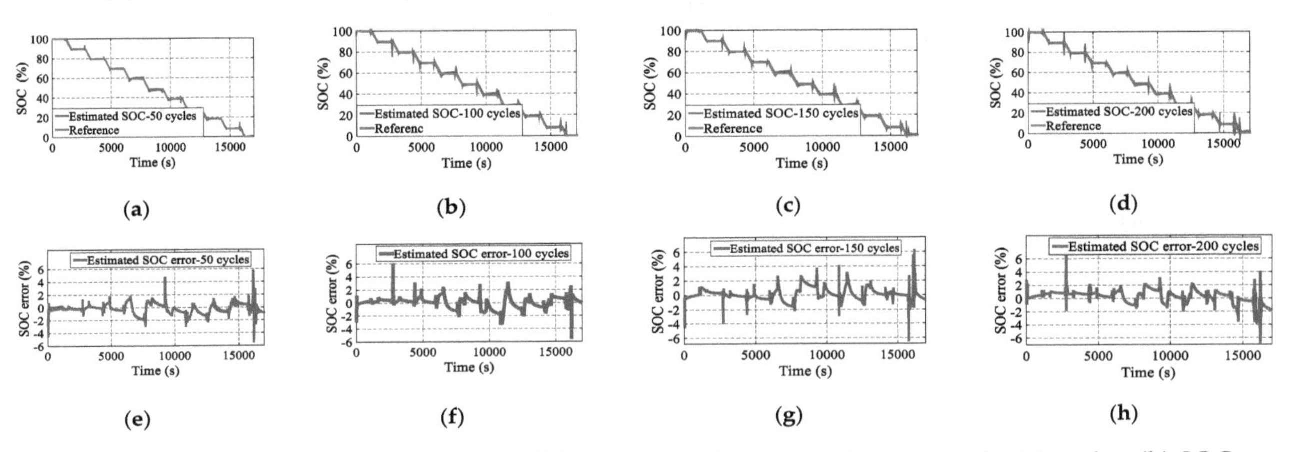

Figure 18. Aging performance for the LiNMC battery: (**a**) SOC estimation under 50 cycles, (**b**) SOC estimation under 100 cycles, (**c**) SOC estimation under 150 cycles, and (**d**) SOC estimation under 200 cycles; (**e**) SOC error under 50 cycles, (**f**) SOC error under 100 cycles, (**g**) SOC error under 150 cycles, and (**h**) SOC error under 200 cycles.

Table 7. Performance assessment of LiNCA and LiNMC batteries under aging cycles.

Aging Cycles	Battery	RMSE (%)	MAE (%)	SOC Error (%)
50	LiNCA	0.933	0.717	[−3.4, 6.7]
	LiNMC	0.821	0.623	[−5.5, 5.7]
100	LiNCA	1.525	0.923	[−3.6, 7.6]
	LiNMC	0.864	0.685	[−5.8, 6]
150	LiNCA	1.878	1.338	[−6.8, 8.8]
	LiNMC	0.927	0.742	[−6.5, 6.2]
200	LiNCA	2.614	1.785	[−7.8, 9.9]
	LiNMC	1.046	0.825	[−6.7, 6.4]

It is observed that the LiNCA battery has higher error rates than the LiNMC battery. For instance, at 50 aging cycles, the LiNCA battery has RMSE of 0.933%, while this is 0.821% in the LiNMC battery. A similar type of results is also noticed in the MAE and SOC error rates. Moreover, the SOC error range is found to be [−3.4%, 6.7%] in the LiNCA battery, while that for LiNMC is [−5.5%, 5.7%]. SOC error rates move up as the aging cycle increases from 50 to 100. For example, in the LiNMC battery, the MAE is calculated to be 0.685% at 100 cycles, while it is 0.623% at 50 cycles. Under 100 aging cycles, the LiNMC battery delivers better solutions than the LiNCA battery, achieving a maximum SOC error and RMSE of 6% and 0.864%, respectively. The assessment of iFA-based TDNN algorithm for SOC estimation is further conducted under 150 aging cycles for LiNCA and LiNMC batteries. The results show that the proposed model tracks the reference SOC precisely for both the LiNCA battery and LiNMC battery. Like previous aging cycles, the LiNMC battery is also dominant over the LiNCA battery in providing a low error rate. The LiNCA battery has a SOC error range of [−6.8%, 8.8%] while that for LiNMC is [−6.5%, 6.2%]. Besides, the LiNMC battery obtains low RMSE and MAE, indicating 0.927% and 0.742%, respectively.

The SOC evaluation of the LiNCA battery and LiNMC battery is further verified under 200 aging cycles. It is observed that the accuracy of the LiNCA battery drops significantly when it is deeply cycled. The results indicate that SOC error range and RMSE is attained to be [−7.8%, 9.9%], and 2.614% respectively, for the LiNCA battery. However, the LiNMC battery delivers outstanding SOC estimation results under 200 aging cycles, with a SOC error limit and RMSE of [−6.7%, 6.4%], and 1.046%, respectively.

To test the LiNCA and LiNMC batteries at a higher number of cycles, the verification of SOC estimation by the proposed algorithm is carried out under 400 and 600 cycles. The results indicate that the cycle life of LiNCA and LiNMC batteries under 400 cycles is 73.685% and 91.597%, respectively. It is reported in the literature that a lithium-ion battery with a cycle life below 80% is declared unsafe and unserviceable [37]. In this regard, LiNCA batteries are no longer usable, and accordingly a replacement is needed. In contrast, the LiNMC battery has a cycle life well above 80% and hence is used for algorithm validation and analysis. At 400 aging cycles, the LiNMC battery achieves RMSE and MAE of 1.327% and 1.128%, which are raised by 21.2% and 26.8%, respectively, compared to the values obtained under 200 aging cycles. Moreover, the LiNMC battery illustrates satisfactory performance under 600 aging cycles, indicating cycle life, RMSE, and MAE values of 88.257%, 1.582%, and 1.368%, respectively. These results show that LiNCA batteries are not appropriate for a high number of aging cycles above 400. Nevertheless, LiNMC batteries are outstanding even if the aging cycles increase to 600, demonstrating high adaptability and robustness for EV operation.

5.6. Comparative Validation with the Existing Methods

Apart from data-driven techniques for SOC validation, the accuracy of the proposed optimized TDNN algorithm is compared with the traditional methods and model-based approaches including OCV, CC, KF, PF, H∞ filter, recursive least square (RLS), and observers, as illustrated in Table 8.

The key implementation factors associated with SOC estimation such as lithium-ion battery chemistry, battery capacity, changing temperature, and validation profiles are taken into consideration to conduct the comparative analysis. The SOC error rates are assessed under a similar type of validation profile to carry out a fair comparative study. For instance, in 1 Coulomb (C) SDT, the proposed method computes MAE of 0.2374% and 0.1452% in the LiNCA battery and LiNMC battery, respectively. Nevertheless, the MAE is found to be more than 2% in unscented Kalman filter (UKF) and H∞ Filter methods under the same load profile. The performance of the iFA-based TDNN for SOC estimation is further examined using different EV drive cycles. It is observed that RMSE is reported below 1% in the proposed algorithm under different EV drive cycles, while that for OCV, the unscented particle filter (UPF), RLS, and proportional integral observer (PIO) is above 1%. In summary, the iFA-based TDNN algorithm has proven to be excellent in terms of accuracy, adaptability, and robustness compared to the existing notable SOC estimation techniques.

Table 8. Comparative performance analysis between the proposed method and the existing methods.

Ref.	Method	Battery Chemistry	Temperature	Experimental Validation Profile	Error Rate
[38]	OCV	1.1 Ah $LiFePO_4$	0 °C to 50 °C at an interval of 10 °C	DST, FUDS	RMSE 5%
[39]	CC	2.3 Ah Lithium-ion cell	Room temperature	C-rates Charging-discharging current	MAE 1.905%
[40]	UKF	24 Ah LiNMC	Room temperature at 25 °C ± 2 °C	1 C SDT	MAE 2.56% Max SOC error 5.36%
[41]	H∞ Filter	2.4 Ah Lithium-ion cell	Constant temperature	1 C SDT	MAE 3.96%
[42]	UPF	10 Ah $LiFePO_4$	−20 °C~50 °C	EV operation condition	RMSE 2.05%
[43]	RLS	90 Ah $LiFePO_4$	−10 °C~50 °C	Urban EV drive cycle	RMSE 2.3% MAE 1.8%
[44]	SMO	5 Ah Lithium polymer battery	Room temperature	1 C SDT	RMSE 1.8%
[45]	PIO	90 Ah Lithium-ion cells	0 °C, 25 °C, 40 °C	DST	RMSE 1.2%
	Proposed Method	3.2 Ah LiNCA	Room Temperature	1 C SDT, HPPC	MAE 0.2374% (SDT) MAE 0.4612% (HPPC)
		2.6 Ah LiNMC			MAE 0.1452% (SDT) MAE 0.3283% (HPPC)
		2.0 Ah LiNCA	0 °C, 25 °C, 45 °C	DST, FUDS, US06	RMSE < 1% MAE < 0.8%

6. Conclusions

An improved data-driven algorithm using TDNN optimized by iFA is proposed to achieve accurate SOC estimation of lithium-ion batteries. The iFA algorithm improves the computational capability of TDNN by choosing the optimum values of UTD and HNs, thus leading to an enhancement in SOC accuracy and robustness. For the verification, the fully developed algorithm is examined with two types of lithium-ion battery cells under two different experiments, namely SDT and HPPC. The TDNN-iFA algorithm achieves excellent SOC estimation results with low RMSE, SD, MSE, MAPE, and MAE, while having a narrow SOC error below 5% in the SDT and HPPC tests. The SOC estimation results under EV drive cycles and variable temperatures also prove the dominance of the iFA-based TDNN algorithm where RMSE is reported under 1%. Besides, the reasonable accuracy under noise and aging impacts illustrates the adaptability and robustness of the TDNN-iFA algorithm against uncertainties. Furthermore, detailed comparative analysis with the existing SOC estimation techniques considering different loads and temperatures demonstrates that the developed optimized data-driven method can provide excellent solutions with respect to the accuracy, efficiency, and robustness. The key information, results, and analysis achieved from this study would be important for the EV automobile industry toward the development of an enhanced SOC estimation approach. Hence, further investigation of SOC estimation via the optimized TDNN method will not only increase the battery life cycle but also confirm a reliable operation of EVs, resulting in a substantial growth of the EV market. Future research work should include the validation of iFA-based TDNN algorithm for SOC estimation of the lithium-ion battery pack in EV applications.

Author Contributions: Conceptualization, M.S.H.L. and M.A.H.; methodology, M.S.H.L.; software, M.S.H.L.; validation, M.S.H.L.; formal analysis, M.S.H.L.; investigation, M.S.H.L. and M.A.H.; resources, M.A.H. and A.H.; data curation, M.S.H.L.; writing—original draft, M.S.H.L.; writing—review and editing, M.A.H., A.H., A.A., M.H.M.S. and K.M.M.; visualization, M.S.H.L.; supervision, M.A.H., A.H., A.A. and M.H.M.S.; project administration, M.A.H.; funding acquisition, M.S.H.L. and M.A.H. All authors have read and agreed to the published version of the manuscript.

References

1. Mongird, K.; Viswanathan, V.; Balducci, P.; Alam, J.; Fotedar, V.; Koritarov, V.; Hadjerioua, B. An Evaluation of Energy Storage Cost and Performance Characteristics. *Energies* **2020**, *13*, 3307. [CrossRef]
2. Stampatori, D.; Raimondi, P.P.; Noussan, M. Li-Ion Batteries: A Review of a Key Technology for Transport Decarbonization. *Energies* **2020**, *13*, 2638. [CrossRef]

3. Huang, B.; Pan, Z.; Su, X.; An, L. Recycling of lithium-ion batteries: Recent advances and perspectives. *J. Power Sources* **2018**, *399*, 274–286. [CrossRef]
4. Yuan, W.-P.; Jeong, S.-M.; Sean, W.-Y.; Chiang, Y.-H. Development of Enhancing Battery Management for Reusing Automotive Lithium-Ion Battery. *Energies* **2020**, *13*, 3306. [CrossRef]
5. Balasingam, B.; Ahmed, M.; Pattipati, K. Battery Management Systems—Challenges and Some Solutions. *Energies* **2020**, *13*, 2825. [CrossRef]
6. Zhang, R.; Xia, B.; Li, B.; Cao, L.; Lai, Y.; Zheng, W.; Wang, H.; Wang, W.; Zhang, R.; Xia, B.; et al. State of the Art of Lithium-Ion Battery SOC Estimation for Electrical Vehicles. *Energies* **2018**, *11*, 1820. [CrossRef]
7. Hussain, S.; Nengroo, S.H.; Zafar, A.; Kim, H.-J.; Alvi, M.J.; Ali, M.U. Towards a Smarter Battery Management System for Electric Vehicle Applications: A Critical Review of Lithium-Ion Battery State of Charge Estimation. *Energies* **2019**, *12*, 446.
8. Hu, X.; Feng, F.; Liu, K.; Zhang, L.; Xie, J.; Liu, B. State estimation for advanced battery management: Key challenges and future trends. *Renew. Sustain. Energy Rev.* **2019**, *114*, 109334. [CrossRef]
9. Zhang, Y.; Song, W.; Lin, S.; Feng, Z. A novel model of the initial state of charge estimation for LiFePO 4 batteries. *J. Power Sources* **2014**, *248*, 1028–1033. [CrossRef]
10. Antonucci, V.; Artale, G.; Brunaccini, G.; Caravello, G.; Cataliotti, A.; Cosentino, V.; Di Cara, D.; Ferraro, M.; Guaiana, S.; Panzavecchia, N.; et al. Li-ion Battery Modeling and State of Charge Estimation Method Including the Hysteresis Effect. *Electronics* **2019**, *8*, 1324. [CrossRef]
11. Lai, X.; Yi, W.; Zheng, Y.; Zhou, L. An All-Region State-of-Charge Estimator Based on Global Particle Swarm Optimization and Improved Extended Kalman Filter for Lithium-Ion Batteries. *Electronics* **2018**, *7*, 321. [CrossRef]
12. Yang, F.; Xing, Y.; Wang, D.; Tsui, K.L. A comparative study of three model-based algorithms for estimating state-of-charge of lithium-ion batteries under a new combined dynamic loading profile. *Appl. Energy* **2016**, *164*, 387–399. [CrossRef]
13. Xiong, R.; Zhang, Y.; He, H.; Zhou, X.; Pecht, M.G. A Double-Scale, Particle-Filtering, Energy State Prediction Algorithm for Lithium-Ion Batteries. *IEEE Trans. Ind. Electron.* **2018**, *65*, 1526–1538. [CrossRef]
14. Liu, C.Z.; Zhu, Q.; Li, L.; Liu, W.Q.; Wang, L.Y.; Xiong, N.; Wang, X.Y. A State of Charge Estimation Method Based on H∞ Observer for Switched Systems of Lithium-Ion Nickel-Manganese-Cobalt Batteries. *IEEE Trans. Ind. Electron.* **2017**, *64*, 8128–8137. [CrossRef]
15. Du, J.; Liu, Z.; Wang, Y.; Wen, C. An adaptive sliding mode observer for lithium-ion battery state of charge and state of health estimation in electric vehicles. *Control Eng. Pract.* **2016**, *54*, 81–90. [CrossRef]
16. Rivera-Barrera, J.; Muñoz-Galeano, N.; Sarmiento-Maldonado, H. SoC Estimation for Lithium-ion Batteries: Review and Future Challenges. *Electronics* **2017**, *6*, 102. [CrossRef]
17. He, W.; Williard, N.; Chen, C.; Pecht, M. State of charge estimation for Li-ion batteries using neural network modeling and unscented Kalman filter-based error cancellation. *Int. J. Electr. Power Energy Syst.* **2014**, *62*, 783–791. [CrossRef]
18. Haddad Zarif, M.; Charkhgard, M.; Alfi, A. Hybrid state of charge estimation for lithium-ion batteries: Design and implementation. *IET Power Electron.* **2014**, *7*, 2758–2764.
19. Lipu, M.S.H.; Hannan, M.A.; Hussain, A.; Saad, M.H.M.; Ayob, A.; Uddin, M. Extreme Learning Machine Model for State of Charge Estimation of Lithium-ion battery Using Gravitational Search Algorithm. *IEEE Trans. Ind. Appl.* **2019**, *55*, 4225–4234. [CrossRef]
20. Cui, D.; Xia, B.; Zhang, R.; Sun, Z.; Lao, Z.; Wang, W.; Sun, W.; Lai, Y.; Wang, M.; Cui, D.; et al. A Novel Intelligent Method for the State of Charge Estimation of Lithium-Ion Batteries Using a Discrete Wavelet Transform-Based Wavelet Neural Network. *Energies* **2018**, *11*, 995. [CrossRef]
21. Liu, K.; Li, Y.; Hu, X.; Lucu, M.; Widanage, W.D. Gaussian Process Regression with Automatic Relevance Determination Kernel for Calendar Aging Prediction of Lithium-Ion Batteries. *IEEE Trans. Ind. Inform.* **2020**, *16*, 3767–3777. [CrossRef]
22. Lipu, M.S.H.; Hannan, M.A.; Hussain, A.; Saad, M.H.M.; Ayob, A.; Blaabjerg, F. State of Charge Estimation for Lithium-ion Battery Using Recurrent NARX Neural Network Model Based Lighting Search Algorithm. *IEEE Access* **2018**, *6*, 28150–28161. [CrossRef]
23. Bian, C.; He, H.; Yang, S. Stacked bidirectional long short-term memory networks for state-of-charge estimation of lithium-ion batteries. *Energy* **2020**, *191*, 116538. [CrossRef]
24. Xiao, B.; Liu, Y.; Xiao, B. Accurate state-of-charge estimation approach for lithium-ion batteries by gated recurrent unit with ensemble optimizer. *IEEE Access* **2019**, *7*, 54192–54202. [CrossRef]

25. Awadallah, M.A.; Venkatesh, B. Accuracy improvement of SOC estimation in lithium-ion batteries. *J. Energy Storage* **2016**, *6*, 95–104. [CrossRef]
26. Chaoui, H.; Ibe-Ekeocha, C.C.; Gualous, H. Aging prediction and state of charge estimation of a LiFePO4 battery using input time-delayed neural networks. *Electr. Power Syst. Res.* **2017**, *146*, 189–197. [CrossRef]
27. Shao, Y.E.; Lin, S.-C. Using a Time Delay Neural Network Approach to Diagnose the Out-of-Control Signals for a Multivariate Normal Process with Variance Shifts. *Mathematics* **2019**, *7*, 959. [CrossRef]
28. Hannan, M.A.; Lipu, M.S.H.; Hussain, A.; Saad, M.H.; Ayob, A. Neural Network Approach for Estimating State of Charge of Lithium-ion Battery Using Backtracking Search Algorithm. *IEEE Access* **2018**, *6*, 10069–10079. [CrossRef]
29. Hagan, M.T.; Menhaj, M.B. Training feedforward networks with the Marquardt algorithm. *IEEE Trans. Neural Netw.* **1994**, *5*, 989–993. [CrossRef] [PubMed]
30. Lv, C.; Xing, Y.; Zhang, J.; Na, X.; Li, Y.; Liu, T.; Cao, D.; Wang, F.-Y. Levenberg–Marquardt Backpropagation Training of Multilayer Neural Networks for State Estimation of a Safety-Critical Cyber-Physical System. *IEEE Trans. Ind. Inform.* **2018**, *14*, 3436–3446. [CrossRef]
31. Xia, X.; Gui, L.; He, G.; Xie, C.; Wei, B.; Xing, Y.; Wu, R.; Tang, Y. A hybrid optimizer based on firefly algorithm and particle swarm optimization algorithm. *J. Comput. Sci.* **2018**, *26*, 488–500. [CrossRef]
32. Xu, S.S.-D.; Huang, H.-C.; Kung, Y.-C.; Lin, S.-K. Collision-Free Fuzzy Formation Control of Swarm Robotic Cyber-Physical Systems Using a Robust Orthogonal Firefly Algorithm. *IEEE Access* **2019**, *7*, 9205–9214. [CrossRef]
33. Ball, A.K.; Roy, S.S.; Kisku, D.R.; Murmu, N.C.; Dos Santos Coelho, L. Optimization of drop ejection frequency in EHD inkjet printing system using an improved Firefly Algorithm. *Appl. Soft Comput. J.* **2020**, *94*, 106438. [CrossRef]
34. Wang, Y.; Zhang, C.; Chen, Z. A method for joint estimation of state-of-charge and available energy of LiFePO4batteries. *Appl. Energy* **2014**, *135*, 81–87. [CrossRef]
35. Hannan, M.A.; Lipu, M.S.H.; Hussain, A.; Ker, P.J.; Mahlia, T.M.I.; Mansor, M.; Ayob, A.; Saad, M.H.; Dong, Z.Y. Toward Enhanced State of Charge Estimation of Lithium-ion Batteries Using Optimized Machine Learning Techniques. *Sci. Rep.* **2020**, *10*, 4687. [CrossRef]
36. CALCE. Lithium-ion Battery Experimental Data. Available online: https://web.calce.umd.edu/batteries/data.htm (accessed on 23 August 2018).
37. Naha, A.; Han, S.; Agarwal, S.; Guha, A.; Khandelwal, A.; Tagade, P.; Hariharan, K.S.; Kolake, S.M.; Yoon, J.; Oh, B. An Incremental Voltage Difference Based Technique for Online State of Health Estimation of Li-ion Batteries. *Sci. Rep.* **2020**, *10*, 9526. [CrossRef] [PubMed]
38. Xing, Y.; He, W.; Pecht, M.; Tsui, K.L. State of charge estimation of lithium-ion batteries using the open-circuit voltage at various ambient temperatures. *Appl. Energy* **2014**, *113*, 106–115. [CrossRef]
39. Wu, T.-H.; Moo, C.-S.; Wu, T.-H.; Moo, C.-S. State-of-Charge Estimation with State-of-Health Calibration for Lithium-Ion Batteries. *Energies* **2017**, *10*, 987.
40. Chen, Y.; Huang, D.; Zhu, Q.; Liu, W.; Liu, C.; Xiong, N. A new state of charge estimation algorithm for lithium-ion batteries based on the fractional unscented kalman filter. *Energies* **2017**, *10*, 1313. [CrossRef]
41. Zhu, Q.; Xiong, N.; Yang, M.L.; Huang, R.S.; Hu, G. Di State of charge estimation for lithium-ion battery based on nonlinear observer: An H ∞ method. *Energies* **2017**, *10*, 679. [CrossRef]
42. He, Y.; Liu, X.; Zhang, C.; Chen, Z. A new model for State-of-Charge (SOC) estimation for high-power Li-ion batteries. *Appl. Energy* **2013**, *101*, 808–814. [CrossRef]
43. Duong, V.H.; Bastawrous, H.A.; See, K.W. Accurate approach to the temperature effect on state of charge estimation in the LiFePO4 battery under dynamic load operation. *Appl. Energy* **2017**, *204*, 560–571. [CrossRef]
44. Chen, X.; Shen, W.; Cao, Z.; Kapoor, A. A novel approach for state of charge estimation based on adaptive switching gain sliding mode observer in electric vehicles. *J. Power Sources* **2014**, *246*, 667–678. [CrossRef]
45. Zheng, L.; Zhang, L.; Zhu, J.; Wang, G.; Jiang, J. Co-estimation of state-of-charge, capacity and resistance for lithium-ion batteries based on a high-fidelity electrochemical model. *Appl. Energy* **2016**, *180*, 424–434. [CrossRef]

3

Study of a Bidirectional Power Converter Integrated with Battery/Ultracapacitor Dual-Energy Storage

Ching-Ming Lai [1], Jiashen Teh [2,*], Yuan-Chih Lin [3] and Yitao Liu [4,*]

1 Department of Electrical Engineering, National Chung Hsing University, Taichung 420, Taiwan; pecmlai@gmail.com
2 School of Electrical and Electronic Engineering, Engineering Campus, Universiti Sains Malaysia (USM), Nibong Tebal 14300, Penang, Malaysia
3 Department of Electrical Engineering, National Taiwan University, Taipei 106, Taiwan; gero.lin1980@gmail.com
4 College of Mechatronics and Control Engineering, Shenzhen University, Shenzhen 518060, China
* Correspondence: jiashenteh@usm.my (J.T.); liuyt@szu.edu.cn (Y.L.)

Abstract: A patented bidirectional power converter was studied as an interface to connect the DC-bus of driving inverter, battery energy storage (BES), and ultracapacitor (UC) to solve the problem that the driving motor damages the battery life during acceleration and deceleration in electric vehicles (EVs). The proposed concept was to adopt a multiport switch to control the power flow and achieve the different operating mode transitions for the better utilization of energy. In addition, in order to improve the conversion efficiency, the proposed converter used a coupled inductor and interleaved-pulse-width-modulation (IPWM) control to achieve a high voltage conversion ratio (i.e., bidirectional high step-up/down conversion characteristics). This study discussed the steady-state operation and characteristic analysis of the proposed converter. Finally, a 500 W power converter prototype with specifications of 72 V DC-bus, 24 V BES, and 48 V UC was built, and the feasibility was verified by simulation and experiment results. The highest efficiency points of the realized prototype were 97.4%, 95.5%, 97.2%, 97.1%, and 95.3% for the UC charge, battery charge, UC discharge, the dual-energy in series discharge, and battery discharge modes, respectively.

Keywords: battery/ultracapacitor; dual-energy; bidirectional power converter; electric vehicles

---

1. Introduction

Electric vehicle (EV) technologies are currently being developed to lessen environmental impact and overcome shortages of fossil fuel [1–10]. The typical power configuration of pure electric vehicle (EV) contains four major parts: the battery energy storage (BES), the power converter, the driving inverter of motor, and the energy management system (EMS) [6,8,10]. Among them, BES is the most critical component, which can directly affect the life and endurance of the EV, driving efficiency, and system performance. In general, the power will be drawn rapidly from the BES during the vehicle acceleration, subsequently causing BES output current and temperature to rise quickly. Moreover, the driving inverter is prone to generate less stable pulse currents for the BES during deceleration [9]. Such long-term use not only causes damage to the external body of the battery but also excessively charges and discharges the BES, which eventually will shorten the lifespan of the BES, specifically in high power applications. Although it is feasible to size up the BES for high power demands, the high price of the overall system still remains an issue. Possible solutions may be to select an ultracapacitor (UC), to assist BES, forming a "hybrid energy" system as shown in Figure 1 for EVs [11,12].

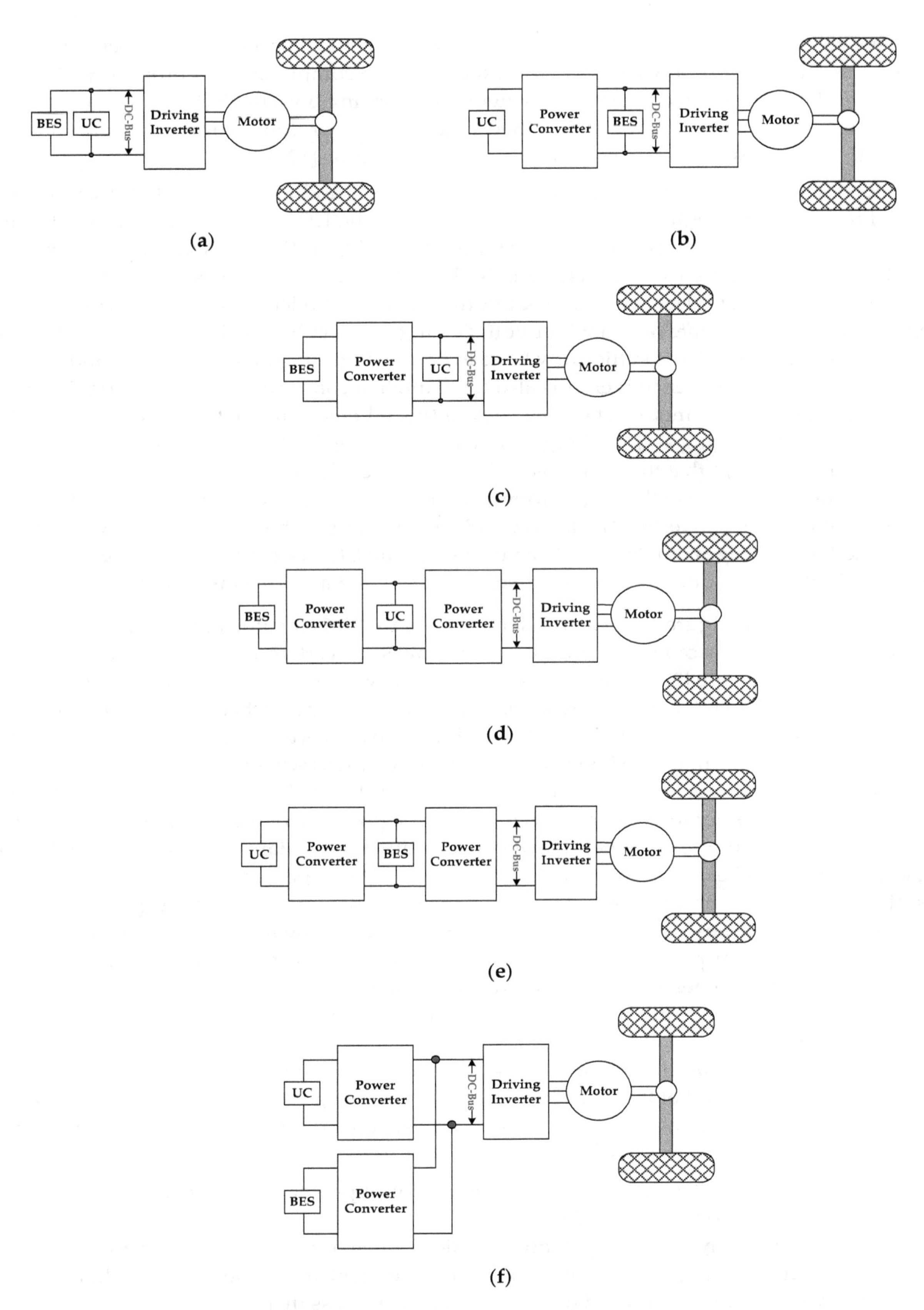

Figure 1. Several schemes of interfacing battery energy storage (BES) and ultracapacitor (UC) to the DC-bus in electric vehicle (EV) power train: (**a**) Directly parallel hybrid scheme; (**b**) UC/BES scheme; (**c**) BES/UC scheme; (**d**) type-I of cascaded scheme; (**e**) type-II of cascaded scheme; (**f**) multiple converter parallel scheme.

UC has high-power density, long cycle life, quick dynamic response, but low-energy-density, which are opposite toward BES. Hence, it should exploit the complementary properties of both the UC and the BES [12]. Several conventional schemes integrating both the BES and UC are shown in Figure 1. These schemes have all been designed to control power flows, supply specific voltages to loads, and to reduce design cost, mass, and power consumptions [12–15].

Figure 1a shows the most basic parallel scheme of the BES and UC, with the latter serving as the low-pass filter [16]. Although simple, the energy stored in the UC is not utilized effectively due to the absence of power converters. The slightly more robust Figure 1b shows that a power converter is added in between the BES and UC [17–20]. In this scheme, the BES is connected directly to the DC-bus instead of the UC. The power output of the UC is controlled by the power converter, and this enables the UC to operate over a wider voltage range than in Figure 1a. Due to this, the power rating of the converter has to be sufficiently large to handle high surges of power demands from the UC. The purpose of the power converter is also to maintain a constant voltage value on the DC-bus during the operation of the motor. The drawback of this scheme is that the BES is exposed to large fluctuations of high charging and discharging current, resulting in its reduced lifetime. Figure 1c is similar to Figure 1b except that the positions of the BES and UC have been swapped [21]. Due to this, the BES is no longer exposed to the large current fluctuations. The power output from the BES is now controlled by the power converter. The main disadvantage of this scheme is that the DC-bus voltage is exposed to large voltages as it is directly connected to the UC. As a result, the power converter is exposed to a high risk of suffering adverse losses, especially in harsh driving conditions.

All the schemes in Figure 1a–c clearly demonstrate that it is insufficient to use only one or no power converter. Hence, cascaded schemes using two power converters, as shown in Figure 1d,e, have also been considered before [22,23]. In these two schemes, two converters decouple the BES and UC from the DC-bus. The circuit, in Figure 1d, is also known as the "type-I scheme" where an extra power converter is added in between the UC and the DC-bus. The converter that is located in between the BES and UC is rated according to the power rating of the BES. This scheme creates more losses for the higher rated converter that is located in between the UC and the DC-bus due to the fluctuations of the UC output voltage. In order to overcome this problem, the positions of the BES and UC are swapped, as shown in Figure 1e (type-II scheme). However, it is difficult to balance the BES cell due to it now being located at the higher voltage terminal. Although both the type-I and type-II schemes are more robust than all the previous designs that use only one or no power converter, the power losses and design costs of the schemes are increased substantially owing to the multi-stage energy conversion processes in the vehicular power train. Besides that, only one power converter is connected to the DC-bus in both of these schemes. An outage in one of the power converters will lead to the loss of the power-control function. An alternative is to employ the scheme in Figure 1f, where the power converters are connected in parallel and directly to the DC-bus [13,24–28]. In this scheme, the power converters have the same output voltage, and the power flow of both the energy sources (BES and UC) are not affected by the output of the other converter. Consequently, this scheme can operate in various modes [28]. But, the fully power-rated converters are needed, and the cost of this scheme is higher than all the aforementioned schemes.

In order to reduce the overall system cost, a multi-input power converter scheme is studied, as shown in Figure 2, into the EV system [29–33].

Multi-input power converters are potential solutions when multiple energy sources with different voltage levels (battery voltage ≠ UC voltage ≠ DC-bus voltage) and/or power capabilities are to be combined and yet maintain a regulated output load voltage across them.

Using multi-input power converters, it is possible to apply a different power control command for each input source. In order to reduce the cost and weight and enhance the overall performance of the hybrid energy storage system, the multi-input power converter scheme was chosen in this paper and further investigated.

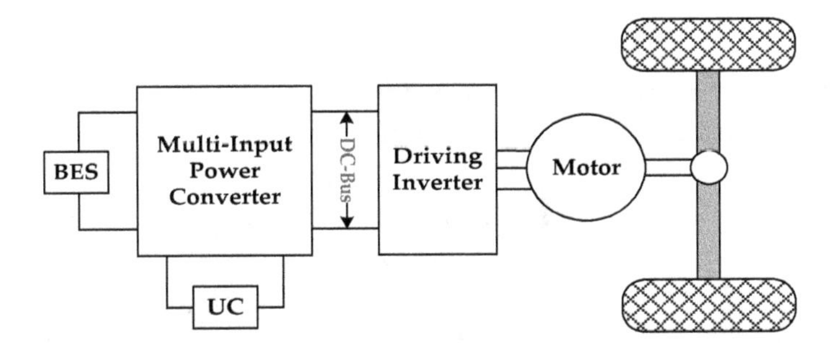

Figure 2. Multi-input power converter interfacing BES and UC to the DC-bus in the EV power train.

In [29], a multi-input power converter topology has been proposed to combine various input energy sources in parallel by using a single-pole triple-throw switch. The major limitations of parallel-connected source topologies are: input source voltage should be asymmetric, and only one input source can supply power to the load at a time to avoid the power coupling effect.

In [30], a single-inductor unidirectional multi-input power converter has been presented, which can operate in buck, boost, or buck-boost modes. To realize the bidirectional power flow mode, all the diodes must be replaced by unidirectional switches, which increase the number of switches.

In [31], a DC-bus interfacing three-port converter with a simple topology and no electrical isolation has been proposed, but it cannot cope with a wide operating voltage ratio; energy storage devices connected to different ports must have a similar operating voltage, and this constricts the application.

In [32], a modular multi-input power converter has been presented to integrate the basic buck-boost circuit and a shared DC-bus. It is a very simple approach to integrate multiple converters into a single unit. However, it has limited static voltage gains, resulting in a narrow voltage range and a low voltage difference between the high- and low-side ports. Besides, since only a few circuit elements are shared among multiple converters, the benefits of the integration are limited.

In [33], a two-phase multi-input converter with a high voltage conversion ratio has been proposed as an interface between dual-energy storage sources. Due to the intrinsic automatic current balance characteristic, the currents of two energy sources are theoretically identical; it indicates that the high power capability of UC cannot be utilized, and the applications of the proposed converter would be limited.

By conducting a research literature review of [29–36], in this paper, a bidirectional power converter integrated BES/UC dual-energy storage was proposed, which had the capability to perform forward power transmission and reverse energy recovery.

First, the proposed converter used a multiport switch to change the different operating modes and to improve the energy utilization of UC and increase battery life.

Second, it was also integrated with interleaved-pulse-width-modulation (IPWM) control to increase power density and reduce bidirectional current ripples, which makes power delivery more reliable.

Third, the proposed converter also used a coupled inductor technique instead of a general single-winding inductor to achieve high voltage conversion ratio and high power density for bidirectional power conversion.

Finally, the steady-state operation and characteristic analysis of the proposed converter were described, validated using simulation and experimentation of a 500 W power converter prototype with specifications of 72 V DC-bus, 24 V BES, and 48 V UC.

The summarized main features of the proposed converter were its ability to:

(1) interface more than two energy sources of different voltage levels,
(2) control power flow between the DC-bus and the two low-voltage energy sources,
(3) control power flow from either the UC or BES or both,
(4) enhance static voltage gain and reduce switch voltage stress, and
(5) possess a reasonable duty cycle and produce a wide voltage difference between its high- and low-side ports.

2. Converter Operating Principles

Figure 3 shows the architecture of the proposed converter integrated with dual-energy storage. The power devices (S_1~S_4) are the multiport switch used to control the power flow between the battery/UC dual-energy and DC-bus. To achieve the high conversion efficiency, the design concept for the converter are based on multi-phase operation and switch stress reduction as (1) the power devices (Q_1~Q_4) are designed to use IPWM control to reduce current stress and ripple on the switch, (2) two-phase coupled inductors T_1 and T_2 are integrated into the bidirectional power converter with high turns ratio to reduce the undesirable duty ratio and conduction loss of metal-oxide-semiconductor field-effect transistors (MOSFETs).

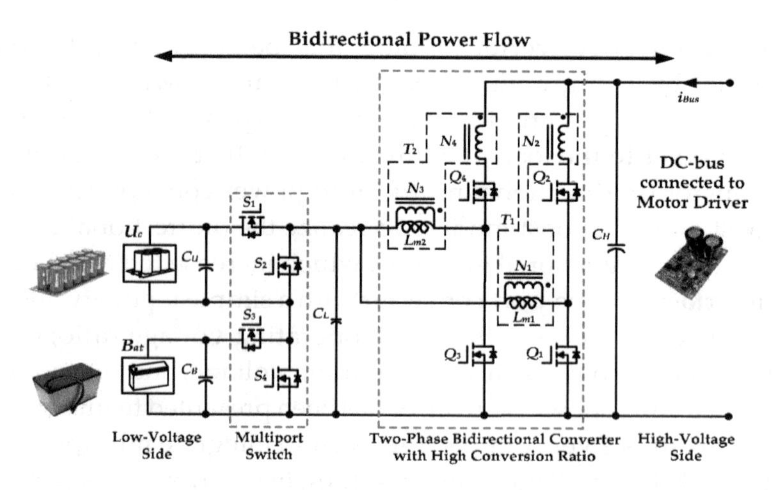

Figure 3. The proposed bidirectional power converter architecture.

2.1. Multiport Switch

Figure 4 shows the equivalent circuits of the multiport switch of the proposed converter under different operating modes. For the converter operating in the UC charge or discharge mode, the multiport switches S_1, S_4 are turned on, and S_2, S_3 are turned off. The equivalent circuit of this condition is shown in Figure 4a. It is shown that the bidirectional energy delivery between the UC and the DC-bus can be achieved. For the converter operating under the battery charge mode or discharge mode, the multiport switches S_2, S_3 are turned on, and S_1, S_4 are turned off. Under this condition, the corresponding equivalent circuit is shown in Figure 4b. The figure shows that the bidirectional energy delivery between the battery and the DC-bus can be achieved. For the converter operating under the dual-energy in series discharge mode, the multiport switches S_1, S_3 are turned on, and S_2, S_4 are turned off. The battery/UC dual-energy delivers the energy to DC-bus, and its equivalent circuit is shown in Figure 4c.

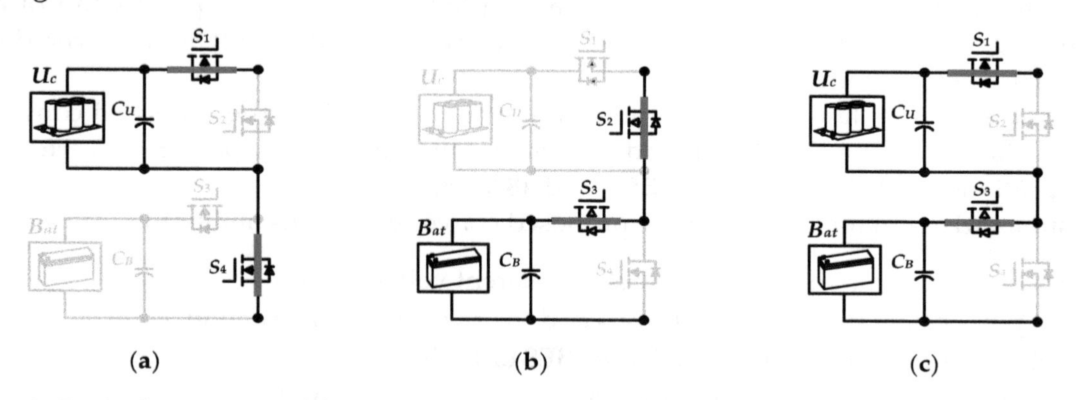

Figure 4. Equivalent circuits of the multiport switch under different operating modes. (**a**) UC charge mode or discharge mode. (**b**) Battery charge mode or discharge mode. (**c**) Battery/UC dual-energy in series discharge mode.

2.2. Operating Principle of the Proposed Converter

Figure 5 shows the equivalent circuits of the different states for the proposed converter, where V_H represents the high-side voltage for the DC-bus, and V_L represents low-side voltage for UC, battery, or battery/UC dual-energy in series modes.

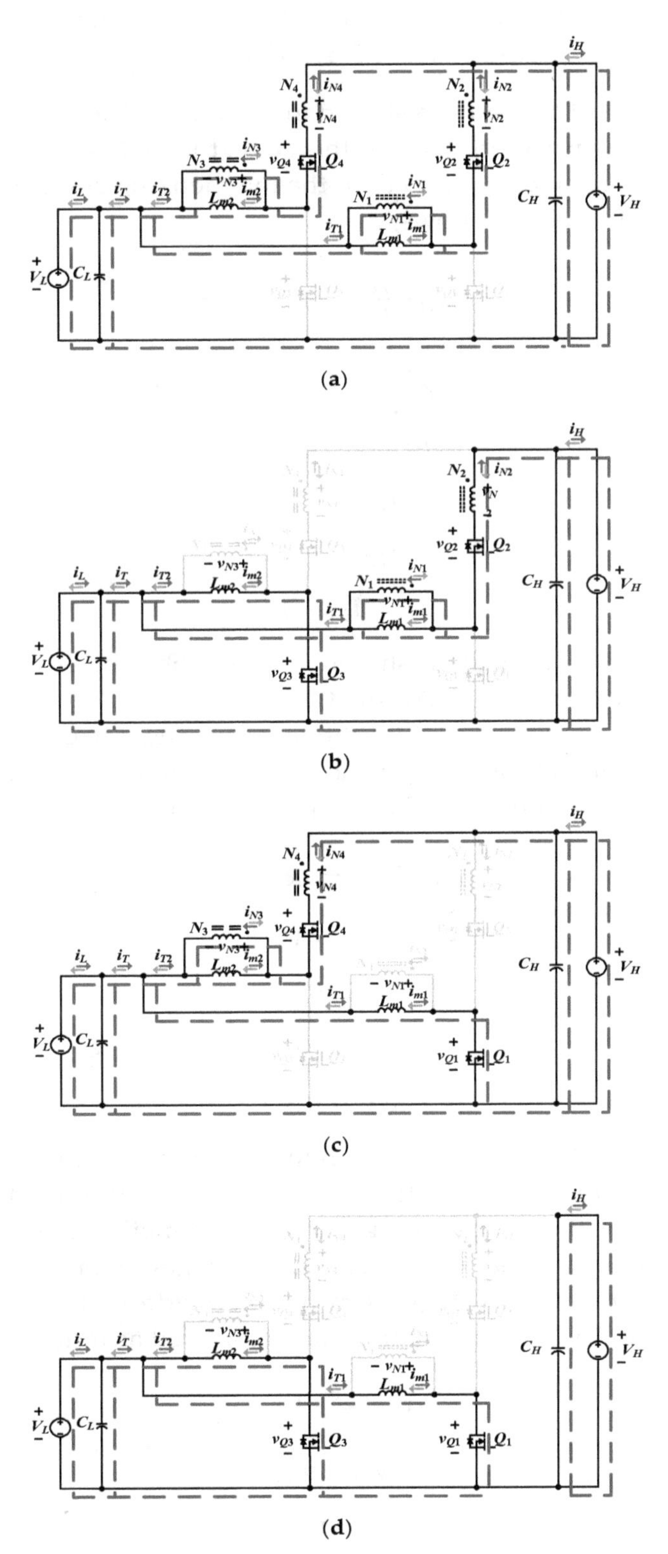

Figure 5. Equivalent circuits of the proposed converter. (**a**) State 1: Q_2, Q_4 on, and Q_1, Q_3 off. (**b**) State 2: Q_2, Q_3 on, and Q_1, Q_4 off. (**c**) State 3: Q_1, Q_4 on, and Q_2, Q_3 off. (**d**) State 4: Q_1, Q_3 on, and Q_2, Q_4 off. (The arrows in **green** indicate the charge mode, and the arrows in **red** indicate the discharge mode.).

The assumptions are made in analyzing the proposed converter:

(1) the converter operates in continuous conduction mode (CCM);
(2) characteristic of the two-phase coupled inductors T_1 and T_2 are the same, i.e., $L_{m1} = L_{m2}$, $i_{m1} = i_{m2}$ and $n = N_2/N_1 = N_4/N_3$;
(3) all voltages and currents in the circuits are periodic in steady-state condition; for simplicity, it is assumed that all the components in Figure 3 are idealized.

State 1. The equivalent circuit of this state is shown in Figure 5a. The power switches Q_2 and Q_4 are turned on, and Q_1 and Q_3 are turned off. During this state, the high-side voltage V_H stores energy to the magnetizing inductance L_{m1} and L_{m2}, and then the magnetizing currents i_{m1}, i_{m2} increase linearly. The circuit equations are expressed as follows,

$$v_{N1} = L_{m1}\frac{di_{m1}}{dt} = V_H - v_{N2} - V_L \quad (1)$$

$$v_{N3} = L_{m2}\frac{di_{m2}}{dt} = V_H - v_{N4} - V_L \quad (2)$$

$$i_H = i_{N2} + i_{N4} \quad (3)$$

$$i_{T1} = i_{N2} \quad (4)$$

$$i_{T2} = i_{N4} \quad (5)$$

$$i_T = i_{T1} + i_{T2} \quad (6)$$

State 2. The equivalent circuit of this state is shown in Figure 5b. The power switches Q_2 and Q_3 are turned on, and Q_1 and Q_4 are turned off. At this time, the high-side voltage V_H continues to store energy to the magnetizing inductance L_{m1}, and the magnetizing current i_{m1} increases linearly. The energy stored in the magnetizing inductor L_{m2} is now released to the low-side energy device, and the magnetizing current i_{m2} decreases linearly. The circuit equations are expressed as follows,

$$v_{N1} = V_H - v_{N2} - V_L \quad (7)$$

$$v_{N3} = -V_L \quad (8)$$

$$i_H = i_{N2} \quad (9)$$

$$i_{T1} = i_{N2} \quad (10)$$

$$i_{T2} = i_{m2} \quad (11)$$

State 3. The equivalent circuit of this state is shown in Figure 5c. The power switches Q_1 and Q_4 are turned on, and Q_2 and Q_3 are turned off. At this time, the energy stored in the magnetizing inductor L_{m1} is now released to the low-side energy storage, and the magnetizing current i_{m1} decreases linearly. The voltage across v_{N1} of the magnetizing inductor L_{m1} is negative of the low-side voltage V_L. The magnetizing inductor L_{m2} draws the energy from the high-side voltage V_H, and the magnetizing current i_{m2} increases linearly. The circuit equations are expressed as follows,

$$v_{N1} = -V_L \quad (12)$$

$$v_{N3} = V_H - v_{N4} - V_L \quad (13)$$

$$i_H = i_{N4} \quad (14)$$

$$i_{T1} = i_{m1} \quad (15)$$

$$i_{T2} = i_{N4}. \quad (16)$$

State 4. The equivalent circuit of this state is shown in Figure 5d. The power switches Q_1 and Q_3 are turned on, and Q_2 and Q_4 are turned off. At this time, the energy stored in the magnetizing inductor L_{m1} and L_{m2} is now released to the low-side energy storage, and the magnetizing currents i_{m1} and i_{m2} decrease linearly. The voltage across v_{N1} and v_{N3} of the magnetizing inductor L_{m1} and L_{m2} is negative of the low-side voltage V_L. The circuit equations are expressed as follows,

$$v_{N1} = -V_L \tag{17}$$

$$v_{N3} = -V_L \tag{18}$$

$$i_H = 0 \tag{19}$$

$$i_{T1} = i_{m1} \tag{20}$$

$$i_{T2} = i_{m2} \tag{21}$$

Considering the different duty ratio conditions in the charge mode and discharge mode, the operating state flow of each condition during a switching period is summarized as follows.

Charge Mode (D_c < 0.5)

State 2 → State 4 → State 3 → State 4

Charge Mode (D_c = 0.5)

State 2 → State 3

Charge Mode (D_c > 0.5)

State 2 → State 1 → State 3 → State 1

Discharge Mode (D_d < 0.5)

State 3 → State 1 → State 2 → State 1

Discharge Mode (D_d = 0.5)

State 3 → State 2

Discharge Mode (D_d > 0.5)

State 3 → State 4 → State 2 → State 4

As mentioned above, D_c is the duty ratio of switch Q_2 and Q_4 for the charge mode, and D_d is the duty ratio of switch Q_1 and Q_3 for the discharge mode.

When the proposed converter operates with the duty ratio of 0.5 in the charge or discharge mode (i.e., $D_c = D_d = 0.5$), the only two operating states of the proposed converter are produced.

When the proposed converter operates in the charge mode with duty ratio $D_c > 0.5$, the operation state in a switching period is the same as the discharge mode with $D_d < 0.5$, and only the reverse current direction is considered.

Similarly, when the proposed converter operates in the discharge mode with the duty ratio $D_d > 0.5$, the operation state in the switching period is the same as the charge mode with $D_c < 0.5$, and only the reverse current direction is considered. Figure 6 shows the key waveforms of the proposed converter in the charge mode with $D_c < 0.5$, and in the discharge mode with $D_d < 0.5$, respectively.

The time intervals of Figure 6 are described as

Charge Mode (D_c < 0.5)
$[t_0 < t \leq t_1]$: state 2; $[t_1 < t \leq t_2]$: state 4; $[t_2 < t \leq t_3]$: state 3; $[t_3 < t \leq t_4]$: state 4.
Discharge Mode (D_d < 0.5)
$[t_0 < t \leq t_1]$: state 3; $[t_1 < t \leq t_2]$: state 1; $[t_2 < t \leq t_3]$: state 2; $[t_3 < t \leq t_4]$: state 1.

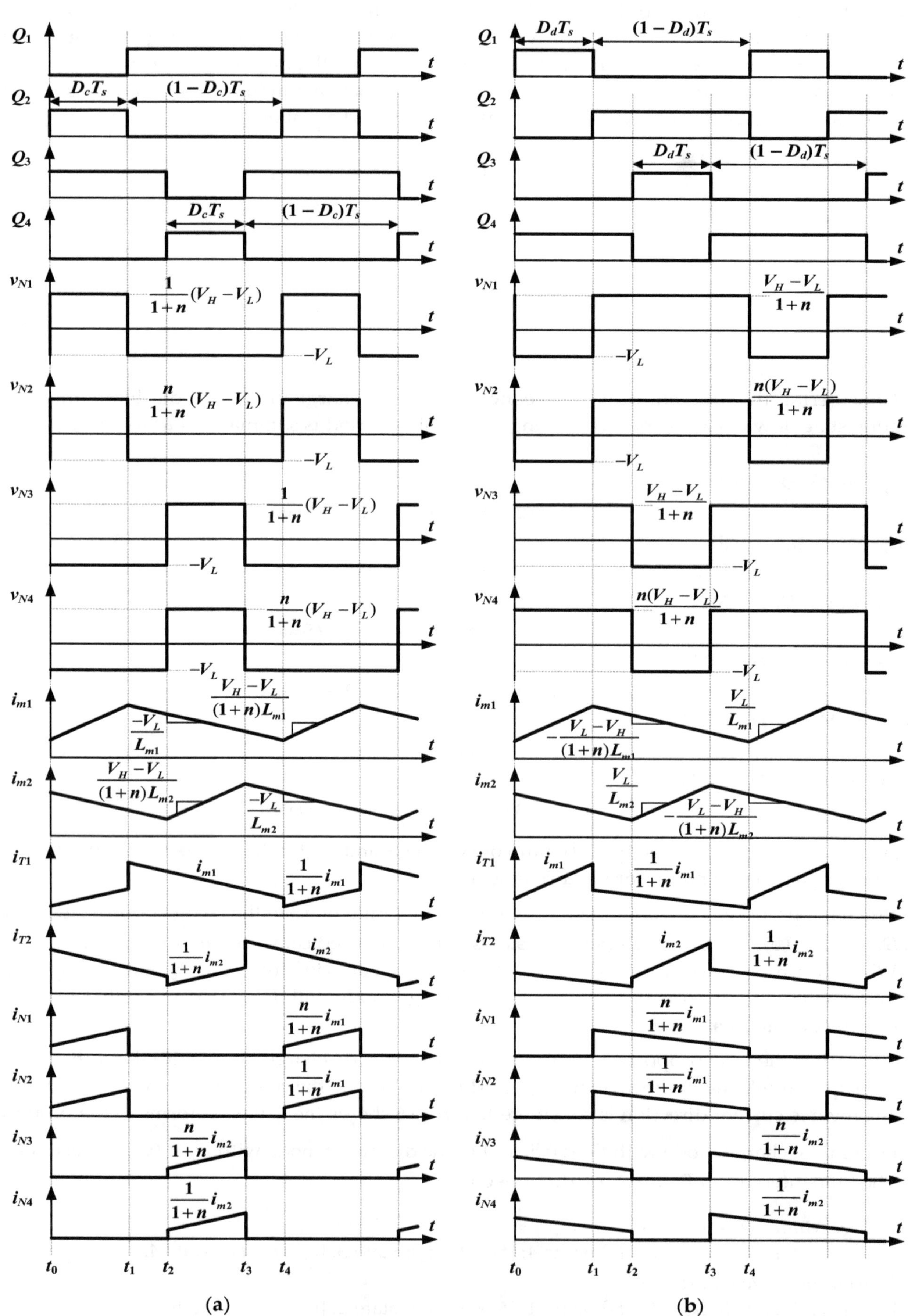

Figure 6. Key waveforms: (**a**) charge mode and $D_c < 0.5$; (**b**) discharge mode and $D_d < 0.5$.

3. Converter Steady-State Analyses

3.1. Static Voltage Conversion Ratio Analysis

Charge Mode (UC Charge; Battery Charge)

During steady-state operation and according to the volt-second balance principle of the magnetizing inductance operating in the charge mode, the static voltage conversion ratio M_c can be derived as from (22)–(25).

The voltage relationship between primary and secondary sides of the coupled inductor is shown as follows

$$v_{N2} = nv_{N1} \quad (22)$$

substituting (22) into (1), it can be rewritten as follows

$$v_{N1} = (V_H - V_L)\frac{1}{1+n} \quad (23)$$

By combining (23) and (12), the average voltage of the primary side for the coupled inductor during a switching period can be expressed as follows

$$\langle v_{N1} \rangle_{T_s} = \int_0^{D_cT_s} \frac{V_H - V_L}{1+n} dt + \int_{D_cT_s}^{T_s} (-V_L)dt = 0 \quad (24)$$

The static voltage conversion ratio of the proposed converter in the charge mode can be derived as follows

$$M_c = \frac{V_L}{V_H} = \frac{D_c}{1 + n(1 - D_c)} \quad (25)$$

Figure 7a shows the relationship between the coupled inductance with different turns ratio and the static voltage conversion ratio M_c of the proposed converter in the charge mode.

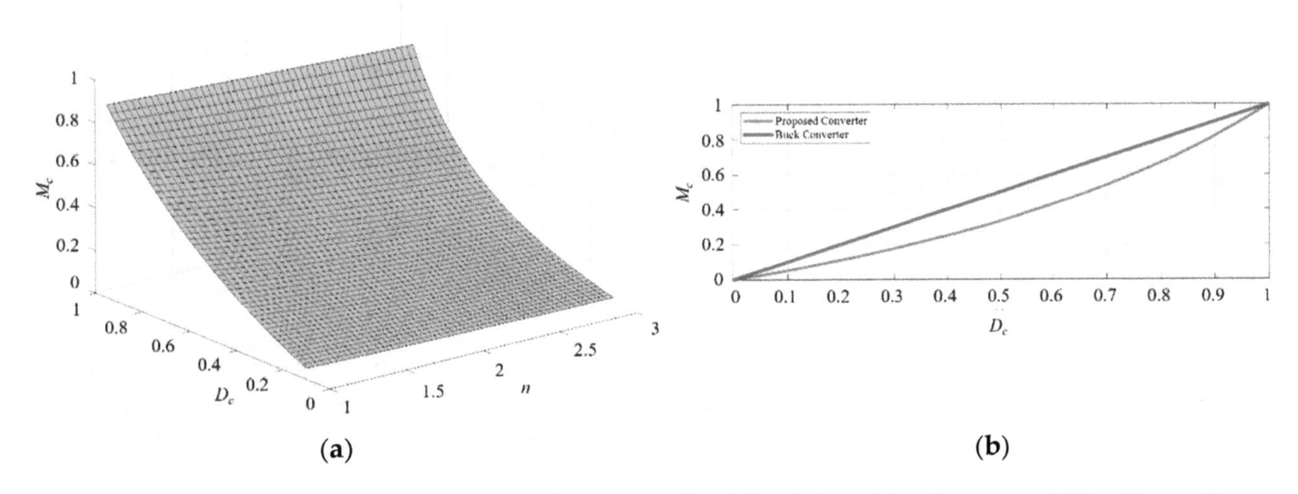

Figure 7. Converter characteristics in charge mode: (**a**) relationship diagram of M_c, D_c, and n; (**b**) relationship diagram of M_c and D_c (n = 1).

For simplicity, assuming that the turns ratio of the coupling inductance is $n = 1$, the relationship between M_c and D_c is shown in Figure 7b.

It can be seen that the static voltage conversion ratio of the proposed converter in the charge mode has a better performance, compared with the conventional buck converter.

Discharge Mode (UC Discharge; Battery Discharge; Dual-Energy in Series Discharge)

The static voltage conversion ratio M_d in the discharge mode can be derived from the average voltage of the magnetizing inductance. According to (23) and (12), and considering the duty ratio D_d of the switch Q_1 and Q_3, the average voltage of the primary side for the coupled inductor during a switching period can be expressed as follows

$$\langle -v_{N1} \rangle_{T_s} = \int_0^{D_d T_s} (-V_L)dt + \int_{D_d T_s}^{T_s} \frac{V_H - V}{1+n} dt = 0 \tag{26}$$

The static voltage conversion ratio of the converter in the discharge mode can be derived as follows

$$M_d = \frac{V_H}{V_L} = \frac{1 + nD_d}{1 - D_d} \tag{27}$$

Figure 8a shows the relationship between the coupled inductance with different turns ratios and the static voltage conversion ratio M_d of the proposed converter in the discharge mode. For simplicity, assuming that the turns ratio of the coupling inductance is $n = 1$, the relationship between M_d and D_d is shown in Figure 8b. It can be seen that the static voltage conversion ratio of the proposed converter in the discharge mode has a better performance, compared with the conventional boost converter.

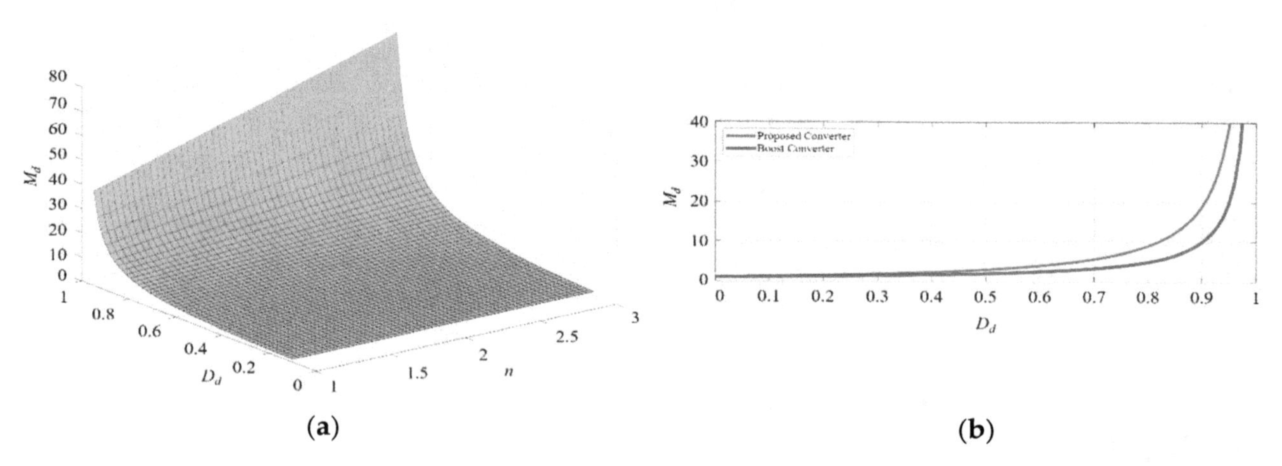

Figure 8. Converter characteristics in discharge mode: (**a**) relationship diagram of M_d, D_d, and n; (**b**) relationship diagram of M_d and D_d (n = 1).

3.2. Boundary Condition Analysis

Charge Mode (UC Charge; Battery Charge)

According to the ampere-second balance principle for the filter capacitor C_L on the low-voltage side, it means that the average current of the filter capacitor should be zero in steady-state, and the sum of the averaged currents I_{T1} and I_{T2} are equal to the low-side current $I_{L,BCM}$ (i.e., UC current or battery current), as described below

$$I_{L,BCM} = \frac{V_L}{R_{L,BCM}} = I_{T1} + I_{T2} \tag{28}$$

$$I_{T1} = \frac{i_{mc1,pk}}{2(1+n)} D_c + \frac{i_{mc1,pk}}{2}(1 - D_c) \tag{29}$$

$$I_{T1} = I_{T2} = i_{mc1,pk}(1 - D_c) \tag{30}$$

In (28), $R_{L,BCM}$ represents the low-side equivalent resistance under boundary-conduction-mode (BCM) condition.

Considering the low-side voltage V_L is constant, the peak value of the magnetizing current $i_{mc1,pk}$ at BCM in the charge mode can be expressed as

$$i_{mc1,pk} = \frac{V_L}{L_m}(1 - D_c)T_s \tag{31}$$

where T_s is the switching period.

Substituting (29), (30), and (31) into (28), the boundary magnetizing inductance $L_{mc,BCM}$ in the charge mode can be derived as follows

$$L_{mc,BCM} = \frac{R_{L,BCM}(1 - D_d)}{f_s}\left[\frac{1 + n(1 - D_c)}{1 + n}\right] \tag{32}$$

The boundary time constant $\tau_{c,BCM}$ of the proposed converter in the charge mode can be derived as (33), and the corresponding relationship curve is depicted as shown in Figure 9.

$$\tau_{c,BCM} = \frac{L_{mc,BCM} f_s}{R_{L,BCM}} = (1 - D_c)\left[\frac{1 + n(1 - D_c)}{1 + n}\right] \tag{33}$$

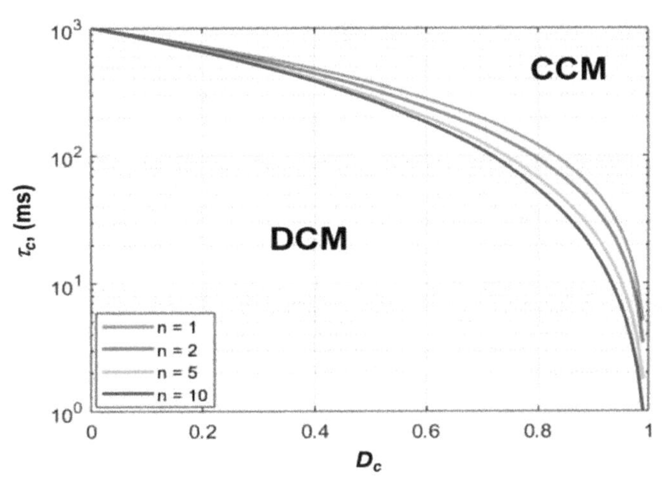

Figure 9. The curve of the boundary time constant $\tau_{c,BCM}$ in the charge mode.

Discharge Mode (UC Discharge; Battery Discharge; Dual-Energy in Series Discharge)

According to the ampere-second balance principle of the filter capacitor C_H on the high-voltage side, it can be shown that the average current on the filter capacitor is zero in steady-state, and the sum of the averaged currents I_{N2} and I_{N4} are equal to the high-side current $I_{H,BCM}$ (i.e., DC-bus current) as described below

$$I_{H,BCM} = \frac{V_H}{R_{H,BCM}} = I_{N2} + I_{N4} \tag{34}$$

$$I_{N2} = I_{N4} = \frac{i_{md1,pk}}{2(1 + n)}(1 - D_d) \tag{35}$$

where $R_{H,BCM}$ represents the high-side equivalent resistance under BCM.

Considering the high-side voltage V_H is constant, the peak value of the magnetizing current $i_{md1,pk}$ at BCM in the discharge mode can be expressed as follows

$$i_{md1,pk} = \frac{V_H T_s}{L_m}\frac{(1 - D_d)D_d}{1 + nD_d} \tag{36}$$

Substituting (35), (36) into (34) and simplifying it, the boundary magnetizing inductance in the discharge mode can be derived as follows

$$L_{md,BCM} = \frac{R_{H,BCM}}{f_s}\left[\frac{(1-D_d)^2 D_d}{(1+n)(1+nD_d)}\right] \tag{37}$$

The boundary time constant $\tau_{d,BCM}$ of the converter in the discharge mode can be derived as (38), and the corresponding relationship curve is depicted as shown in Figure 10.

$$\tau_{d,BCM} = \frac{L_{md,BCM} f_s}{R_{H,BCM}} = \frac{(1-D_d)^2 D_d}{(1+n)(1+nD_d)} \tag{38}$$

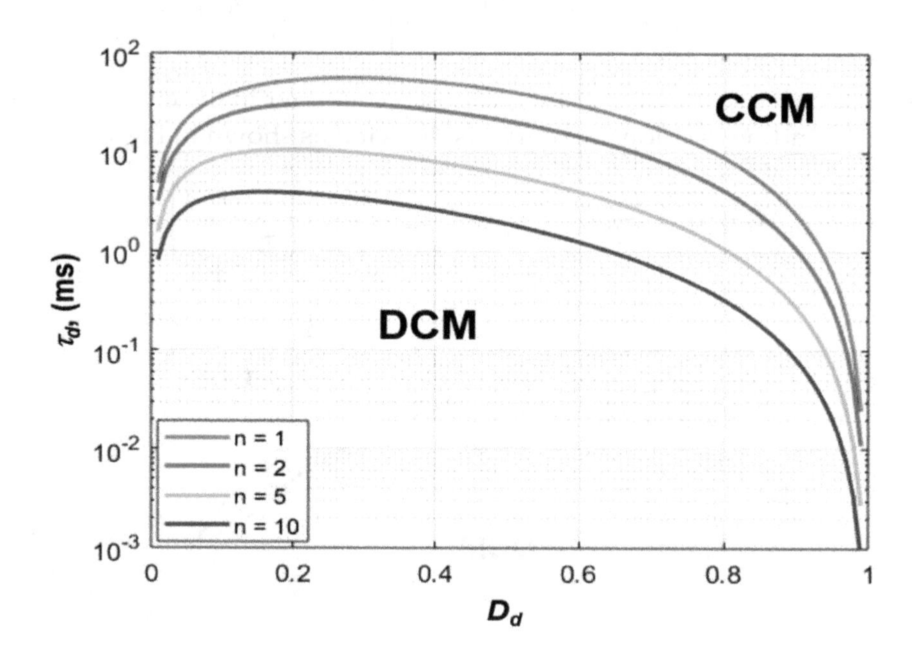

Figure 10. The curve of the boundary time constant $\tau_{d,BCM}$ in the discharge mode.

3.3. Voltage and Current Stresses Analyses of Power Devices

3.3.1. Voltage Stress Derivations

The power switching device is the main design considerations when implementing the power converter. The voltage and current stresses of the power device for the converter circuit are analyzed, and then the appropriate components are selected as below.

The multiport switches S_1~S_4 are used as the pre-stage for the discharge mode or post-stage for the charge mode. The voltage stress of the multiport switches S_1 and S_2 is equal to the UC voltage V_U, and the voltage stress of S_3 and S_4 is equal to the battery voltage V_B, as follows

$$V_{S1,\max} = V_{S2,\max} = V_U \tag{39}$$

$$V_{S3,\max} = V_{S4,\max} = V_B \tag{40}$$

The voltage stress of the power switches Q_1 to Q_4 for the converter can be expressed as follows

$$V_{Q1,\max} = V_{Q3,\max} = V_H - v_{N2} = \frac{V_H + nV_L}{1+n} \tag{41}$$

$$V_{Q2,\max} = V_{Q4,\max} = V_H - v_{N2} = V_H + nV_L \tag{42}$$

3.3.2. Current Stress Derivations

The root mean square (RMS) current of the magnetizing inductances L_{m1} and L_{m2} are derived based on the operating state of the proposed converter, as follows

$$I_{m1,rms} = I_{m2,rms} = \sqrt{I_{m1}{}^2 + \left(\frac{\Delta i_{m1}}{2\sqrt{3}}\right)^2} \tag{43}$$

where I_{m1} and I_{m2} are the DC value of the magnetizing current i_{m1} and i_{m2}, respectively; Δi_{m1} and Δi_{m2} are the magnetizing ripple currents, as follows

$$I_{m1} = I_{m2} = \frac{I_L}{2} \cdot \frac{1+n}{1+n(1-D_c)} \tag{44}$$

$$\Delta i_{m1} = \Delta i_{m2} = \frac{V_L}{L_m}(1-D_c)T_s \tag{45}$$

The RMS current of the power switches Q_1~Q_4 of the proposed converter in the charge mode can be derived as follows

$$I_{Q1,rms} = I_{Q3,rms} = I_{m1,rms}\sqrt{1-D_c} \tag{46}$$

$$I_{Q2,rms} = I_{Q4,rms} = \frac{I_{m1,rms}}{1+n}\sqrt{D_c} \tag{47}$$

The RMS current of the filter capacitors C_L and C_H of the proposed converter in the charge mode can be derived as follows

$$I_{CL,rms} = \sqrt{I_{T1,rms}{}^2 + I_{T2,rms}{}^2 - I_L{}^2} \tag{48}$$

$$I_{CH,rms} = \sqrt{I_{Q2,rms}{}^2 + I_{Q4,rms}{}^2 - I_H{}^2} \tag{49}$$

where

$$I_{T1,rms} = I_{T2,rms} = I_{m1,rms}\sqrt{\frac{D_c}{(1+n)^2} + (1-D_c)} \tag{50}$$

4. Simulated and Experimented Results

The realized converter prototype is shown in Figure 11, and Table 1 shows the electrical specifications and the circuit parameters of the realized power converter. For the convenience of the experiments, in the charge mode, the power supply (ITECH IT6726G) was used as the DC-bus on the high-voltage side, and the electronic load (ITECH IT8814B) was used as the UC or the battery on the low-voltage side. Conversely, in the discharge mode, the power supply was used as the UC, the battery, or the dual-energy storage in series.

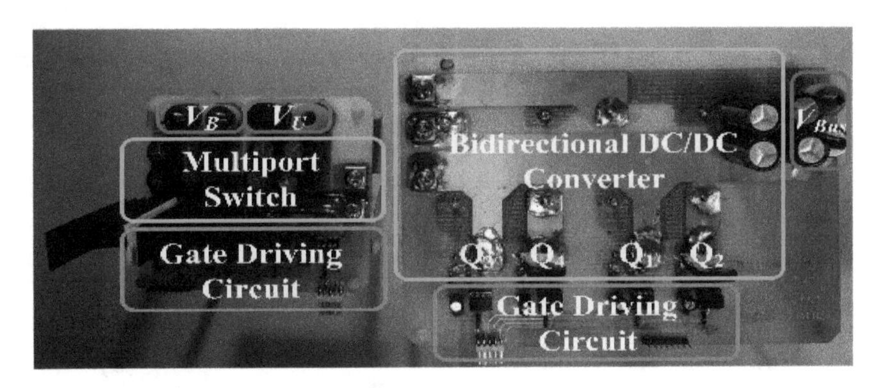

Figure 11. Prototype circuit of the proposed converter.

Table 1. Specifications and circuit parameters of the realized power converter.

Symbol	Descriptions	Specifications
V_H (V_{bus})	high-side voltage (DC-bus voltage)	72 V
V_L	low-side voltage	
V_B	battery voltage	20 V~26 V
V_U	UC voltage	0 V~48 V
P_o	rated output power	500 W
f_s	switching frequency	20 kHz
Symbol	**Descriptions**	**Parameters**
L_{m1}, L_{m2}	magnetizing inductances of the coupled inductors	250 µH
n	turns ratio of the coupled inductors	1
C_H	high-side capacitor	2400 µF
C_L	low-side capacitor	800 µF

UC Charge Mode

Figures 12 and 13 show the waveforms of the gate signals of Q_2 and Q_4, the primary-side currents of the coupled inductor (i_{T1}, i_{T2}), the secondary-side currents of the coupled inductor (i_{N2}, i_{N4}), and the low-side voltage V_U in the UC charge mode with full load condition, respectively. In this mode, the UC voltage was about 48 V, the duty ratio of the switches Q_2 and Q_4 was set to 80% (i.e., D_c = 0.8), the DC values of the primary currents (i_{T1}, i_{T2}) and secondary currents (i_{N2}, i_{N4}) of the coupled inductance were about 5.2 A and 3.5 A, respectively.

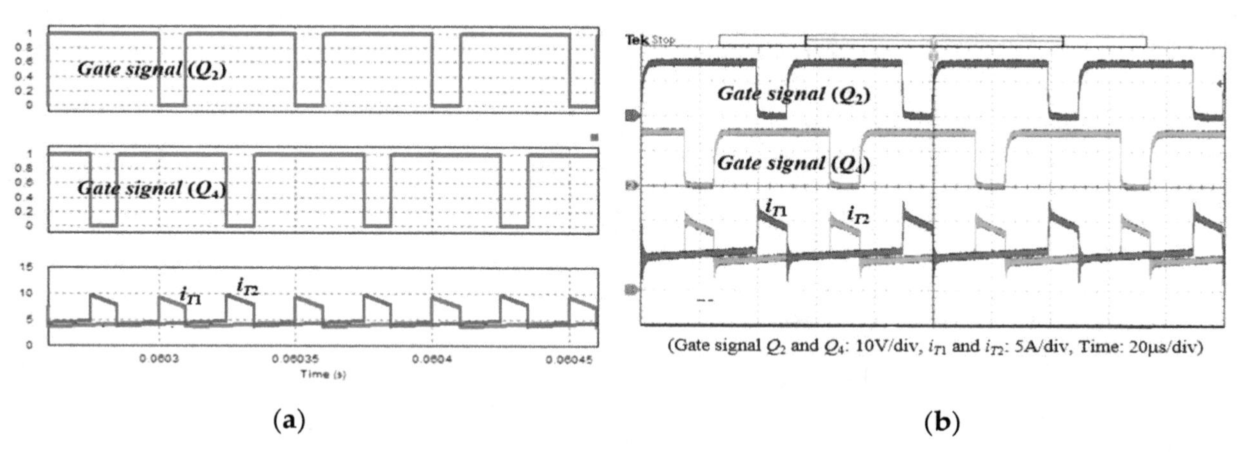

Figure 12. Waveforms of the switching gate signals and the primary-side currents of the coupled inductor in the UC recharging mode with D_c = 0.8: (**a**) simulated and (**b**) experimental.

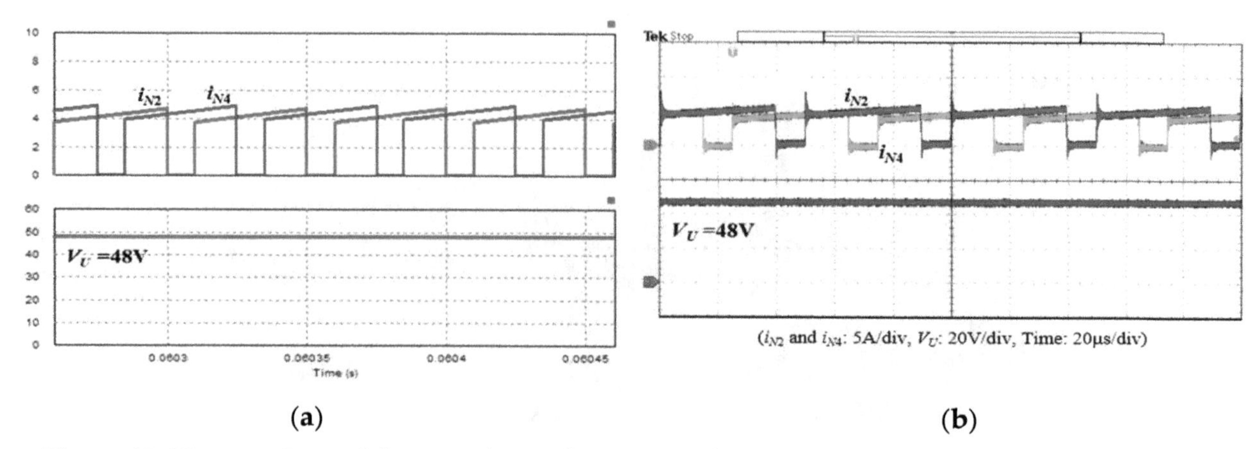

Figure 13. The waveform of the secondary-side currents of the coupled inductor and UC voltage in the UC charge mode with D_c = 0.8: (**a**) simulated and (**b**) experimental.

Figure 14 shows the waveforms of the steady-state switching voltages across the power devices in the UC charge mode. The results showed that the steady-state switching voltages across the lower-leg MOSFETs Q_1 and Q_3 were about 60 V, and the steady-state switching voltages across the upper-leg MOSFETs Q_2 and Q_4 were about 120 V. It could be seen that in Figure 14, the simulation and the experimental results were consistent and corresponded to (41) and (42).

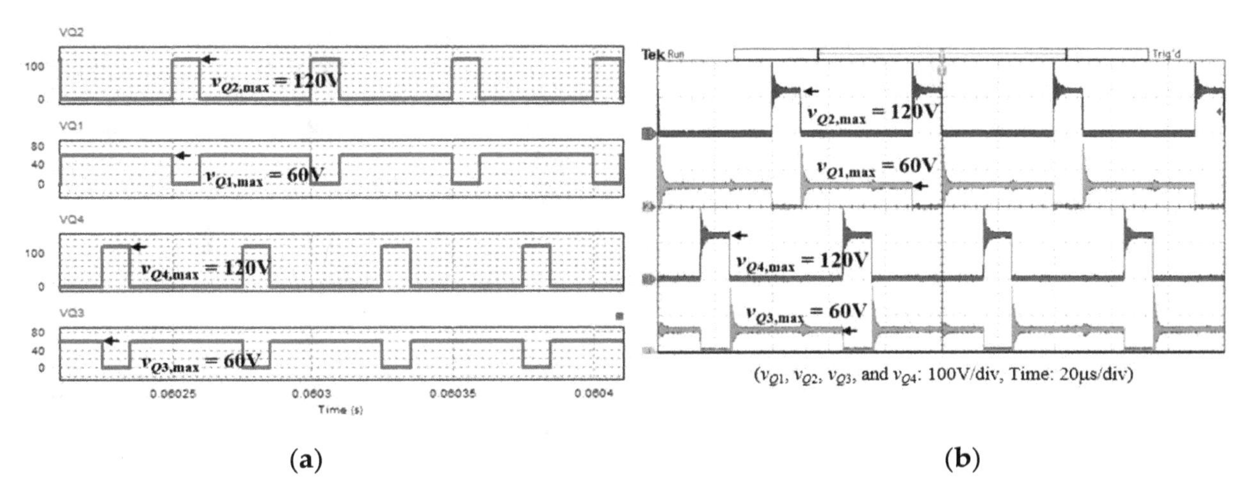

Figure 14. The waveform of switching voltage across the power devices in the UC charge mode with D_c = 0.8: (**a**) simulated and (**b**) experimental.

Battery Charge Mode

Figures 15 and 16 show the waveforms of the gate signals of Q_2 and Q_4, the primary-side currents of the coupled inductor (i_{T1}, i_{T2}), the secondary-side currents of the coupled inductor (i_{N2}, i_{N4}), and the low-side voltage V_U in the battery charge mode with full load condition, respectively.

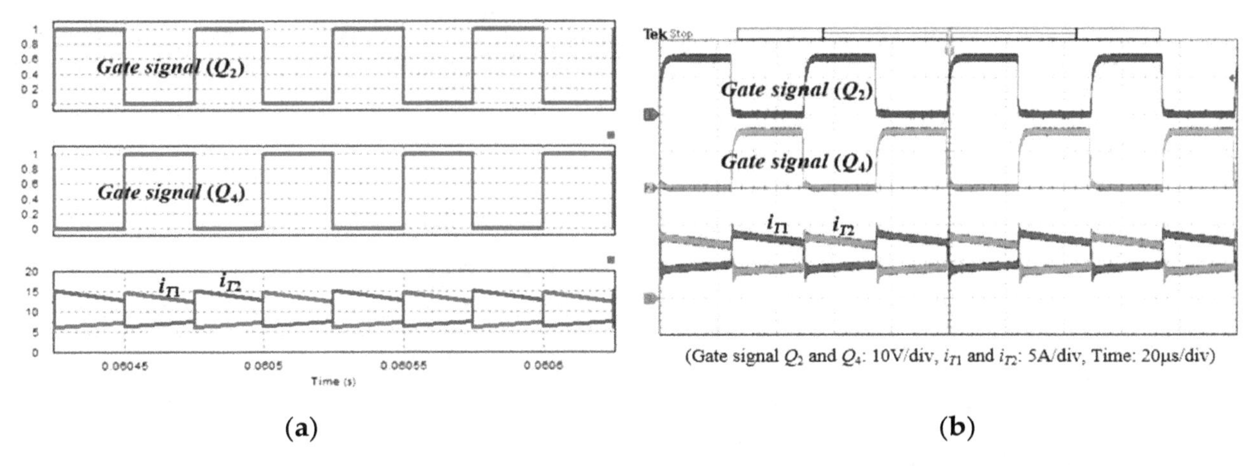

Figure 15. Waveforms of the switching gate signals and the primary-side currents of the coupled inductor in the battery charge mode with D_c = 0.5: (**a**) simulated and (**b**) experimental.

In this mode, the battery voltage was about 24 V, the duty ratio of the switches Q_2 and Q_4 was set to 50% (i.e., D_c = 0.5), and the DC values of the primary currents (i_{T1}, i_{T2}) and secondary currents (i_{N2}, i_{N4}) of the coupled inductance were about 10.4 A and 3.5 A, respectively. It could be seen that in Figures 15 and 16, the simulation and the experimental results were consistent.

Figure 17 shows the waveforms of the steady-state switching voltages across the power devices in the battery charge mode. The results showed that the steady-state switching voltages across the lower-leg MOSFETs Q_1 and Q_3 were about 48 V, and the steady-state switching voltages across the upper-leg MOSFETs Q_2 and Q_4 were about 96 V. It could be seen that in Figure 17, the simulation and the experimental results were consistent and corresponded to (41) and (42).

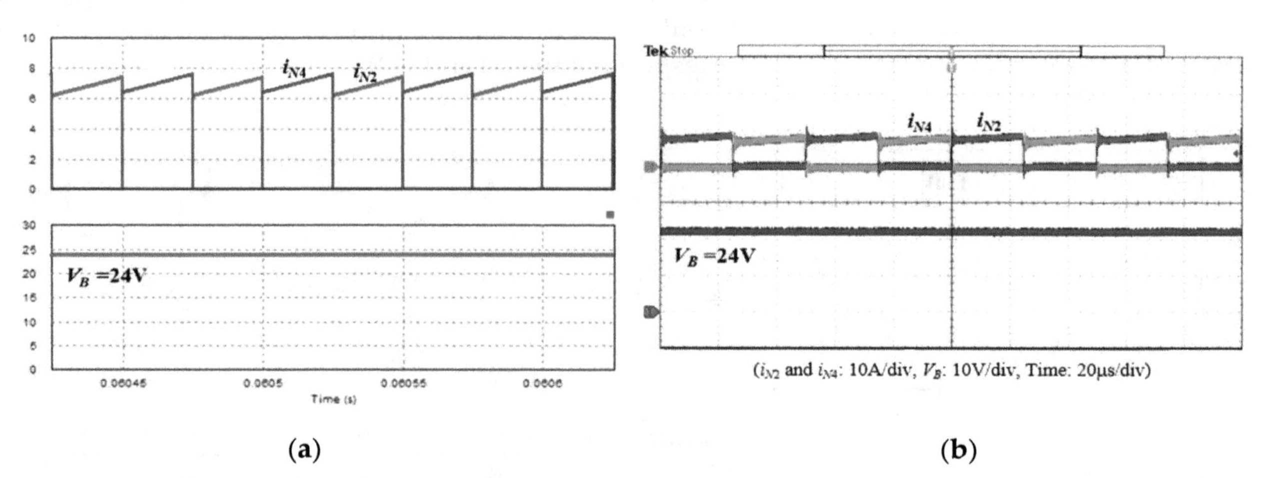

(a) **(b)**

Figure 16. The waveform of the secondary-side currents of the coupled inductor and UC voltage in the battery charge mode with D_c = 0.5: (**a**) simulated and (**b**) experimental.

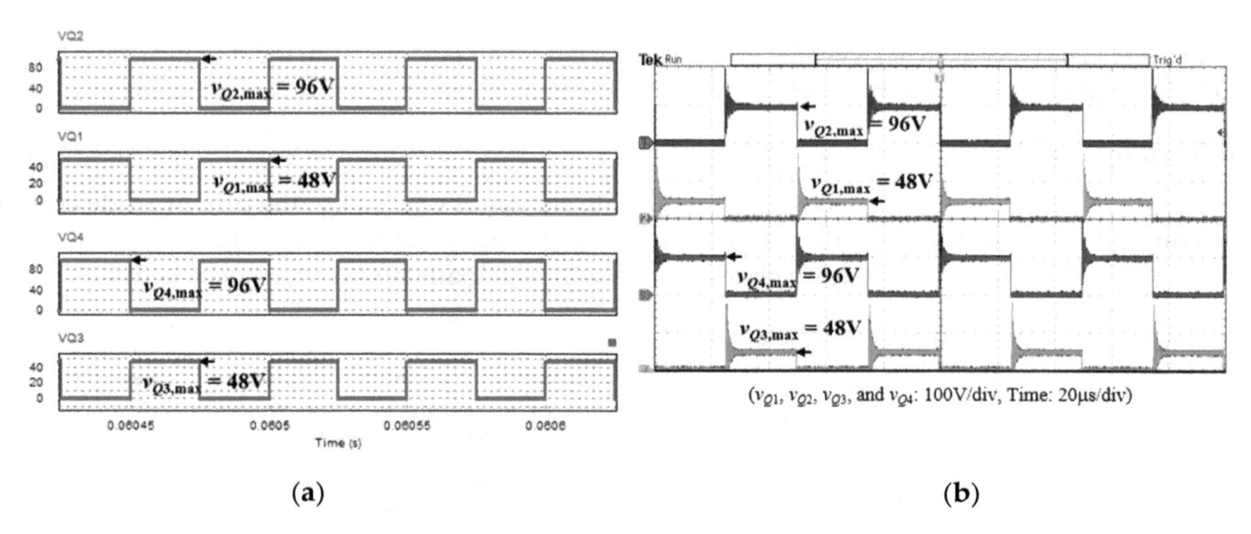

(a) **(b)**

Figure 17. The waveform of switching voltage across the power devices in the battery charge mode with D_c = 0.5: (**a**) simulated and (**b**) experimental.

UC Discharge Mode

Figures 18 and 19 show the waveforms of the gate signals of Q_2 and Q_4, the primary-side currents of the coupled inductor (i_{T1}, i_{T2}), the secondary-side currents of the coupled inductor (i_{N2}, i_{N4}), and the high-side voltage V_H in the UC discharge mode with full load condition, respectively.

In this mode, the DC-bus voltage was about 72 V, the duty ratio of the switches Q_1 and Q_3 was set to 20% (i.e., D_d = 0.2), and the DC values of the primary currents (i_{T1}, i_{T2}) and secondary currents (i_{N2}, i_{N4}) of the coupled inductance were about 5.2 A and 3.5 A, respectively. It could be seen that in Figures 18 and 19, the simulation and the experimental results were consistent.

Figure 20 shows the waveforms of the steady-state switching voltages across the power devices in the UC discharge mode. The results showed that the steady-state switching voltages across the lower-leg MOSFETs Q_1 and Q_3 were about 60 V, and the steady-state switching voltages across the upper-leg MOSFETs Q_2 and Q_4 were about 120 V. It could be seen that in Figure 20, the simulation and the experimental results were consistent and corresponded to (41) and (42).

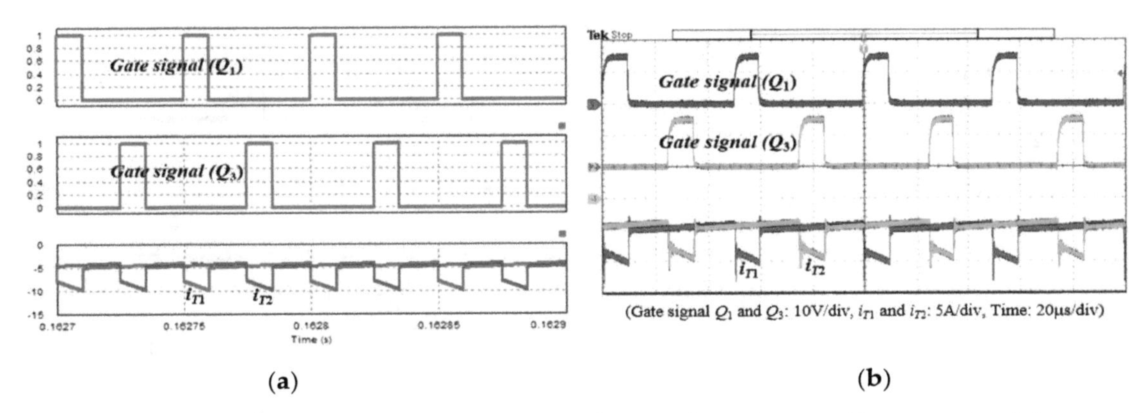

(a) (b)

Figure 18. Waveforms of the switching gate signals and the primary-side currents of the coupled inductor in the UC discharging mode with D_d = 0.2: (**a**) simulated and (**b**) experimental.

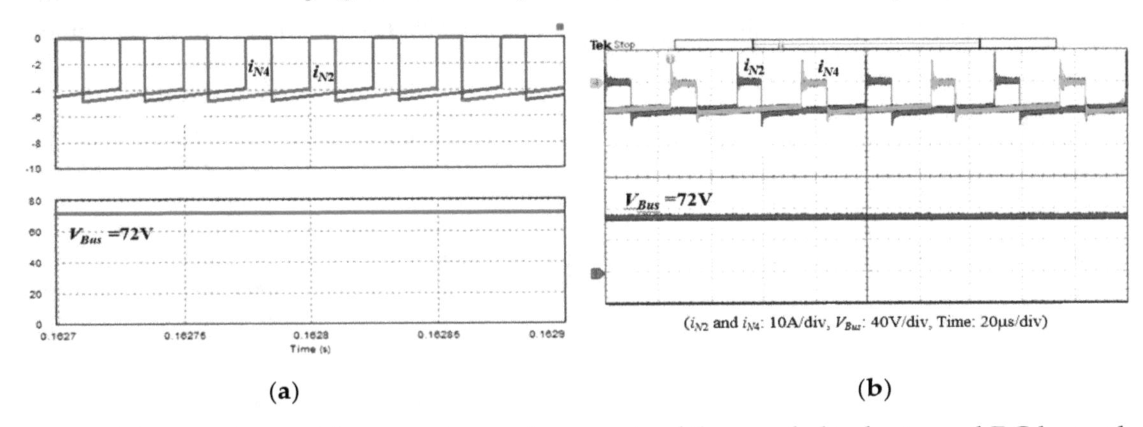

(a) (b)

Figure 19. The waveform of the secondary-side currents of the coupled inductor and DC-bus voltage in the UC discharging mode with D_d = 0.2: (**a**) simulated and (**b**) experimental.

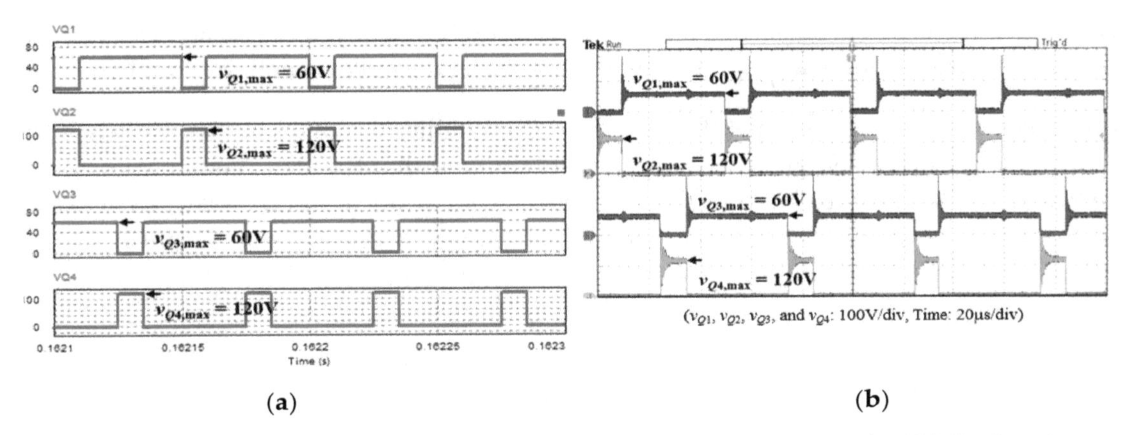

(a) (b)

Figure 20. The waveform of switching voltage across the power devices in the UC discharging mode with D_d = 0.2: (**a**) simulated and (**b**) experimental.

Dual-Energy in Series Discharge Mode

Figures 21 and 22 show the waveforms of the gate signals of Q_2 and Q_4, the primary-side currents of the coupled inductor (i_{T1}, i_{T2}), the secondary-side currents of the coupled inductor (i_{N2}, i_{N4}), and the high-side voltage V_H in the dual-energy discharge mode with full load condition, respectively.

In this mode, the DC-bus voltage was about 72 V, the low-side voltage V_L was 44 V, the duty ratio of the switches Q_1 and Q_3 was set to 25% (i.e., D_d = 0.25), and the DC values of the primary currents (i_{T1}, i_{T2}) and secondary currents (i_{N2}, i_{N4}) of the coupled inductance were about 5.8 A and 3.5 A, respectively. It could be seen that in Figures 21 and 22, the simulation and the experimental results were consistent.

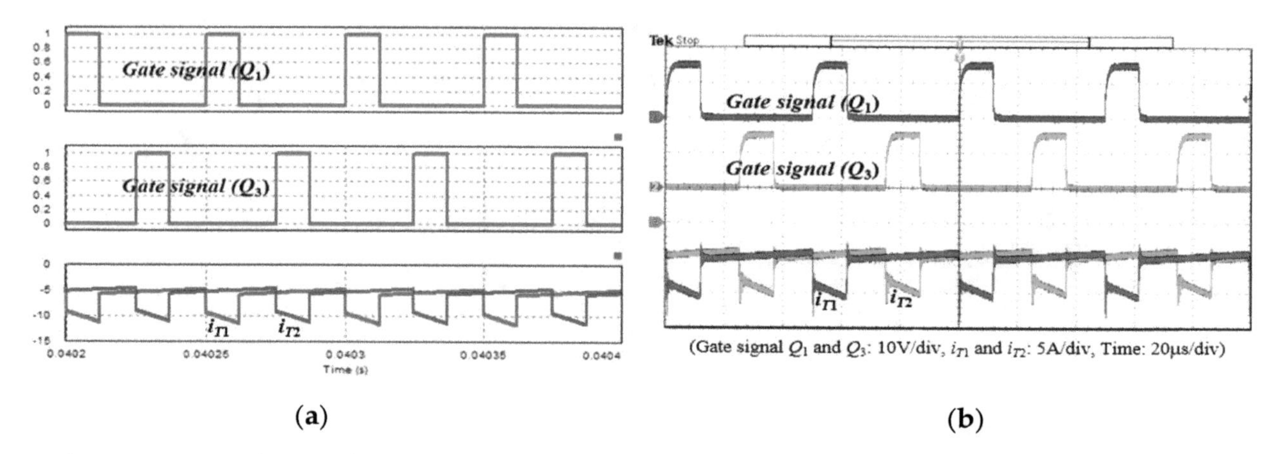

(a) (b)

Figure 21. Waveforms of the switching gate signals and the primary-side currents of the coupled inductor in the dual-energy in series discharge mode with $D_d = 0.25$: (**a**) simulated and (**b**) experimental.

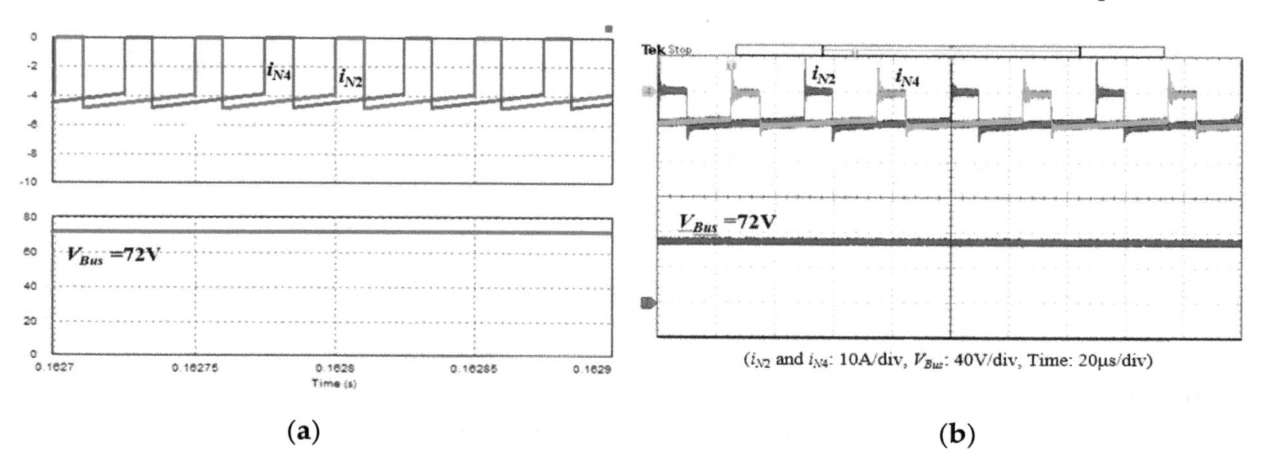

(a) (b)

Figure 22. The waveform of the secondary-side currents of the coupled inductor and DC-bus voltage in the dual-energy in series discharge mode with $D_d = 0.25$: (**a**) simulated and (**b**) experimental.

Figure 23 shows the waveforms of the steady-state switching voltages across the power devices in the dual-energy in series discharge mode. The results showed that the steady-state switching voltages across the lower-leg MOSFETs Q_1 and Q_3 were about 58 V, and the steady-state switching voltages across the upper-leg MOSFETs Q_2 and Q_4 were about 116 V. It could be seen that in Figure 23, the simulation and the experimental results were consistent and corresponded to (41) and (42).

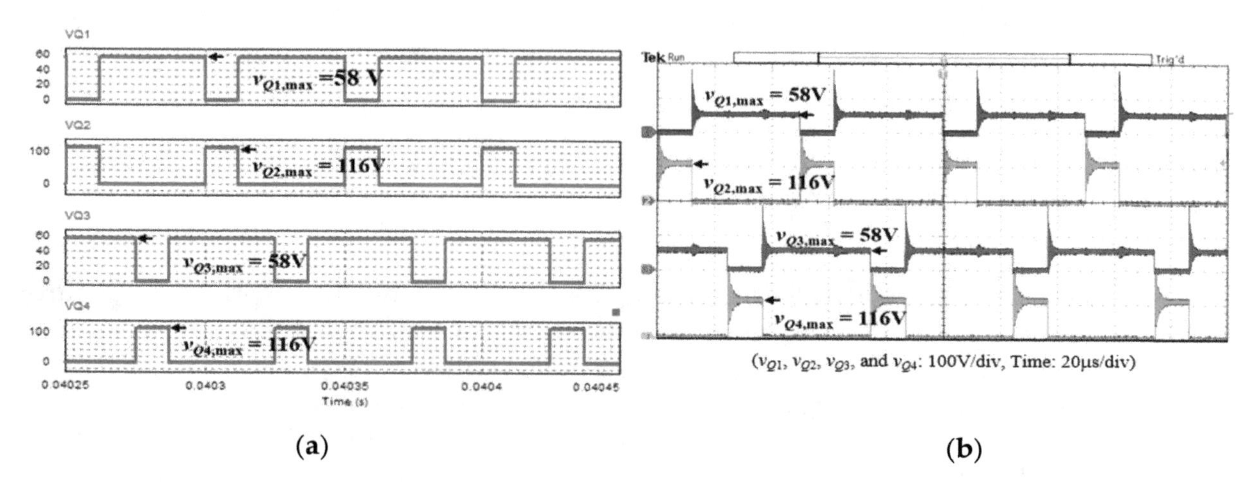

(a) (b)

Figure 23. The waveform of switching voltage across the power devices in the dual-energy in series discharge mode with $D_d = 0.25$: (**a**) simulated and (**b**) experimental.

Efficiency Measurement

The system used two power analyzers (YOKOGAWA WT310) connected to the input and output of the realized converter prototype. As could be seen in Figure 24, in the UC charge mode, the highest efficiency point was 97.4%; in the battery charge mode, the highest efficiency point was 95.5%; in the UC discharge mode, the highest efficiency point was 97.2%; in the dual-energy in series discharge mode, the highest efficiency point was 97.1%; in the battery discharge mode, the highest efficiency point was 95.3%.

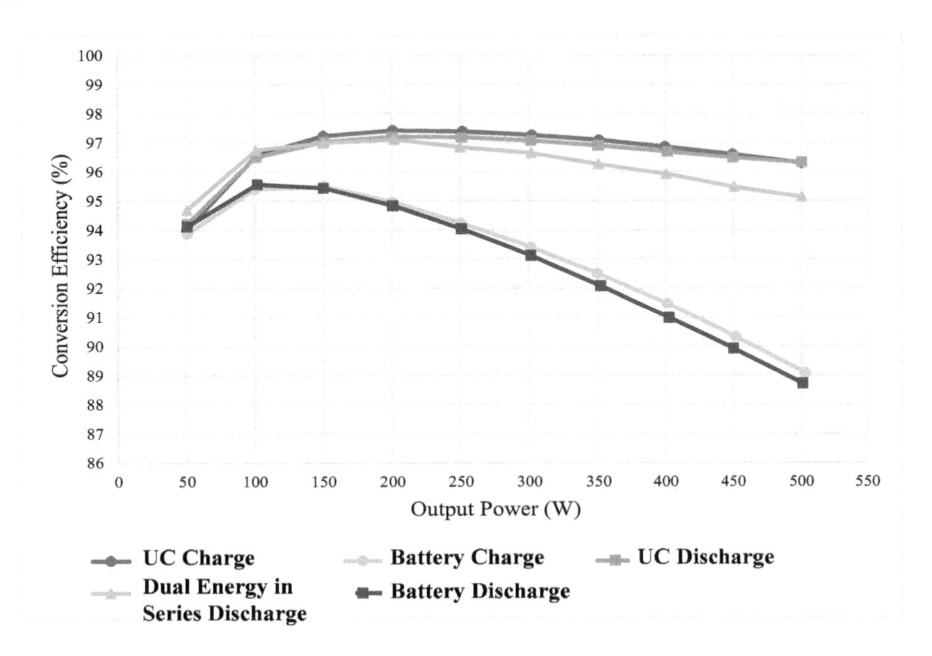

Figure 24. The measured efficiency of the proposed converter for the different operating modes.

5. Conclusions

This study proposed a patented bidirectional power converter that used dual-energy storage as input sources and incorporated a coupled inductor to obtain a higher voltage conversion ratio. The converter control used IPWM control to achieve low current ripple, dissipate low side current stress, and reduce the conduction loss of the power MOSFET. Moreover, the proposed bidirectional power converter in this study also discussed the steady-state operation in the charge mode and discharge mode, respectively. The voltage conversion ratio, boundary conditions, and voltage and current stress of each power component of the converter were analyzed. Finally, this study implemented a converter prototype with a 500 W power rating for verification. The simulation results and the experimental results were consistent; the highest efficiency points of the realized prototype were 97.4%, 95.5%, 97.2%, 97.1%, and 95.3% for the UC charge mode, battery charge mode, UC discharge mode, the dual-energy in series discharge mode, and battery discharge mode, respectively. In summary, this paper demonstrated that the proposed bidirectional power converter could be potentially applied to produce hybrid power architecture (has been patented [37]).

Author Contributions: C.-M.L. substantially contributed to the examination and interpretation of the results, development of the overall system, and review and proofreading of the manuscript. J.T. and Y.L. substantially contributed to the review and proofreading of the manuscript. Y.-C.L. substantially contributed to literature search, control strategy design, and production and analysis of the results. All authors have read and agreed to the published version of the manuscript.

Acknowledgments: The authors would like to express their appreciation to the Chih-Yu Yang (guided by Prof. Ching-Ming Lai) for the experimental bench setup.

Nomenclature

T_1, T_2	Two-phase coupled inductors
L_{m1}, L_{m2}	Magnetizing inductors of the two-phase coupled inductors
$L_{mc,BCM}$	Boundary magnetizing inductance in the charge mode
$L_{md,BCM}$	Boundary magnetizing inductance in the discharge mode
n	Turns ratio of the two-phase coupled inductors ($n = N_2/N_1 = N_4/N_3$)
N_1	Primary winding of T_1
N_2	Secondary winding of T_1
N_3	Primary winding of T_2
N_4	Secondary winding of T_2
k	Coupling coefficient
C_U	Input capacitor paralleled with UC
C_B	Input capacitor paralleled with BES
S_1~S_4	Power devices of the multiport switch
Q_1~Q_4	Power devices of the two-phase bidirectional power converter
V_H	High-side voltage for the DC-bus
V_L	Low-side voltage for UC, BES, or BES/UC dual-energy in series
V_U	UC voltage
V_B	BES voltage
i_{Bus}	DC-bus current
i_{Uc}	UC current
i_{Bat}	BES current
i_H	High voltage side current
i_L	Low voltage side current
$I_{L,BCM}$	Low voltage side current under BCM condition
$I_{H,BCM}$	High voltage side current under BCM condition
$I_{m1,rms}, I_{m2,rms}$	RMS value of the magnetizing currents of the coupled inductors
$I_{T1,rms}, I_{T2,rms}$	RMS value of the primary-side currents of the coupled inductors
v_{N1}	Voltage of the winding N_1 of the T_1
v_{N2}	Voltage of the winding N_2 of the T_1
v_{N3}	Voltage of the winding N_3 of the T_2
v_{N4}	Voltage of the winding N_4 of the T_2
$V_{S1,\mathrm{max}}$~$V_{S4,\mathrm{max}}$	Switch voltage stress of the multiport switch
$V_{Q1,\mathrm{max}}$~$V_{Q4,\mathrm{max}}$	Switch voltage stress of the two-phase bidirectional power converter
i_{T1}, i_{T2}	The primary-side currents of the two-phase coupled inductors
i_T	The sum of the primary-side currents i_{T1} and i_{T2}
i_{N2}, i_{N4}	The secondary-side currents of the two-phase coupled inductors
i_{m1}, i_{m2}	Magnetizing inductor currents of the coupled inductors T_1 and T_2
I_{m1}, I_{m2}	DC value of the magnetizing currents
$i_{mc,pk}$	peak value of the magnetizing inductor current under BCM in the charge mode
$i_{md,pk}$	peak value of the magnetizing inductor current under BCM in the discharge mode
$\Delta i_{m1}, \Delta i_{m2}$	Magnetizing ripple currents
$I_{Q1,\mathrm{rms}}$~$I_{Q4,\mathrm{rms}}$	RMS current of the power switches Q_1~Q_4
$I_{CH,\mathrm{rms}}$~$I_{CL,\mathrm{rms}}$	RMS current of the filter capacitors C_L and C_H
D_c, D_d	Duty ratio of charge mode and discharge mode
T_s	Switching period
$\tau_{c,BCM}$	Boundary time constant in the charge mode
$\tau_{d,BCM}$	Boundary time constant in the discharge mode
$R_{L,BCM}$	Low-side equivalent resistance under BCM condition
$R_{H,BCM}$	High-side equivalent resistance under BCM condition
M_c	Static voltage conversion ratio in the charge mode
M_d	Static voltage conversion ratio in the discharge mode

References

1. Lai, J.S.; Nelson, D.J. Energy management power converters in hybrid electric and fuel cell vehicles. *Proc. IEEE* **2007**, *95*, 766–777. [CrossRef]
2. Bauman, J.; Kazerani, M. A comparative study of fuel-cell-battery, fuel-cell-ultracapacitor, and fuel-cell-battery-ultracapacitor vehicles. *IEEE Trans. Veh. Technol.* **2008**, *57*, 760–769. [CrossRef]
3. Khaligh, A.; Li, Z. Battery ultracapacitor fuel cell and hybrid energy storage systems for electric hybrid electric fuel cell and plug-in hybrid electric vehicles: State of the art. *IEEE Trans. Veh. Technol.* **2010**, *59*, 2806–2814. [CrossRef]
4. Chan, C.C.; Bouscayrol, A.; Chen, K. Electric, hybrid, and fuel-cell vehicles: Architectures and modeling. *IEEE Trans. Veh. Technol.* **2010**, *59*, 589–598. [CrossRef]
5. Rajashekara, K. Present status and future trends in electric vehicle propulsion technologies. *IEEE J. Emerg. Sel. Top. Power Electron.* **2013**, *1*, 3–10. [CrossRef]
6. Zhang, Y.; Meng, D.; Zhou, M.; Li, S. Energy management of an electric city bus with battery/ultra-capacitor HESS. In Proceedings of the 2016 IEEE Vehicle Power and Propulsion Conference (VPPC), Hangzhou, China, 17–20 October 2016.
7. Cheng, Y.H.; Lai, C.M. Control strategy optimization for parallel hybrid electric vehicles using memetic algorithm. *Energies* **2017**, *10*, 305. [CrossRef]
8. Cheng, L.; Acuna, P.; Aguilera, R.P.; Jiang, J.; Flecther, J.; Baier, C. Model predictive control for energy management of a hybrid energy storage system in light rail vehicles. In Proceedings of the 2017 11th IEEE International Conference on Compatibility, Power Electronics and Power Engineering (CPE-POWERENG), Cadiz, Spain, 4–6 April 2017; pp. 683–688.
9. Un-Noor, F.; Padmanaban, S.; Mihet-Popa, L.; Mollah, M.N.; Hossain, E.A. Comprehensive study of key electric vehicle (EV) components, technologies, challenges, impacts, and future direction of development. *Energies* **2017**, *10*, 1217. [CrossRef]
10. Serpi, A.; Porru, M. Modelling and design of real-time energy management systems for fuel cell/battery electric vehicles. *Energies* **2019**, *12*, 4260. [CrossRef]
11. Schaltz, E.; Khaligh, A.; Rasmussen, P.O. Influence of battery/ultracapacitor energy-storage sizing on battery lifetime in a fuel cell hybrid electric vehicle. *IEEE Trans. Veh. Technol.* **2009**, *58*, 3882–3891. [CrossRef]
12. Cao, J.; Emadi, A. A new battery/ultracapacitor hybrid energy storage system for electric, hybrid, and plug-in hybrid electric vehicles. *IEEE Trans. Power Electron.* **2012**, *27*, 122–132.
13. Momayyezan, M.; Hredzak, B.; Agelidis, V.G. A new multiple converter topology for battery/ultracapacitor hybrid energy system. In Proceedings of the Annual Conference of the IEEE Industrial Electronics Society, Yokohama, Japan, 9–12 November 2015; pp. 464–468.
14. Juned, S.; Mohammad, S.; Bhanabhagvanwala, D. Simulation analysis of battery/ultracapacitor hybrid energy storage system for electric vehicle. In Proceedings of the International Conference on Intelligent Sustainable Systems, Palladam, India, 21–22 February 2019.
15. Chakraborty, S.; Vu, H.-N.; Hasan, M.M.; Tran, D.-D.; Baghdadi, M.E.; Hegazy, O. DC-DC converter topologies for electric vehicles, plug-in hybrid electric vehicles and fast charging stations: State of the art and future trends. *Energies* **2019**, *12*, 1569. [CrossRef]
16. Ding, S.; Wei, B.; Hang, J.; Zhang, P.; Ding, M. *A Multifunctional Interface Circuit for Battery-Ultracapacitor Hybrid Energy Storage System*; NSW: Sydney, Australia, 2017.
17. Ortúzar, M.; Moreno, J.; Dixon, J. Ultracapacitor-based auxiliary energy system for an electric vehicle: Implementation and evaluation. *IEEE Trans. Ind. Electron.* **2007**, *54*, 2147–2156. [CrossRef]
18. Machado, F.; Antunes, C.H.; Dubois, M.R.; Trovao, J.P. Semi-active hybrid topology with three-level DC-DC converter for electric vehicle application. In Proceedings of the 2015 IEEE Vehicle Power and Propulsion Conference (VPPC), Montreal, QC, Canada, 19–22 October 2015; pp. 1–6.
19. Shen, J.; Khaligh, A. A supervisory energy management control strategy in a battery/ultracapacitor hybrid energy storage system. *IEEE Trans. Transp. Electrif.* **2015**, *1*, 223–231. [CrossRef]
20. Castaings, A.; Lhomme, W.; Trigui, R.; Bouscayrol, A. Practical control schemes of a battery/supercapacitor system for electric vehicle. *IET Electr. Syst. Transp.* **2016**, *6*, 20–26. [CrossRef]
21. Kuperman, A.; Aharon, I.; Malki, S.; Kara, A. Design of a semiactive battery-ultracapacitor hybrid energy source. *IEEE Trans. Power Electron.* **2013**, *28*, 806–815. [CrossRef]

22. Onar, O.; Khaligh, A. Dynamic modeling and control of a cascaded active battery/ultra-capacitor based vehicular power system. In Proceedings of the 2008 IEEE Vehicle Power and Propulsion Conference (VPPC), Harbin, China, 3–5 September 2008.
23. Jing, W.; Lai, C.H.; Wong, S.H.W.; Wong, M.L.D. Battery-supercapacitor hybrid energy storage system in standalone DC microgrids: Areview. *IET Renew. Power Gener.* **2017**, *11*, 461–469. [CrossRef]
24. Allegre, A.L.; Bouscayrol, A.; Trigui, R. Flexible real-time control of a hybrid energy storage system for electric vehicles. *IET Electr. Syst. Transp.* **2013**, *3*, 79–85. [CrossRef]
25. Trovão, J.P.F.; Pereirinha, P.G. Control scheme for hybridised electric vehicles with an online power follower management strategy. *IET Electr. Syst. Transp.* **2015**, *5*, 12–23. [CrossRef]
26. Trovao, J.P.; Silva, M.A.; Dubois, M.R. Coupled energy management algorithm for MESS in urban EV. *IET Electr. Syst. Transp.* **2017**, *7*, 125–134. [CrossRef]
27. Livreri, P.; Castiglia, V.; Pellitteri, F.; Miceli, R. Design of a battery/ultracapacitor energy storage system for electric vehicle applications. In Proceedings of the IEEE International Forum on Research and Technologies for Society and Industry, Palermo, Italy, 10–13 September 2018; pp. 1–5.
28. Lu, X.; Wang, H. Optimal sizing and energy management for cost-effective PEV hybrid energy storage systems. *IEEE Trans. Ind. Inform.* **2020**, *16*, 3407–3416. [CrossRef]
29. Gummi, K.; Ferdowsi, M. Double-input DC–DC power electronic converters for electric-drive vehicles-Topology exploration and synthesis using a single-pole triple-throw switch. *IEEE Trans. Ind. Electron.* **2010**, *57*, 617–623. [CrossRef]
30. Kumar, L.; Jain, S. Multiple-input DC/DC converter topology for hybrid energy system. *IET Power Electron.* **2013**, *6*, 1483–1501. [CrossRef]
31. Lai, C.M.; Yang, M.J. A high-gain three-port power converter with fuel cell, battery sources and stacked output for hybrid electric vehicles and DC-microgrids. *Energies* **2016**, *9*, 180. [CrossRef]
32. Hintz, A.; Prasanna, U.R.; Rajashekara, K. Novel modular multiple-input bidirectional DC-DC power converter (MIPC) for HEV/FCV application. *IEEE Trans. Ind. Electron.* **2015**, *62*, 3163–3172. [CrossRef]
33. Lai, C.M.; Cheng, Y.H.; Hsieh, M.H.; Lin, Y.C. Development of a bidirectional DC/DC converter with dual-battery energy storage for hybrid electric vehicle system. *IEEE Trans. Veh. Technol.* **2018**, *67*, 1036–1052. [CrossRef]
34. Hernándeza, J.C.; Ruiz-Rodriguezb, F.J.; Juradoc, F. Modelling and assessment of the combined technical impact of electric vehicles and photovoltaic generation in radial distribution systems. *Energy* **2017**, *141*, 316–332. [CrossRef]
35. Hernándeza, J.C.; Sanchez-Sutila, F.; Muñoz-Rodríguezb, F.J. Design criteria for the optimal sizing of a hybrid energy storage system in PV household-prosumers to maximize self-consumption and self-sufficiency. *Energy* **2019**, *186*, 115827. [CrossRef]
36. Gomez-Gonzaleza, M.; Hernandezb, J.C.; Veraa, D.; Juradoa, F. Optimal sizing and power schedule in PV household-prosumers for improving PV self-consumption and providing frequency containment reserve. *Energy* **2020**, *191*, 116554. [CrossRef]
37. Lai, C.-M.; Yang, C.-Y.; Cheng, Y.-H. Power Supply System and Power Supply Method for Electric Vehicle. Taiwan Patent No. I642575, 1 December 2018.

Electric Vehicles Plug-In Duration Forecasting using Machine Learning for Battery Optimization

Yukai Chen [1], Khaled Sidahmed Sidahmed Alamin [1], Daniele Jahier Pagliari [1,*], Sara Vinco [1], Enrico Macii [2] and Massimo Poncino [1]

1 Department of Control and Computer Engineering (DAUIN), Politecnico di Torino, 10129 Turin, Italy; yukai.chen@polito.it (Y.C.); khaled.alamin@polito.it (K.S.S.A.); sara.vinco@polito.it (S.V.); massimo.poncino@polito.it (M.P.)

2 Interuniversity Department of Regional and Urban Studies and Planning (DIST), Politecnico di Torino, 10129 Turin, Italy; enrico.macii@polito.it

* Correspondence: daniele.jahier@polito.it

Abstract: The aging of rechargeable batteries, with its associated replacement costs, is one of the main issues limiting the diffusion of electric vehicles (EVs) as the future transportation infrastructure. An effective way to mitigate battery aging is to act on its charge cycles, more controllable than discharge ones, implementing so-called battery-aware charging protocols. Since one of the main factors affecting battery aging is its average state of charge (SOC), these protocols try to minimize the standby time, i.e., the time interval between the end of the actual charge and the moment when the EV is unplugged from the charging station. Doing so while still ensuring that the EV is fully charged when needed (in order to achieve a satisfying user experience) requires a "just-in-time" charging protocol, which completes exactly at the plug-out time. This type of protocol can only be achieved if an estimate of the expected plug-in duration is available. While many previous works have stressed the importance of having this estimate, they have either used straightforward forecasting methods, or assumed that the plug-in duration was directly indicated by the user, which could lead to sub-optimal results. In this paper, we evaluate the effectiveness of a more advanced forecasting based on machine learning (ML). With experiments on a public dataset containing data from domestic EV charge points, we show that a simple tree-based ML model, trained on each charge station based on its users' behaviour, can reduce the forecasting error by up to 4× compared to the simple predictors used in previous works. This, in turn, leads to an improvement of up to 50% in a combined aging-quality of service metric.

Keywords: electric vehicles; light gradient boosting; battery charging; intelligent charging; optimal charging behavior; battery aging

---

1. Introduction

Given the environmental impact of petroleum-based transportation and the recent developments of renewable energy production technologies, electric vehicles (EVs) are gaining traction as the most promising transportation infrastructure for the future [1]. EVs are considered environmental-friendly because the electric power they consume can be generated from a wide variety of sources including various renewable ones [2]. Using renewables for transportation has the potential to massively reduce fuel consumption and gas emissions as well as increase the security level of energy usage, via geographic diversification of the available sources. In addition to protecting the environment, EVs also have advantages in terms of higher energy efficiency and lower noise than internal combustion engines [3].

Although EVs are currently ordinary in most sectors of public and private transportation, with rapidly growing market demands, one critical issue that still limits their adoption is related to batteries. There is an urgent demand for advanced battery optimization strategies that can satisfy customers' demands, such as increasing the driving range and reducing the charge time, while also containing the costs associated with battery replacement [4]. In fact, the battery is the main contributor to the total cost of an EV, accounting for more or less 75% of the total capital cost of the full vehicle [5]. Thus, prolonging the lifetime of the battery becomes a crucial issue in the development of EVs.

Since two decades, Lithium-ion has become the dominant battery chemistry adopted in EVs due to its relatively high energy density and power delivery ability [6]. Among the weaknesses of Lithium-ion technology, capacity loss is one of the essential aspects that influences EVs' widespread adoption [7]. The aging of a Lithium-ion battery, intended as the loss of usable capacity over time, strongly affects the battery replacement cost. The aging of the battery depends on several quantities such as temperature, depth of discharge (DOD), average state of charge (SOC), and magnitude of the charge/discharge currents [8–10]. The values of these quantities during the discharge phase (in particular the DOD and the discharge current) cannot be controlled because they depend on the motor power demand, which in turn is a function of the EV driving profile, and therefore is determined by user habits. In contrast, the charge phase is more controllable and as such provides some space for battery aging optimization.

EV battery charging is usually constrained to use standardized schemes, based on pre-defined current and voltage profiles. Because of the low cost and straightforward implementations, constant current-constant voltage (CC-CV) charge protocol is the most common one for Lithium-ion batteries. The CC-CV protocol effectively limits the risk of overcharging, which has to be carefully managed in Lithium-ion batteries. Although constraining to CC-CV does limit the space for optimizations, this protocol still offers some degrees of freedom (namely, the charge starting time and the charge current) that can be leveraged to mitigate aging [10–17].

As a matter of fact, vehicles are often connected to a charge station for a time much longer than what is needed to charge their battery, that is, the plug-in duration is often much longer than the actual charge duration. The default charge scheme starts the CC phase as soon as the vehicle is plugged in, using a large current to obtain a fast charge. While this solution yields a 100% charged battery as fast as possible, it can significantly degrade the battery capacity. This fast capacity loss is due to a twofold effect. The first direct cause is that battery aging worsens with large charge currents. Moreover, according to the default charge protocol, charging is typically completed well before the actual plug-out time. This early completion implies a higher average SOC stored in the battery, which also negatively affects the aging [8]. Therefore, there exists a trade-off between (fast) charge time and battery aging: the former requires larger charge currents and an immediate start of the charging protocol at plug-in time, both of which degrade battery capacity.

Several works in the literature have proposed aging-aware charging protocols that try to overcome the limitations of this default solution [10–17]. To do so, these protocols jointly optimize the final charge state of the battery in each charge cycle and its aging. Both objectives have to be considered since, paradoxically, the ideal policy for aging-only optimization would otherwise coincide with leaving the battery fully discharged (0 charge current and SOC), which is clearly not meaningful. In contrast, the charge state of the battery at plug-out time is fundamental for the EV driver's user experience. In practice, these approaches optimize the two free parameters of the CC-CV protocol (start time and charge current) in such a way that the battery becomes fully charged exactly when it is plugged out of the charge station.

To this end, an estimate of the plug-in duration is needed. The accuracy of such an estimate is fundamental, as pointed out in [17]. Indeed, overestimating the plug-in duration would cause the battery to have a low SOC at plug-out time, dramatically degrading user experience. On the other hand, overestimating it would worsen the battery aging as charging would complete before the actual plug-out time, increasing the average SOC. Despite the importance of accurate plug-in duration estimation, previous works have only relied on elementary models. Several works [10,11,13] assume that the plug-in duration is set directly by the user when plugging the EV into the station. In those approaches, the responsibility of providing an accurate estimate is entirely left to the user, who mostly cares about the driving experience rather than battery aging. Other approaches, although not targeting EV batteries specifically, attempt an automatic estimate [12,17], but only using basic models such as fixed-time predictions or moving averages.

In this work, we assess the effectiveness of using a machine learning (ML) model in improving the accuracy of plug-in duration estimation. Our approach is based on the observation that, especially for domestic stations, plug-in and plug-out instants depend on the habits of a single user (or a small group). Therefore, we envision a system in which each charge station autonomously learns its users' plug-in behaviour from history. With experiments on a public dataset containing records from domestic charge stations in the UK [18], we show that a simple tree-based model (light gradient boosting or LightGBM) can reduce the prediction error compared to all straightforward policies considered in previous work. Using this model on top of an aging-aware charging protocol, in turn, yields an improvement of up to 54% on a combined quality-of-service/aging metric. Moreover, the accuracy of the ML predictor is strongly related to the number of records present in the dataset for a given charge station, suggesting that even better results could be achieved with more available data.

In summary, our main contributions are the following:

- We evaluate for the first time the effectiveness of a ML-based approach for predicting the plug-in duration of EVs in domestic charge stations.
- We show that this method is superior in terms of prediction accuracy with respect to the straightforward policies used by previous works.
- Finally, we show that this reduction of the prediction error actually translates into an improvement in terms of quality-of-service and battery aging, when the prediction is used within an aging-aware EV charging protocol.

The rest of the paper is organized as follows. Section 2 provides the required background on battery charging and aging models, and discusses related works; Section 3 describes the different plug-in duration estimates considered in our experiments. Section 4 reports the results, while Section 5 concludes the paper.

2. Background and Related Works

2.1. EV Battery Capacity Aging Degradation

The capacity loss of rechargeable Lithium-ion batteries depends on four main factors [8,13,19,20]: (i) temperature, (ii) DOD at each cycle (also referred to as deviation of the SOC), (iii) average SOC, and (iv) charge/discharge current. Aging worsens with an increase in any of these quantities.

Among these four main factors, given that temperature cannot be easily controlled and that DOD and discharge currents depend on the power demand and duration of the discharge phase, only the charging current and the average SOC can be managed during the charging process for optimization.

The average SOC for a generic time interval from t_0 to t_1, is given by:

$$SOC_{avg} = \int_{t_0}^{t_1} SOC(t)dt \, / \, (t_1 - t_0) \tag{1}$$

For instance, [19] reports that, for $LiFePO_4$ batteries, SOC should be less than 60% on average for maintaining battery life acceptable.

Aging is usually evaluated through the state of health (SOH) aggregate metric, defined as the ratio of the capacity of an aged battery and its nominal capacity. In this work, since we focus on multiple charge cycles, we need a model that expresses the aging for each cycle. To this purpose we use the classical model of [19] augmented as in [13] to account for charge and discharge current. The following expression determines battery capacity loss (L) in the i-th cycle:

$$L_i = L_{0,i} \cdot e^{(K_{ic,i} \cdot I_{ch,i} + K_{id,i} \cdot I_{dis,i})} \quad (2)$$

where $L_{0,i}$ is the battery aging factor provided by the Millner's model [19], which accounts for temperature, deviation of the SOC, and SOC_{avg} in the i-th cycle; and L_i is the battery aging computed by the model provided by [13], which strengthen the Millner's model by adding aging dependence on discharge and charge current values in the i-th cycle, $I_{ch,i}$ and $I_{dis,i}$. $K_{ic,i}$ and $K_{id,i}$ are empirical coefficients extracted from battery datasheet information [13] or from experimental measurements. By summing L_i over M cycles, we get the total loss of capacity L_M. L_M and SOH are both normalized, so they are related as $SOH = 1 - L_M$. The SOH after M-th cycle is therefore indicated by:

$$SOH_M = 1 - L_M = 1 - \sum_{i=0}^{M} (L_i) \quad (3)$$

This model can be applied to any device powered with Lithium-ion type batteries, and supports battery aging estimation after multiple cycles. Therefore, it can be used to identify charging protocols that best fit a specific user behavior from the point of view of both battery aging and quality of service (QoS).

In this work, we use the average SOC at plug-out time as a metric of QoS, since the available residual capacity at the plug-out time determines the quality of user experience. Mathematically:

$$QoS = \frac{1}{n} \sum_{i=1}^{n} SOC_{plug-out,i} \quad (4)$$

This metric is commonly used [10,17], since a higher SOC at the plug-out time guarantees a longer driving range and a better driving experience (e.g., allowing the enabling of auto-auxiliary driving system, on-board multi-media systems, etc.), and vice-versa.

2.2. EV Battery Charging with CC-CV

Charging a battery is an operation that may significantly impact its lifetime, even more than discharging when comparing the effects of the same absolute current rate in both phases [9]. Selecting the appropriate charge protocol is thus a critical step, not only for keeping as much as possible unaltered the battery performance, but also for avoiding dangerous side effects, like overheating and overcharging, which besides creating obvious hazards, can also worsen the battery SOH, defined in (3) and accelerate the battery aging.

The selection of the proper charge scheme depends on the battery chemistry. For Lithium-ion batteries (the majority of EVs are equipped with this kind of battery cells), the standard is to adopt the CC-CV protocol [21]. Although various optimized charging methods were explored and reported in the literature (e.g., [15,22]), CC-CV is still adopted in the great majority of Lithium-ion battery-based systems, due to the simplicity of chargers implementations from an electrical point of view, and because it guarantees battery safety against over-voltage and over-current.

The CC-CV protocol operates in three phases, as shown in Figure 1. In the first phase (CC), the battery is charged at a constant current until its voltage reaches a pre-determined limit; in the second phase (CV), the battery is charged at a constant voltage until the current drops to a pre-defined value. This second phase effectively manages the risk of overcharging, which is quite dangerous in Lithium-ion batteries. The time interval starting after the end of the second phase until to the unplug time is called standby time. As shown in Figure 1, the SOC remains fixed at 100% in this third phase, which is detrimental for the SOH of Lithium-ion batteries, since the length of standby period significantly increases the average SOC of the battery. An analytical macro-model of CC-CV charge time based on a subset of all relevant parameters, namely, average SOC, deviation of the SOC (here considered as depth-of-discharge), discharge/charge current, and temperature is presented in [23].

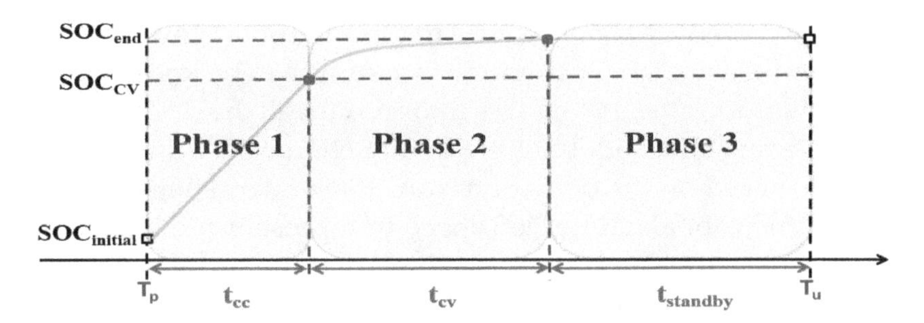

Figure 1. Typical CC-CV charging protocol scenario.

In this work, we evaluate the impact of an accurate plug-in duration estimation on aging-aware charging protocols that stick to the CC-CV scheme, and only act on the free parameters made available by it. As explained in [10,17] these free parameters are the starting time and the slope (which depends on the current) of the first phase of Figure 1, i.e., the CC phase.

Solutions that remain compliant with CC-CV do so in order to maintain its electrical simplicity, low cost and safety properties, while still adapting the standard to consider battery aging. In our work, considering these solutions allows us to assess the impact of our plug-in duration forecasting on realistic charge stations most commonly used nowadays. However, our method is actually orthogonal to a specific charging profile, and could therefore be used also in conjunction with advanced aging-aware schemes that do not follow the CC-CV standard [9,14–16].

2.3. *Aging-Aware Charging Protocols*

As anticipated, most of the previous works on aging-aware charging of Lithium-ion batteries focus on altering two main CC-CV protocol variables, namely the average SOC (shortening length of Phase 3 in Figure 1) and the constant charging current (decreasing the slope of the line in Phase 1 of Figure 1). The two approaches are schematized in Figure 2.

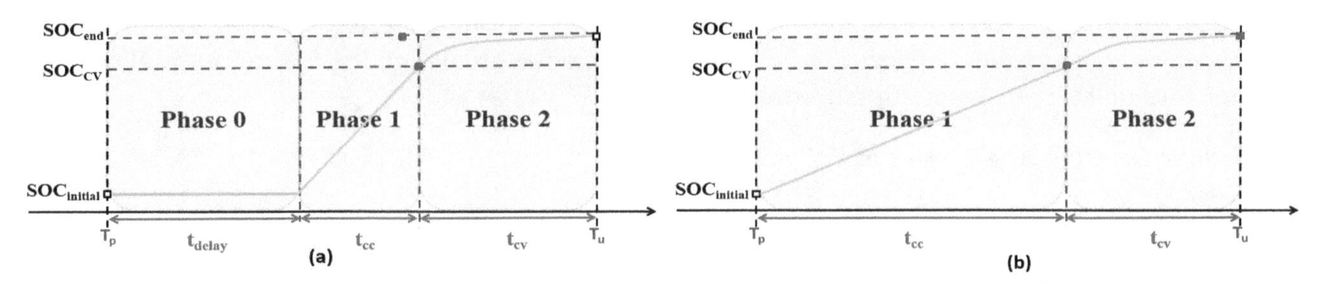

Figure 2. Aging-aware charging protocols: (**a**) Delay the charge starting time to reduce the average SOC; (**b**) Decrease the charging current in CC phase.

In terms of the SOC effect on aging, the charge phase should ideally reach 100% exactly at unplug time (see Figure 2a); this would yield the smallest possible average SOC, minimizing the length of the standby period, while still guaranteeing a fully charged battery, which corresponds to the best QoS [12]. In contrast, when charging is started immediately at the plug-in time, batteries are often left fully charged for a long time, as indicated by the $t_{standby}$ in Figure 1, with a significant impact on battery aging [8]. The work in [12] was one of the firsts to consider delayed start time for charging batteries as late as possible, thus minimizing the average SOC.

Concerning the charge current, Ref. [11] mitigates battery aging by considering only this parameter, and calculating a minimum current that ensures a fully charged battery at the end of a predicted plug-in period, whenever it is higher than the charge time needed in standard CC-CV. In [16], the non-linear relation between charge current and charging time is analyzed.

In [13], both charge current and average SOC are taken into account. The authors show that the aging-optimal charge current is more related to battery usage rather than plug-in time, and that the capacity loss vs. charge current characteristic is not monotonic. However, their aging analysis is limited to a single cycle, and the actual plug-in time is assumed to be known.

A CC-CV compliant charge protocol that takes into account all the relevant parameters is proposed in [17]. Because of the opposing goals of obtaining a fully charged EV and optimizing battery aging, Ref. [17] uses a QoS metric based on the deviation from 100% charge level at plug-out time. Results are then reported in a 2-dimensional aging/QoS space to represent the trade-off between these two quantities. With this multi-objective analysis, it is shown that the proposed protocol obtains a better trade-off under various user charge/discharge pattern statistics compared to [13], which only considers aging at the expense of low QoS. Although both [13,17] focus on the Lithium-ion batteries found in mobile devices (such as smartphones), the proposed charging protocols are general and can be applied to any battery-powered device, including EVs.

Finally, other works have proposed aging-aware charge protocols that however do not stick to the CC-CV scheme [14–16].

Importantly, many recent works [10–13,17] analyze the need for predicting the plug-in duration by extracting the data from battery usage history; however, no accurate prediction mechanisms are proposed. For example, Ref. [17] calculates the optimal charge current based on a simple prediction of the plug-in duration to achieve a "just-in-time" charge. Although this work gives a detailed analysis of the effect of plug-in duration forecasting, it only uses a simple predictor based on an exponential moving average (EMA) to estimate the current cycle's plug-in duration. The few simple predictors proposed in literature are used as baselines for comparison in our experiment, and are detailed in Section 3.3.

The recent work of [24] proposed an intelligent charging solution to prolong the lifetime of battery-powered devices. Also this work underlines the importance of plug-out time prediction (Notice that, since the plug-in time is known at the beginning of each charging phase, estimating the plug-in duration or the plug-out time is equivalent, and we use both terms interchangeably in the paper) accuracy for alleviating the aging of battery during charging phase. It suggests using multiple data sources and connecting multiple battery-powered intelligent devices to increase the accuracy of predictions and to allocate charging power intelligently among different devices. Unfortunately, this work only lists the need of accurate prediction as an open challenge for designing intelligent EV chargers, and no specific forecasting model is proposed.

2.4. *Machine Learning Applications in EVs*

Machine learning is used in many other applications related to EVs and to battery-related optimizations. Among the most relevant ones is driving range estimation [4,25,26], which has been addressed, among others, using ML models based on linear regression [26] and self-organizing maps [25]. ML based on deep neural networks has also been used to optimize the energy requested by EVs in a demand-side management framework [27]. Finally, other applications of ML models to EVs include engine faults diagnosis [28] and estimation of the battery State-of-Charge (SOC) [29] and State-of-Health (SOH) [30]. Notice that the latter are different and orthogonal to the goal of this work. In fact, our aim is estimating the plug-in time, which is then used to guide a battery-aware charging protocol. In turn this determines the SOC and the SOH of the battery, which can be estimated either with the analytical models described in Section 2.1 or with the ML methods of [31,32]. To the best of our knowledge, ML-based approaches have never been used for plug-in duration estimation.

3. Methods

The objective of our work is to build a reliable predictor of the plug-in duration of an EV onto a domestic charging station. Specifically, we want to assess whether a ML solution does yield superior prediction accuracy compared to the basic prediction policies used by previous works, which we take as comparison baselines. Then, we want to assess whether this superior accuracy corresponds to a sizeable improvement in the QoS/aging space for the EV battery. To this end, we apply our ML-based prediction, as well as the baselines, on top of the As Soon As Possible (ASAP) and Aging-Optimal battery-aware CC-CV protocols described in [13,17]. We briefly describe the two protocols in Section 4.1 and refer the readers to the original papers for more details. As anticipated, our predictor is independent of the underlying charge protocol; nonetheless, testing with these two state-of-the-art protocols allows us to assess its real impact on battery aging and QoS.

In the rest of this section, we first describe the scenario in which we envision to deploy our ML-based forecasting in Section 3.1. We then detail the selected ML algorithm in Section 3.2 and the comparison baselines in Section 3.3.

3.1. Continuous Charge Behaviour Learning with Edge Computing

In the target scenario, plug-in time forecasting (i.e., ML inference) must be performed at the beginning of each EV charging phase, in order to let the charge station implement an optimized protocol which needs this information as input. Moreover, once the charge cycle ends, the real plug-in duration becomes known. In case of a ML approach, this new information should be used to update the prediction model, so that the system continuously learns from the users' charging behaviours. This strategy can allow significant improvements in prediction accuracy over time [33]. However, it also implies that, at the end of each charge cycle, a re-training of the ML model should be executed.

Modern EV charge stations include either a central processing unit (CPU) or a micro-controller unit (MCU) [34], normally used to monitor the charging, provide alerts and feedback to the user, etc. One option could be to use this processing device just for collecting historical data of past charging phases (e.g., a record containing plug-in time, plug-out time, day of the year, etc. for each charge cycle), and then offload all the processing to a cloud server [35]. However, this solution requires the availability of a cloud infrastructure, and that the charge station is constantly connected to the Internet. Moreover, transmitting EV charging records over the Internet may also raise security concerns, as these data might be intercepted by malicious third parties and used to infer private information about users' behaviours (e.g., when they are at home or not) [36]. As an alternative to this approach, we propose a solution based on edge computing, where the ML algorithm is directly executed in the charge station processing hardware [35,36]. The two alternatives are schematized in Figure 3.

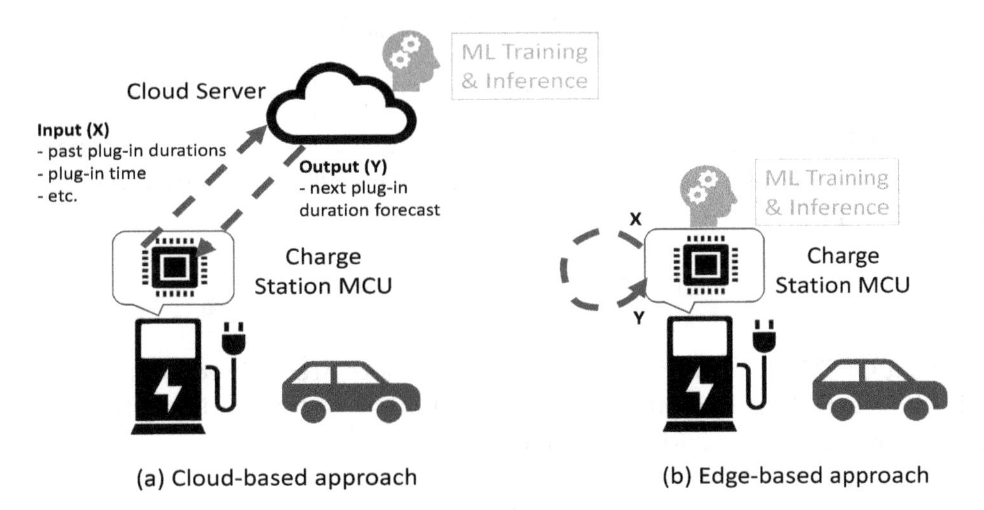

Figure 3. Alternative scenarios for the deployment of ML-based plug-in duration forecasting.

As shown in the Figure, the edge-based solution solves all aforementioned problems, but it introduces new limitations to the characteristics of the selected ML algorithm. In fact, processors present in EV charging stations (e.g., from the ARM Cortex family) are normally designed for embedded applications. While these devices are sufficiently powerful to implement simple ML-based predictors, their compute power (number of cores, clock frequency, available memory) is significantly more limited than what is available on a cloud server [35,37]. This is especially critical since, as anticipated, both training and inference should be performed repeatedly. Therefore, computational complexity for the training and inference phases of the model becomes a key design metric. For this reason, complex models such as deep neural networks [38], although possibly very accurate, are not a viable option. In contrast, we select light gradient boosting (LightGBM) [39] to implement our ML-based predictor. LightGBM models are based on tree learning, and are explicitly designed to limit the computational complexity for training (i.e., tree growing), as detailed in the following section.

3.2. *Light Gradient Boosting*

LightGBM is a type of ML model belonging to the gradient boosting family, and it is based on ensembles of trees. It was first proposed by [39] in 2017.

In our work, we decided to focus on tree-based learning since it is inherently one of the most efficient types of ML in terms of both inference and training complexity. In fact, performing inference (i.e., classification or regression) on an input datum simply requires performing a set of comparisons with thresholds, one for each visited tree node [40]. Therefore, the number of operations performed in an inference is $O(\sum_{m=1}^{M} d_m)$, where d_i is the depth of the m-th tree and M the number of trees in the ensemble. Similarly, the number of parameters (i.e., thresholds) of the model is $O(\sum_{m=1}^{M} w_m)$, where w_m is the total number of nodes in the m-th tree. Comparisons are very simple operations from a hardware point of view, which makes this type of inference easy to implement even on micro-controllers [40]. In addition to maintaining this inference simplicity, LightGBM extends the standard gradient boosting decision tree (GBDT) algorithm [41] with several optimizations that speed-up its training time up to 20 times. Moreover, tree-based learning is also effective when training set sizes are not extremely large [42], as in the case of the dataset described in Section 4. In such a "small-data" setting, other models, such as deep learning ones, which are extremely accurate in general, would likely lead to over-fitting and therefore to poor results [43].

Considering the training set of a supervised learning problem $(X, Y) = \{(x_i, y_i)\}_{i=1}^{n}$, both GBDT and LightGBM build predictions in the form:

$$\hat{y} = \hat{F}_M(X) = \sum_{m=1}^{M} \gamma_m \hat{f}_m(X) \tag{5}$$

where $\hat{f}_m(X)$ is a decision tree, also called weak learner and γ_m is a weight. The goal of training is to minimize the expected value of a loss function $L(y_i, \hat{F}_M(x_i))$, such as the squared error loss in the case of regression problems [44]:

$$L(y_i, \hat{F}_M(x_i)) = \frac{1}{2}(y_i - \hat{F}_M(x_i))^2 \tag{6}$$

As in standard GBDT, also in LightGBM weak learners are trained sequentially one after the other, and each tree learns to predict the negative residual error (or gradient) from previous models. In mathematical terms, negative residuals are computed as:

$$r_{i,m} = -\frac{\delta L(y_i, \hat{F}_{m-1}(x_i))}{\hat{F}_{m-1}(x_i)} = y_i - \hat{F}_{m-1}(x_i) \tag{7}$$

where the second equality comes from the squared error loss in (6). The m-th tree is then grown using the training set $(X, R_m) = \left\{(x_i, r_{i,m})\right\}_{i=1}^{n}$. Finally, the corresponding weight γ_m is computed as:

$$\gamma_m = \underset{\gamma}{\text{argmin}} \sum_{i=1}^{n} L(y_i, \hat{F}_{m-1}(x_i) + \gamma \hat{f}_m(x_i)) \tag{8}$$

In terms of complexity, the most critical step of this training procedure is the growth of individual trees. In LightGBM, trees are grown leaf-wise, as shown in Figure 4. As explained in [39], this requires $O(n{\cdot}\phi)$ operations for each expansion, where ϕ is the number of features in each input sample x_i.

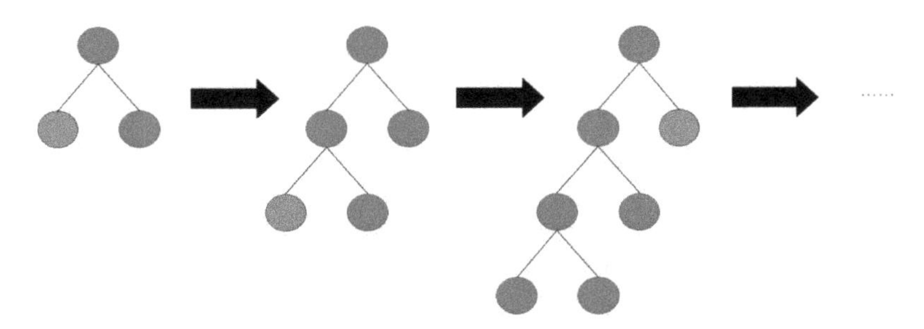

Figure 4. Leaf-wise tree growth in LightGBM.

LightGBM reduces this computational burden by sub-sampling training data and bundling features. The first objective is achieved thanks to a technique called gradient-based one-side sampling (GOSS): when growing the m-th tree, GOSS samples the training instances considered for split-point selection based on the magnitude of their gradient $r_{i,m}$, computed as in (7). Specifically, the $a \times 100\%$ training instances with largest gradient are always selected, whereas $b \times 100\%$ of the remaining instances are randomly sampled. The values of a and b are hyper-parameters of the algorithm. The authors of [39] show that $a, b < 0.1$ are sufficient to obtain good accuracy, while significantly speeding up training time. LightGBM combines GOSS with so-called exclusive feature bundling (EFB), an optimization which further reduces the training time by "bundling" mutually-exclusive features (i.e., those that are never simultaneously $\neq 0$). While EFB is very effective in general [39,44], it is less relevant than GOSS for our work, since we train our LightGBM predictor using few and not mutually exclusive features (see Section 4).

In our experiments, we retrain the LightGBM model after each EV charge cycle, using as training data all past history until that moment. For simplicity, the model is currently retrained from scratch every time. However, we plan to also experiment with incremental learning techniques for GBDT-like models, such as those proposed in [45] in our future work, to further speed-up training time.

3.3. *Baseline Algorithms*

Previous works on aging-aware Li-ion battery charging schemes have almost always assumed the plug-in duration as a known input. The few exceptions, such as [12,17] have used simple time-series predictors such as the exponential moving average (EMA) and the historical average (HA). Therefore, to the best of our knowledge, ours is the first work to utilize a proper ML algorithm for plug-in duration forecasting. Accordingly, in our experiments, we have compared the proposed LightGBM predictor with the following four simple baselines.

3.3.1. Fix Duration and Fix Time

It is not easy to compare the ML-based solution against a scenario in which the EV users directly indicate the predicted unplug time when they leave the car at the charge station (as assumed in [10,11,13]). In fact, the behaviour of each vehicle user is different. Nonetheless, we can use two simple predictors called Fix Duration and Fix Time to mimic two common categories of behaviours.

The Fix Duration prediction always assumes a fixed plug-in duration $\overline{y}_{fd}$, regardless of any other condition (plug-in time, day of the week, etc.). In our experiments, we have tried setting $\overline{y}_{fd}$ to 6 h, 8 h and 12 h. The best results have been obtained with 6.

The slightly more complex Fix Time predictor, instead, assumes that the unplug time of the EV always occurs at one of two possible fixed times of the day $\bar{t}_{ft,1}$ and $\bar{t}_{ft,2}$. This is similar to a user-specified "alarm-like" unplug prediction. Specifically, if the EV is plugged in at time t where

$\bar{t}_{ft,1} \leq t < \bar{t}_{ft,2}$ this predictor assumes that the unplug time is $\bar{t}_{ft,2}$. Conversely, if $t < \bar{t}_{ft,1} \vee t \geq \bar{t}_{ft,2}$ it assumes that the EV will be unplugged at $\bar{t}_{ft,1}$, where this value refers to the next day if the second condition is verified. The predicted plug-in duration $\hat{y}_{ft}$ is then computed as the difference between the predicted unplug time and t. Mathematically:

$$\hat{y}_{ft} = \begin{cases} \bar{t}_{ft,1} - t & \text{if } t < \bar{t}_{ft,1} \\ \bar{t}_{ft,2} - t & \text{if } \bar{t}_{ft,1} \leq t < \bar{t}_{ft,2} \\ \bar{t}_{ft,1} + 24h - t & \text{if } t \geq \bar{t}_{ft,2} \end{cases} \quad (9)$$

This predictor takes as its only input the latest plug-in time t. Using two different fixed times allows to account for EV charges happening both during the day and during the night, a common pattern as pointed out in [10]. In our experiments, we tried different combinations of $\bar{t}_{ft,1}$ and $\bar{t}_{ft,2}$, and achieved the best results using 7.00 a.m. and 7.00 p.m. respectively.

3.3.2. Exponential Moving Average

This baseline is taken from the works of [12,17] which, although targeting smartphone batteries rather than EVs', proposed to estimate the unplug time using an exponential moving average (EMA). The estimated plug-in duration at cycle i is therefore computed as:

$$\hat{y}_{ema}[i] = w * y[i-1] + (1-w) * \hat{y}_{ema}[i-1] \quad (10)$$

where $y[i-1]$ is the actual (measured) plug-in duration at the previous charge cycle, and $w = 0.6$ is a smoothing weight. EMA-based forecasting assumes that consecutive charge cycles have a similar duration; as such, most of the estimate depends on the latest measurement, while the previous history is accounted for by the second addend. This prediction does not depend on the plug-in instant, like (9), and only uses the previous plug-in duration as an input.

3.3.3. Historical Average

The historical average (HA) is different from the EMA in that it gives equal importance to the entire past history of plug-in duration measurements.In this case, the predicted duration is simply computed as the average of all past measurements, i.e.,:

$$\hat{y}_{ha}[i] = \frac{1}{i-1} \sum_{j=1}^{i-1} y[j] \quad (11)$$

This predictor is used in previous work by [12]. Again, the only required input for this forecasting strategy is the set of past plug-in duration.

4. Results

4.1. Experimental Setup

The methods described in Section 3, have been applied to the publicly available "Electric Chargepoint Analysis: Domestics" dataset, collected by the United Kingdom's office for low emission vehicles (OLEV) [18]. This dataset contains records of charging events from approximately 25,000 domestic charge-points across the UK, collected during the year 2017, for a total of 3.2 million charging events. Each charging record contains the dates and times of the start and end of the plug-in, as well as the acquired energy in KW, the plug-in duration, the charge point identifier, and the charge event identifier. Our experiments have been performed on 5 charging points with identifiers AN05770, AN10157, AN23533, AN08563, and AN03003, selected based on the large number of available charging events (see Table 1). We did not generalize our results to the entire dataset because the great majority of the charging points contain a very limited number of charging event records, which are not sufficient to train the proposed LightGBM model. Before training the predictors, all charge events longer than 40 h have been filtered out as outliers (e.g., holidays), since we have found that they worsened the training results.

Table 1. Number of charge events in each considered station.

Station ID	Events (Cycles)
AN05770	326
AN10157	196
AN23533	186
AN08563	159
AN03003	141

Both the LightGBM predictor and the comparison baselines are trained and tested on individual charge points, in order to simulate the scenario described in Section 3.1. The first 65% of the total events in each charge station have been used as initial training set, and the predictors have been evaluated on the remaining 35%.

As reported in Table 2, as inputs for the LightGBM model, we used the following features for each charging event: the plug-in date and time, expressed as day of the year, hour and minute; the day of the week of plug-in, encoded using a one-hot format, in order to account for different charging patterns during the week and on the weekend; the plug-in duration of the previous charge cycle; the plug-out date and time of the previous charge cycle. LightGB algorithm parameters have been tuned using grid search on each charge station. We also tried feeding the LightGBM model with a longer past history, but we empirically found out that this led to overfitting (i.e., a reduction of the forecasting error on the training set but an increase on the test set) and therefore worsened the performance of the model when used within an aging-aware battery charging protocol.

Table 2. Input features for the LightGBM model.

Feature	Description
Plug-in Instant	Day $\in$ 0–366, Hour $\in$ 0–23, Minute $\in$ 0–59
Plug-in day of the week	One-hot encoded $\{0, 1\}^7$ vector
Prev. Plug-out Instant	Day $\in$ 0–366, Hour $\in$ 0–23, Minute $\in$ 0–59
Prev. Plug-in Duration	In hours (possibly fractional)

Plug-in duration predictors have been written in Python, using the LightGBM package [46] for the proposed model. Battery discharge and charge cycles simulations have been performed in MATLAB. We selected the A123 Systems ANR26650M1A automotive Lithium-ion battery in our simulations. All the necessary parameters of its aging model to compute the L_0, i in (2) are provided in [19], whereas the related discharge and charge current rate coefficients, K_{ic} and K_{id} in (2) are extracted from [13]. The environment temperature in our simulations is set as 25 °C and the battery operating temperature is assumed as a constant value equal to 35 °C. To simulate the charging phase, we adopted the model of [36], which supports changing the current values and computing the length of CC and CV phases. The maximum and minimum CC phase charging currents have been set to 0.1C rate and 2C rate in the aging-aware charging protocols.

The EV discharge behaviour cannot be inferred from the dataset. So, it has been modeled using the same method of [10]. For each charge station, the simulations relative to different forecasting methods have been fed with the exact same sequence of discharge profiles, so that the comparison among them is fair. For what concerns the charge phase, which is the main focus of this work, as anticipated in Section 3, we inserted the different plug-in duration forecasting algorithms in two aging-aware CC-CV-based charging protocols. The first one, which we refer to as Aging-Optimal, was proposed in [13] and works by delaying the charging start time and simultaneously reducing the charging current. Moreover, we also consider the as soon as possible (ASAP) protocol of [17], which starts charging immediately and only reduces the charging current. Both of them need an accurate prediction of the plug-out time to determine the optimal charging current in the CC phase or its starting time. We remark that designing an aging and/or QoS optimal charging protocol is not the target in this work. We test on these two aging-aware charging protocols just to illustrate the importance of plug-in duration prediction accuracy.

4.2. Forecasting Error

As a first experiments, the proposed model and the baselines have been compared in terms of pure prediction error. For this, the mean square error (MSE) has been used as a target metric, defined as:

$$\mathrm{MSE} = \frac{1}{n}\sum_{i=1}^{n}(y_i - \hat{y}_i)^2 \quad (12)$$

where y is the actual plug-in duration, $\hat{y}$ is the forecast one and n is the number of plug-in events at a given charge point.

The results of this experiment are presented in Figure 5, which reports the MSE obtained by all forecasting methods for each of the five considered charge points. As shown, the LightGBM forecasting obtains the lowest error on all five stations.

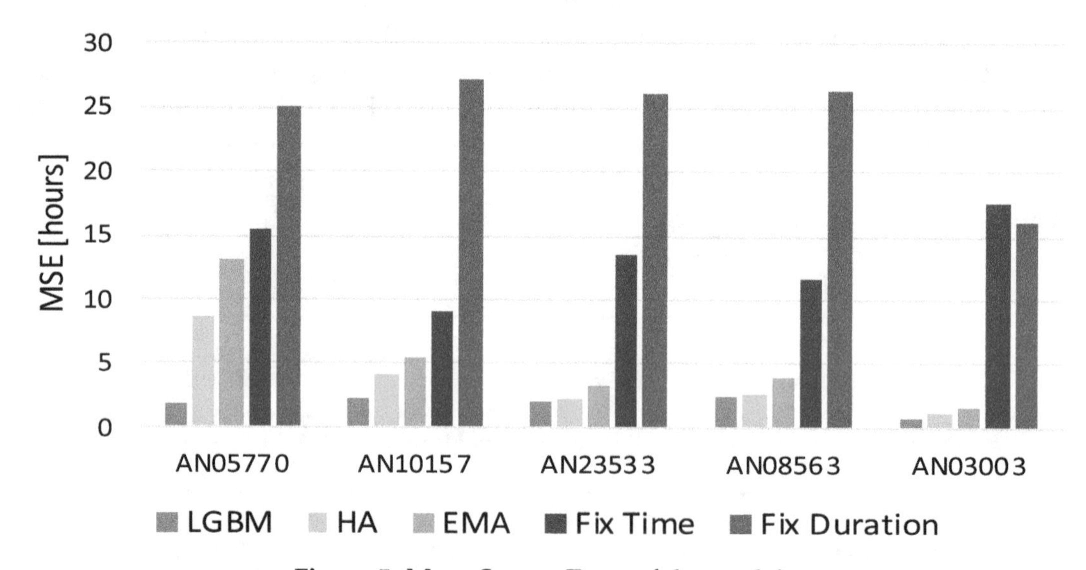

Figure 5. Mean Square Error of the models.

Interestingly, the HA predictor is a close second for some stations, showing that the average plug-in duration is approximately constant over long periods. However, on average, LightGBM reduces the prediction error by 34% compared to HA. Moreover, the difference between the two methods is the largest for the charge point labeled AN05770, which is also the one with the most numerous charge cycles in the dataset (see Table 1). For that station, the error reduction of LightGBM with respect to HA is >4×. This suggests that having more data for training allows the ML-based method to learn the subtleties of user EV charging behaviours, and consequently improve the forecasting accuracy compared to a simple predictor like HA. Figure 6a shows the actual plug-in duration in hours and the corresponding LightGBM prediction for the plug-in events of station AN05770; Figure 6b indicates the absolute error between our prediction and the real plug-in duration for each charge event; Figure 6c displays the histogram of the LightGBM prediction error. This example visually shows that the proposed ML-based predictor is able to provide an accurate estimate in most cases, with the largest errors happening in correspondence of outliers.

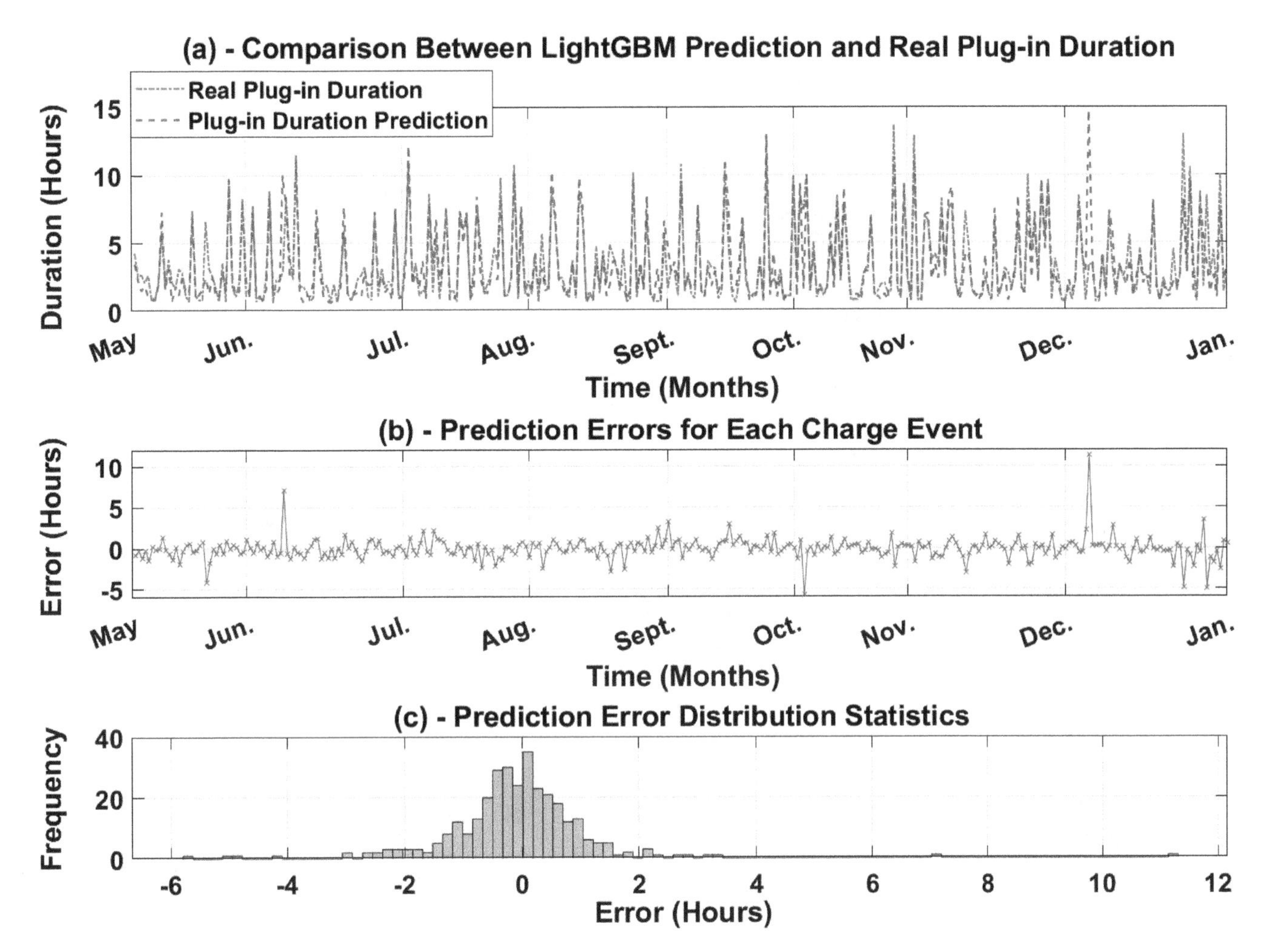

Figure 6. Comparison between LightGBM prediction and actual plug-in duration (in hours) for charging point AN05770.

Aging and QoS Results

After verifying the performance of LightGBM forecasting, we inserted our proposed plug-in duration estimator into the two aging-aware charging protocols described in Section 4.1. According to previous works [17], we measured the performance of the charge protocols enhanced with the estimators on a 2-dimensional space. The impact of forecasting on battery aging has been measured using the SOH metric expressed in (3), whereas (4) was used to measure the QoS. Both SOH and QoS metrics are normalized between 0 and 1, with 1 corresponding to the ideal value.

Figures 7 and 8 show the results of these simulations on the 2D metrics space. Optimality corresponds to the top-right corner of each chart, where both metrics have value 1. Table 3 reports the numerical results of these experiments. As shown, the proposed LightGBM always achieves the best QoS among all forecasting methods, for all stations and for both charging protocols. In terms of SOH, the ML-based prediction achieves slightly worse results than those obtained with the other methods, although still significantly better than those obtained by a standard (i.e., not aging-aware) CC-CV solution with no forecast. However, this demonstrates that pure aging-optimization without considering the QoS is meaningless. Indeed, looking only at the SOH axis, the Fix Duration prediction would be the best. However, this approach yields extremely low QoS values, ranging from 30% to less than 10%, which means that, using this solution, drivers would find their EV still almost fully discharged when they take it from the station.

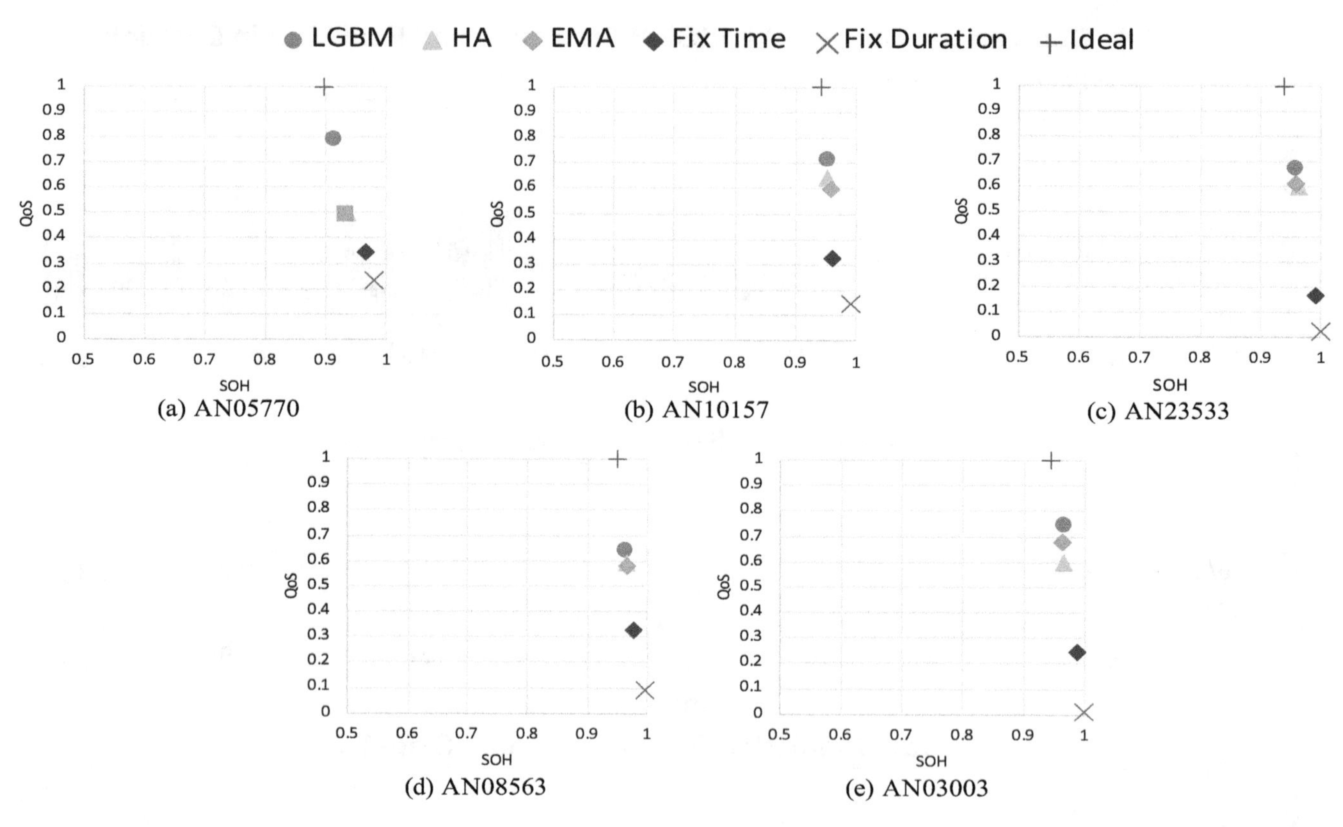

Figure 7. Battery aging (SOH) versus QoS using different plug-in duration predictors and the Aging Optimal charge protocol from [13].

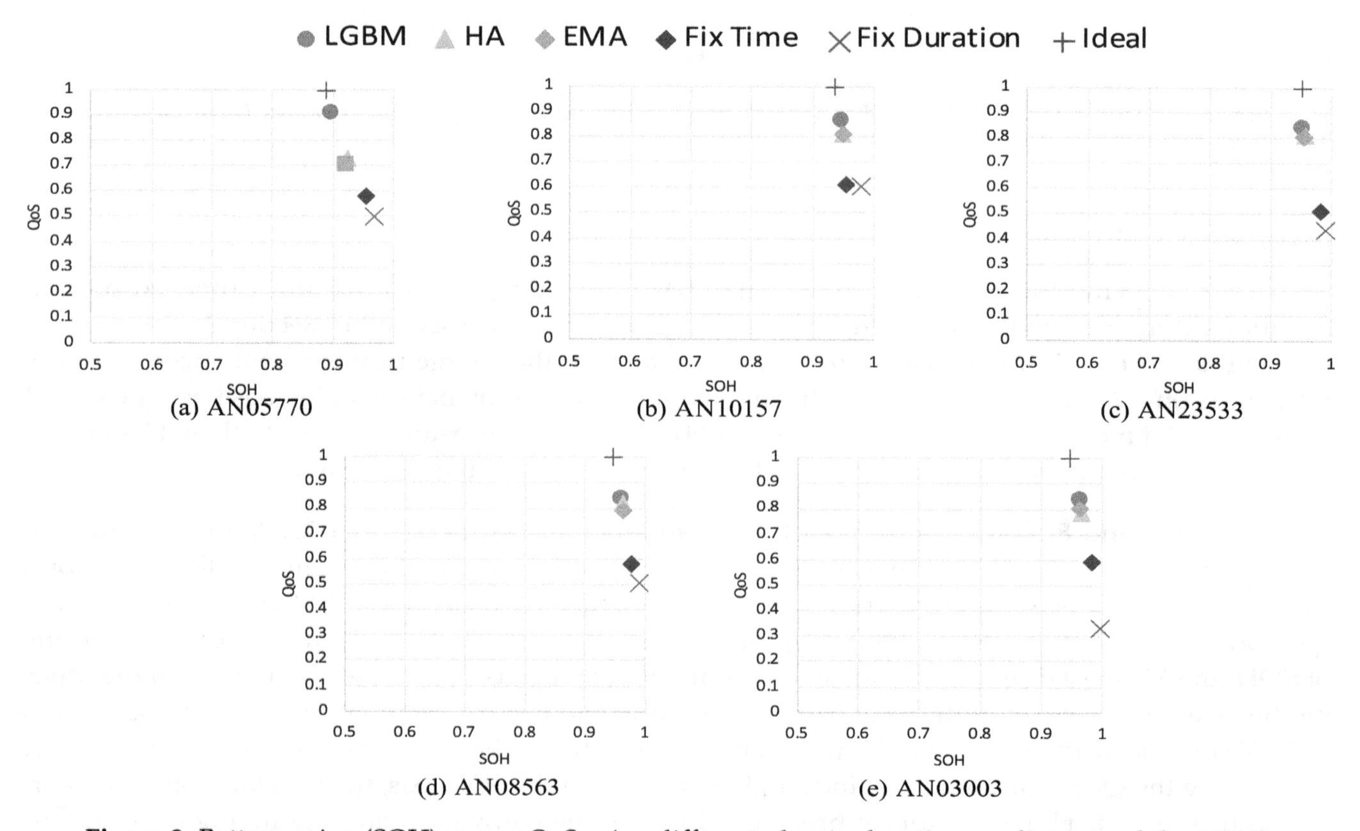

Figure 8. Battery aging (SOH) versus QoS using different plug-in duration predictors and the ASAP charge protocol from [17].

Table 3. MSE, QoS and SOH results using different plug-in duration predictors and the ASAP and Optimal charging protocols.

Station	Model	MSE	SOH		QoS		SOH * QoS	
			Aging Optimal	ASAP	Aging Optimal	ASAP	Aging Optimal	ASAP
AN05770	Ideal	0	0.89	0.88	1	1	0.89	0.88
	LightGBM	1.75	0.91	0.89	0.79	0.91	0.72	0.81
	HA	8.62	0.93	0.92	0.50	0.72	0.46	0.67
	EMA	12.99	0.93	0.92	0.49	0.70	0.46	0.65
	Fix Time	15.43	0.96	0.95	0.34	0.58	0.33	0.55
	Fix duration	25.02	0.98	0.96	0.23	0.50	0.23	0.48
AN10157	Ideal	0	0.94	0.93	1	1	0.94	0.93
	LightGBM	2.18	0.95	0.94	0.71	0.87	0.68	0.82
	HA	4.15	0.95	0.95	0.63	0.81	0.60	0.77
	EMA	5.40	0.95	0.95	0.59	0.80	0.57	0.77
	Fix Time	9.07	0.96	0.95	0.32	0.61	0.31	0.59
	Fix duration	27.21	0.99	0.98	0.14	0.60	0.14	0.59
AN23533	Ideal	0	0.93	0.93	1	1	0.93	0.93
	LightGBM	2.00	0.95	0.95	0.67	0.84	0.64	0.80
	HA	2.14	0.96	0.95	0.59	0.80	0.57	0.77
	EMA	3.32	0.95	0.95	0.60	0.80	0.58	0.77
	Fix Time	13.50	0.99	0.98	0.16	0.51	0.16	0.50
	Fix duration	25.99	0.99	0.99	0.02	0.43	0.02	0.43
AN08563	Ideal	0	0.94	0.94	1	1	0.94	0.94
	LightGBM	2.35	0.96	0.95	0.64	0.83	0.62	0.80
	HA	2.55	0.96	0.96	0.58	0.82	0.56	0.78
	EMA	3.93	0.96	0.96	0.58	0.79	0.58	0.79
	Fix Time	11.54	0.97	0.97	0.32	0.58	0.31	0.56
	Fix duration	26.26	0.99	0.99	0.09	0.50	0.08	0.49
AN03003	Ideal	0	0.94	0.94	1	1	0.94	0.94
	LightGBM	0.75	0.96	0.96	0.74	0.84	0.72	0.81
	HA	8.62	0.96	0.96	0.59	0.78	0.57	0.75
	EMA	1.58	0.96	0.96	0.67	0.80	0.65	0.77
	Fix Time	17.52	0.98	0.98	0.24	0.59	0.24	0.58
	Fix duration	16.06	0.99	0.99	0.01	0.33	0.01	0.33

To better show this aspect, in Figures 7 and 8 we have also plotted the result achieved by an Ideal predictor, i.e., an oracle algorithm that always knows the actual plug-in duration. As shown in the figures, this perfect predictor would also yield a slightly worse battery aging, in exchange for a perfect QoS (always 100% SOC at unplug time). The proposed LightGBM is therefore closer to an ideal forecast in all experiments. Interestingly, the HA method is significantly worse than LightGBM in terms of average QoS for some stations (e.g., AN05770 in Figure 7). This is probably due to the fact that, contrarily to LightGBM, once many training data become available, HA tends towards a constant prediction, and cannot adapt to changes in the user behaviour.

A quantitative way to compare the actual effectiveness of the different predictors using a single number is to measure how close they are to the optimal point $(1, 1)$ in Figures 7 and 8 or, equivalently, compute the product of the QoS and SOH metrics. This result is shown in Figures 9 and 10, and in the rightmost columns of Table 3.

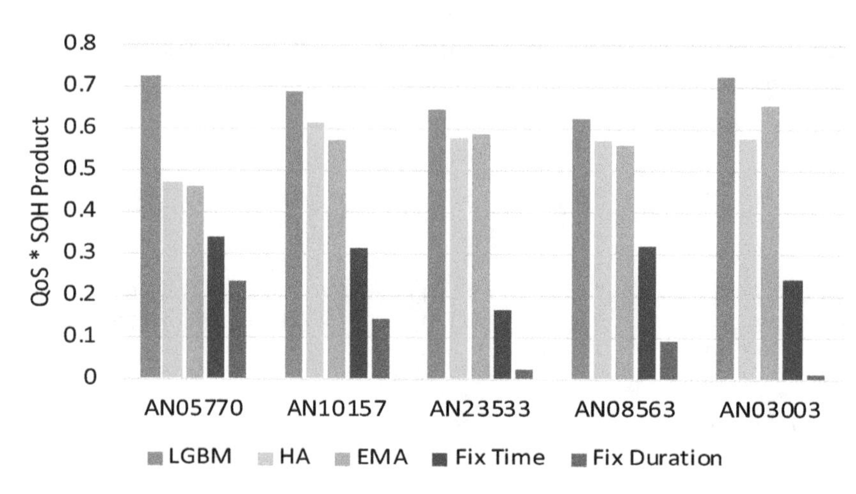

Figure 9. QoS-SOH product using different plug-in duration predictors and the Aging Optimal charge protocol from [13].

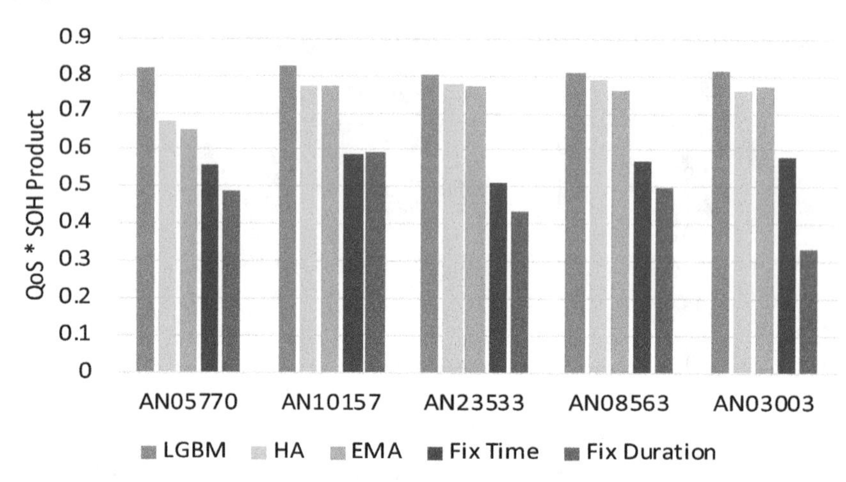

Figure 10. QoS-SOH product using different plug-in duration predictors and the ASAP charge protocol from [17].

The graphs and table clearly show that the ML-based solution is superior to all baselines. On average, LightGBM improves the QoS-SOH product by 20% and 8% using the Aging Optimal and ASAP protocols respectively. Due to the lower prediction MSE, which in turn is a consequence of the larger amount of training data available, AN05770 is always the station for which the improvement is maximum, i.e., 54% and 21% respectively.

5. Conclusions

We have proposed a novel ML-based forecasting method to estimate the plug-in duration of EVs, and consequently improve the effectiveness of aging-aware charging protocols that need this information to implement a "just-in-time" charge. With experiments on a dataset containing real EV plug-in measurements from domestic charge stations, we have shown that this forecasting is superior to the basic approaches used in previous work, reducing the prediction error by 34% on average and up to 4×. Using our proposed predictor on top of two state-of-the-art aging-aware charging protocols, causes an improvement of their effectiveness, measured with a combined aging and user-experience metric, of 20% and 8% on average, and up to 54% and 21% respectively. Moreover, the fact that the largest improvements are always achieved for the charge station with the most records in the dataset suggests that even better results could be obtained with more data available. However, the availability of large public datasets to experiment on remains one of the biggest open challenges for this research. ML-based prediction of EVs plug-in duration may lead to a more cost-efficient use of batteries, and therefore help speed up the mass adoption of electric transportation. Moreover, these predictions could also be used to optimize load distribution and vehicle-to-grid solutions, to further improve the efficiency of smart grids. This analysis will be part of our future work.

Author Contributions: Conceptualization, Y.C., D.J.P., E.M. and M.P.; Data curation, K.S.S.A.; Investigation, Y.C. and D.J.P; Methodology, Y.C., D.J.P. and S.V.; Software, K.S.S.A.; Supervision, E.M. and M.P.; Writing—original draft, Y.C., K.S.S.A. and D.J.P; Writing—review & editing, S.V., E.M. and M.P. All authors have read and agreed to the published version of the manuscript.

Abbreviations

The following abbreviations are used in this manuscript:

EVs	Electric Vehicles
SOC	State Of Charge
ML	Machine Learning
DOD	Depth Of Discharge
CC-CV	Constant Current-Constant Voltage
LightGBM	Light Gradient Boosting
SOH	State Of Health
QoS	Quality of Service
EMA	Exponential Moving Average
HA	Historical Average
CPU	Central Processing Unit
MCU	Micro-controller Unit
GBDT	Gradient Boosting Decision Tree
GOSS	Gradient-based One-Side Sampling
EFB	Exclusive Feature Bundling
MSE	Mean Square Error

References

1. Zhang, Q.; Ou, X.; Yan, X.; Zhang, X. Electric vehicle market penetration and impacts on energy consumption and CO2 emission in the future: Beijing case. *Energies* **2017**, *10*, 228. [CrossRef]
2. Eberle, U.; Von Helmolt, R. Sustainable transportation based on electric vehicle concepts: A brief overview. *Energy Environ. Sci.* **2010**, *3*, 689–699. [CrossRef]
3. Yang, Z.; Shang, F.; Brown, I.P.; Krishnamurthy, M. Comparative study of interior permanent magnet, induction, and switched reluctance motor drives for EV and HEV applications. *IEEE Trans. Transp. Electrif.* **2015**, *1*, 245–254. [CrossRef]
4. Baek, D.; Chen, Y.; Bocca, A.; Bottaccioli, L.; Cataldo, S.D.; Gatteschi, V.; Pagliari, D.J.; Patti, E.; Urgese, G.; Chang, N.; et al. Battery-Aware Operation Range Estimation for Terrestrial and Aerial Electric Vehicles. *IEEE Trans. Veh. Technol.* **2019**, *68*, 5471–5482. [CrossRef]
5. Berckmans, G.; Messagie, M.; Smekens, J.; Omar, N.; Vanhaverbeke, L.; Van Mierlo, J. Cost projection of state of the art lithium-ion batteries for electric vehicles up to 2030. *Energies* **2017**, *10*, 1314. [CrossRef]
6. Hannan, M.; Hoque, M.; Mohamed, A.; Ayob, A. Review of energy storage systems for electric vehicle applications: Issues and challenges. *Renew. Sustain. Energy Rev.* **2017**, *69*, 771–789. [CrossRef]
7. Jafari, M.; Gauchia, A.; Zhang, K.; Gauchia, L. Simulation and analysis of the effect of real-world driving styles in an EV battery performance and aging. *IEEE Trans. Transp. Electrif.* **2015**, *1*, 391–401. [CrossRef]
8. Barré, A.; Deguilhem, B.; Grolleau, S.; Gérard, M.; Suard, F.; Riu, D. A review on lithium-ion battery ageing mechanisms and estimations for automotive applications. *J. Power Sources* **2013**, *241*, 680–689. [CrossRef]
9. Bashash, S.; Moura, S.J.; Forman, J.C.; Fathy, H.K. Plug-in hybrid electric vehicle charge pattern optimization for energy cost and battery longevity. *J. Power Sources* **2011**, *196*, 541–549. [CrossRef]
10. Bocca, A.; Chen, Y.; Macii, A.; Macii, E.; Poncino, M. Aging and Cost Optimal Residential Charging for Plug-In EVs. *IEEE Design Test* **2017**, *35*, 16–24. [CrossRef]

11. Matsumura, N.; Otani, N.; Hamaji, K. Intelligent Battery Charging Rate Management. U.S. Patent Application 12/059,967, 1 October 2009.
12. Pröbstl, A.; Kindt, P.; Regnath, E.; Chakraborty, S. Smart2: Smart charging for smart phones. In Proceedings of the 2015 IEEE 21st International Conference on Embedded and Real-Time Computing Systems and Applications, Hong Kong, China, 19–21 August 2015; pp. 41–50.
13. Bocca, A.; Sassone, A.; Macii, A.; Macii, E.; Poncino, M. An aging-aware battery charge scheme for mobile devices exploiting plug-in time patterns. In Proceedings of the 2015 33rd IEEE International Conference on Computer Design (ICCD), New York, NY, USA, 18–21 October 2015; pp. 407–410.
14. Klein, R.; Chaturvedi, N.A.; Christensen, J.; Ahmed, J.; Findeisen, R.; Kojic, A. Optimal charging strategies in lithium-ion battery. In Proceedings of the 2011 American Control Conference, San Francisco, CA, USA, 29 June–1 July 2011; pp. 382–387.
15. Shen, W.; Vo, T.T.; Kapoor, A. Charging algorithms of lithium-ion batteries: An overview. In Proceedings of the 2012 7th IEEE Conference on Industrial Electronics and Applications (ICIEA), Singapore, 18–20 July 2012; pp. 1567–1572.
16. Guo, Z.; Liaw, B.Y.; Qiu, X.; Gao, L.; Zhang, C. Optimal charging method for lithium ion batteries using a universal voltage protocol accommodating aging. *J. Power Sources* **2015**, *274*, 957–964. [CrossRef]
17. Chen, Y.; Bocca, A.; Macii, A.; Macii, E.; Poncino, M. A li-ion battery charge protocol with optimal aging-quality of service trade-off. In Proceedings of the 2016 International Symposium on Low Power Electronics and Design, San Francisco, CA, USA, 8–10 August 2016; pp. 40–45.
18. Electric Chargepoint Analysis 2017: Domestics. Available online: https://www.gov.uk/government/statistics/electric-chargepoint-analysis-2017-domestics (accessed on 12 August 2020).
19. Millner, A. Modeling lithium ion battery degradation in electric vehicles. In Proceedings of the 2010 IEEE Conference on Innovative Technologies for an Efficient and Reliable Electricity Supply, Waltham, MA, USA, 27–29 September 2010; pp. 349–356.
20. Hoke, A.; Brissette, A.; Smith, K.; Pratt, A.; Maksimovic, D. Accounting for lithium-ion battery degradation in electric vehicle charging optimization. *IEEE J. Emerg. Sel. Top. Power Electron.* **2014**, *2*, 691–700. [CrossRef]
21. Hussein, A.A.H.; Batarseh, I. A review of charging algorithms for nickel and lithium battery chargers. *IEEE Trans. Veh. Technol.* **2011**, *60*, 830–838. [CrossRef]
22. Wang, Y.; Lin, X.; Xie, Q.; Chang, N.; Pedram, M. Minimizing state-of-health degradation in hybrid electrical energy storage systems with arbitrary source and load profiles. In Proceedings of the 2014 Design, Automation & Test in Europe Conference & Exhibition (DATE), Dresden, Germany, 24–28 March 2014; pp. 1–4.
23. Shin, D.; Sassone, A.; Bocca, A.; Macii, A.; Macii, E.; Poncino, M. A compact macromodel for the charge phase of a battery with typical charging protocol. In Proceedings of the 2014 International Symposium on Low Power Electronics and Design, La Jolla, CA, USA, 11–13 August 2014; pp. 267–270.
24. Pröbstl, A.; Islam, B.; Nirjon, S.; Chang, N.; Chakraborty, S. Intelligent Chargers Will Make Mobile Devices Live Longer. *IEEE Des. Test* **2020**. [CrossRef]
25. Zheng, B.; He, P.; Zhao, L.; Li, H. A Hybrid Machine Learning Model for Range Estimation of Electric Vehicles. In Proceedings of the 2016 IEEE Global Communications Conference (GLOBECOM), Washington, DC, USA, 4–8 December 2016; pp. 1–6.
26. Sun, S.; Zhang, J.; Bi, J.; Wang, Y. A Machine Learning Method for Predicting Driving Range of Battery Electric Vehicles. *J. Adv. Transp.* **2019**, *2019*, 4109148. [CrossRef]
27. López, K.L.; Gagné, C.; Gardner, M. Demand-Side Management Using Deep Learning for Smart Charging of Electric Vehicles. *IEEE Trans. Smart Grid* **2019**, *10*, 2683–2691. [CrossRef]
28. Murphey, Y.L.; Masrur, M.A.; Chen, Z.; Zhang, B. Model-based fault diagnosis in electric drives using machine learning. *IEEE/ASME Trans. Mechatron.* **2006**, *11*, 290–303. [CrossRef]
29. Chemali, E.; Kollmeyer, P.J.; Preindl, M.; Emadi, A. State-of-charge estimation of Li-ion batteries using deep neural networks: A machine learning approach. *J. Power Sources* **2018**, *400*, 242–255. [CrossRef]
30. Vidal, C.; Malysz, P.; Kollmeyer, P.; Emadi, A. Machine Learning Applied to Electrified Vehicle Battery State of Charge and State of Health Estimation: State-of-the-Art. *IEEE Access* **2020**, *8*, 52796–52814. [CrossRef]
31. Hu, X.; Li, S.E.; Yang, Y. Advanced Machine Learning Approach for Lithium-Ion Battery State Estimation in Electric Vehicles. *IEEE Trans. Transp. Electrif.* **2016**, *2*, 140–149. [CrossRef]
32. Hu, X.; Jiang, J.; Cao, D.; Egardt, B. Battery Health Prognosis for Electric Vehicles Using Sample Entropy and Sparse Bayesian Predictive Modeling. *IEEE Trans. Ind. Electron.* **2016**, *63*, 2645–2656. [CrossRef]

33. Xiao, J.; Xiong, Z.; Wu, S.; Yi, Y.; Jin, H.; Hu, K. Disk Failure Prediction in Data Centers via Online Learning. In Proceedings of the ICPP 2018 47th International Conference on Parallel Processing, Eugene, OR, USA, 13–16 August 2018; Association for Computing Machinery: New York, NY, USA, 2018. [CrossRef]
34. Pate, M.; Ho, M.; Texas Instruments. Charging ahead toward an EV Support Infrastructure. Available online: https://www.ti.com/lit/wp/swpy030/swpy030.pdf (accessed on 12 August 2020).
35. Chen, J.; Ran, X. Deep Learning With Edge Computing: A Review. *Proc. IEEE* **2019**, *107*, 1655–1674. [CrossRef]
36. Shi, W.; Cao, J.; Zhang, Q.; Li, Y.; Xu, L. Edge Computing: Vision and Challenges. *IEEE Internet Things J.* **2016**, *3*, 637–646. [CrossRef]
37. Jahier Pagliari, D.; Poncino, M.; Macii, E. Energy-Efficient Digital Processing via Approximate Computing. In *Smart Systems Integration and Simulation*; Bombieri, N., Poncino, M., Pravadelli, G., Eds.; Springer International Publishing: Cham, Switzerland, 2016; Chapter 4; pp. 55–89. [CrossRef]
38. LeCun, Y.; Bengio, Y.; Hinton, G. *Deep Learning*; The MIT Press: Cambridge, MA, USA, 2006.
39. Ke, G.; Meng, Q.; Finley, T.; Wang, T.; Chen, W.; Ma, W.; Ye, Q.; Liu, T.Y. LightGBM: A Highly Efficient Gradient Boosting Decision Tree. In *Advances in Neural Information Processing Systems 30*; Guyon, I., Luxburg, U.V., Bengio, S., Wallach, H., Fergus, R., Vishwanathan, S., Garnett, R., Eds.; Curran Associates, Inc.: Nice, France, 2017; pp. 3146–3154.
40. Kumar, A.; Goyal, S.; Varma, M. Resource-efficient Machine Learning in 2 KB RAM for the Internet of Things. In Proceedings of the 34th International Conference on Machine Learning, Sydney, Australia, 6–11 August 2017; Precup, D., Teh, Y.W., Eds.; PMLR, International Convention Centre: Sydney, Australia, 2017; Volume 70, pp. 1935–1944.
41. Li, F.; Zhang, L.; Chen, B.; Gao, D.; Cheng, Y.; Zhang, X.; Yang, Y.; Gao, K.; Huang, Z.; Peng, J. A light gradient boosting machine for remainning useful life estimation of aircraft engines. In Proceedings of the 2018 21st International Conference on Intelligent Transportation Systems (ITSC), Maui, HI, USA, 4–7 November 2018; pp. 3562–3567.
42. Jiang, J.; Wang, R.; Wang, M.; Gao, K.; Nguyen, D.D.; Wei, G.W. Boosting Tree-Assisted Multitask Deep Learning for Small Scientific Datasets. *J. Chem. Inf. Model.* **2020**, *60*, 1235–1244. [CrossRef] [PubMed]
43. Goodfellow, I.; Bengio, Y.; Courville, A. *Deep Learning*; The MIT Press: Cambridge, MA, USA, 2016.
44. Sun, X.; Liu, M.; Sima, Z. A novel cryptocurrency price trend forecasting model based on LightGBM. *Financ. Res. Lett.* **2020**, *32*, 101084. [CrossRef]
45. Zhang, C.; Zhang, Y.; Shi, X.; Almpanidis, G.; Fan, G.; Shen, X. On Incremental Learning for Gradient Boosting Decision Trees. *Neural Process. Lett.* **2019**, *50*, 957–987. [CrossRef]
46. Python LightGBM Package. Available online: https://lightgbm.readthedocs.io/en/latest/ (accessed on 12 August 2020).

5

Operating and Investment Models for Energy Storage Systems

Marija Miletić [1], Hrvoje Pandžić [1,*] and Dechang Yang [2]

[1] Faculty of Electrical Engineering and Computing, University of Zagreb, Unska ulica No. 3, 10000 Zagreb, Croatia; marija.miletic@fer.hr

[2] College of Information and Electrical Engineering, China Agricultural University, No. 17 Qinghuadonglu, Haidian, Beijing 100083, China; yangdechang@cau.edu.cn

* Correspondence: hrvoje.pandzic@fer.hr

Abstract: In the context of climate changes and the rapid growth of energy consumption, intermittent renewable energy sources (RES) are being predominantly installed in power systems. It has been largely elucidated that challenges that RES present to the system can be mitigated with energy storage systems (ESS). However, besides providing flexibility to intermittent RES, ESS have other sources of revenue, such as price arbitrage in the markets, balancing services, and reducing the cost of electricity procurement to end consumers. In order to operate the ESS in the most profitable way, it is often necessary to make optimal siting and sizing decisions, and to determine optimal ways for the ESS to participate in a variety of energy and ancillary service markets. As a result, many publications on ESS models with various goals and operating environments are available. This paper aims at presenting the results of these papers in a structured way. A standard ESS model is first outlined, and that is followed by a literature review on operational and investment ESS models at the transmission and distribution levels. Both the price taking and price making models are elaborated on and presented in detail. Based on the examined body of work, the paper is concluded with recommendations for future research paths in the analysis of ESS.

Keywords: mathematical modelling; energy storage systems; electricity markets; power system planning; power system operation

---

1. Introduction

Liberalisation of the power sector caused electricity to become commodified and traded in the markets. However, unlike other commodities, electrical energy cannot be stored in its original form and the power systems are operated with the goal of maintaining the balance between the consumption and the production of electricity at all times. As our society recognised human influences on the environment and started to require more renewable energy sources (RES), maintaining power balance became a much harder task and consideration of the uncertainties caused by the intermittent energy sources became imperative. Several solutions for addressing RES intermittency exist: installing new, fast ramping generators such as gas power plants, building new transmission lines to secure power supply in the events of renewable energy shortage, designing demand response programs in which the demand is managed to meet the production and using energy storage systems (ESS) to store the surplus and supply the shortage of electricity. While the term ESS can generally represent a larger set of energy storing technologies, in this paper we use it to describe a set of technologies that enable storing of electricity in some other form: potential energy in pumped-hydro plants, kinetic in flywheels, electrochemical in batteries, etc.

ESS as a market participant changes its role from the generating unit to the consumer depending on the market conditions. This puts the ESS in a unique position and gives it an opportunity to strategically choose its market position in order to maximise profit more efficiently than producers and consumers, who can only sell or buy in the markets. Figure 1 presents a concise overview of the models representing the ESS either as a non-strategic player (price-taker) or a strategic player (price-maker). A price-taker has no influence on market prices and bids competitively. On the other hand, a price-maker is a strategic player that exercises market power by bidding over its marginal price or by withholding capacity. The two terms are not completely accurate because even a non-strategic ESS can influence market prices, as was shown in [1]. For this reason, we adopt terms "strategic" and "non-strategic" instead of "price-maker" and "price-taker." In the investments phase, a strategic ESS, as opposed to a non-strategic one, tends to install larger ESS facilities. In the operational phase, strategic ESS, compared to the non-strategic one, is generally signified by higher profits earned through market participation.

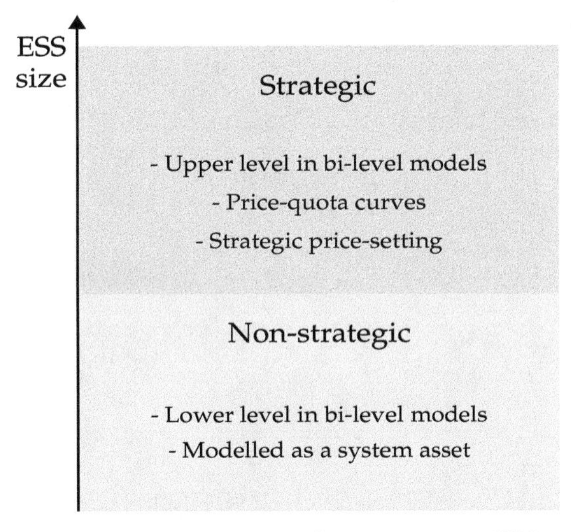

Figure 1. Modelling approaches to strategic and non-strategic ESSpresented in this paper.

Based on the system connection and size of storage, technical literature divides the ESS in two groups: transmission-level and distribution-level. Transmission-level ESS are large-scale installations connected to the transmission network such as pumped-hydropower stations, compressed air energy storage plants and large-scale battery storage plants. Sizes of these facilities range from a couple of megawatts to a couple of gigawatts. Distribution-level ESS are smaller systems connected to the distribution network which can be placed at the consumers' premises (behind the meter) or be a part of a microgrid, virtual power plant or distribution grid operator's (DSO's) assets. The size of such facilities depends on the distribution system operator's grid rules and is usually less than one megawatt. We adopt this approach as well, analysing transmission- and distribution-level ESS separately. The analysed body of literature consists of 57 articles on investments and 77 on the operating of ESS in transmission and distribution systems. Various markets in both investment and operating phases were considered for ESS participation in these papers, which is outlined in Figure 2.

Figure 3 shows a large gap between the number of papers dealing with transmission- and distribution-level ESS in operational phase. The gap is understandable if we take into account the fact that large numbers of papers on ESS operating in real-time markets employ optimal control algorithms, which are out of scope of this review.

Technological background for ESS can be found in [2–4]. Luo et al. [2] presented an overview of ESS technologies and listed possible applications for ESS in electrical power systems. A more recent paper by Koohi-Fayegh and Rosen [3] also focused on technologies and listed a smaller number of potential applications, concluding with a list of technological issues that researchers are facing, such as the need to increase the cycling ability for electrochemical storage and to discover new materials for all types of energy storage. The paper singled out hydrogen storage as the most promising technology for the future. A comparison of conclusions from [2,3], indicates that the majority of the earlier issues have been solved, but some still remain open, such as seasonal storage, especially at the distribution level. Khan et al. [4] presented not only an overview of ESS technologies but also the potential for storing primary energy sources, i.e., natural gas and coal. This is interesting in the context of multi-energy systems where ESS could be displaced by primary energy source storage.

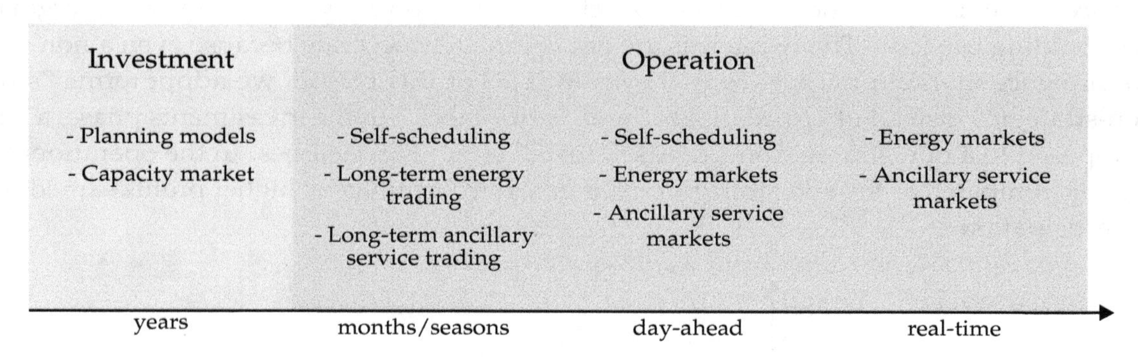

Figure 2. Investment and operational phases represented in the ESS models.

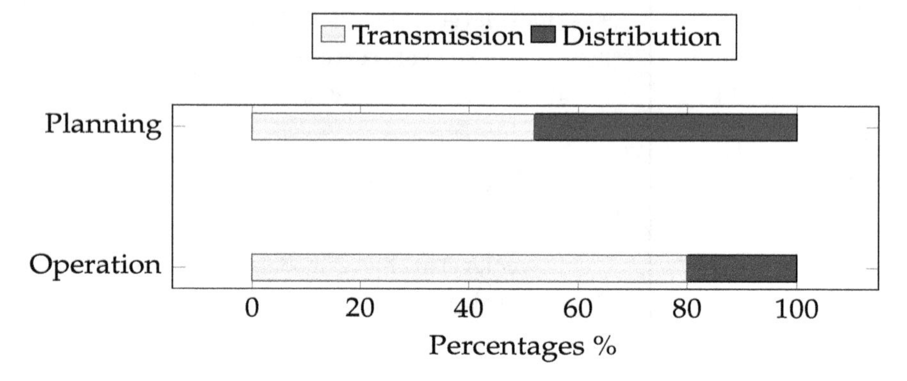

Figure 3. Shares of the papers for transmission- and distribution-level ESS in operational and investment phases.

ESS planning is a widely reviewed area. Awadallah and Venkatesh [5] presented a modelling framework and reviewed literature dealing with ESS planning and operation in distribution networks but did not address uncertainties in models. As a conclusion, they expressed the need for more general studies in the distribution systems—the reason being that the body of literature contains results that are hard to generalise. They also concluded that more research on market participation of the ESS in distribution network is needed and that development of techniques for long-term large-capacity ESS operation are necessary to enable seasonal price arbitrage. Lorente et al. [6] performed a short review of research papers on ESS planning published between 2016 and 2018 and concluded that the ESS impact on prices is commonly ignored in the planning models. They also stated that more research on ESS siting is needed. After a thorough literature review wherein they categorised ESS expansion planning models via different modelling approaches and listed objective functions and constraints, Sheibani et al. [7] recognised open issues in ESS expansion planning. These included the necessity of risk assessments for investors, determination of the optimal financial support for ESS expansion and consideration of different services to system operators. The authors stated, based on conclusions from several case studies, that it is not profitable for the ESS to participate in only one market and the investors must consider more revenue streams. This is in accordance with conclusion drawn by Zidar et al. [8], who reviewed solving methods for ESS siting and sizing in distribution grids, categorising them into: mathematical programming, exhaustive search, analytical methods and heuristic methods.

Optimal financial support for ESS was addressed by Miller and Carriveau [9], who gave a review of financing opportunities for ESS investors. They presented the state-of-the art of the financing schemes categorised as: governmental incentives, partnering with the renewable technologies and innovative finance models. Koohi-Fayegh and Rosen [3] confirmed that governmental policy support will play a large role in development of ESS technologies. However, the latest EU energy legislation, except for some special circumstances, forbids the transmission system operators (TSOs) and DSOs from owning ESS [10]. It remains to be seen whether this will slow down ESS integration in the European power system or will just shift the focus to the privately owned ESS.

Mejia and Kajikawa [11] performed data analysis of a large number of papers and patents. Their findings show an overlap in topics covered by the research community and industry that contains optimisation techniques for ESS operation and planning and various topics in the area of materials science.

Traditionally, ESS have been used for peak-shaving and energy arbitrage, but nowadays they are considered for balancing, congestion management and other purposes as well. As many ESS technologies are reaching maturity and their investment costs are decreasing, the following questions have arisen: how can they operate profitably in today's markets; what capacity and which storage technology should be installed; and where should they be placed? To answer these questions, the scientific community uses mathematical models for simulation and optimisation. In this paper, we present the work aimed at answering these questions. We concentrate on the research of market participation of the ESS during the operational and investment planning phases. Our contribution to the body of literature is a detailed survey of mathematical models for the analysis of ESS used for said purposes. Based on the survey, we provide recommendations for future research in the area of market-participating ESS.

The paper is outlined as follows. The standard mathematical model of ESS is given in Section 2. Section 3 presents a detailed literature survey on ESS market participation, and Section 4 presents a literature survey on expansion planning. Section 5 describes the ways of dealing with the computational complexity of ESS models. We conclude the paper in Section 6.

2. Energy Storage System Models

This section presents a standard model that represents any type of ESS mathematically, without assuming any technological details. As a generic mathematical model, the measuring units associated with the variables and parameters are there for illustration purposes and can be scaled up or down. Depending on the modelling objective, some of the constraints from the following set can be left out or modified:

$$0 \leq p_t^{\text{ch}} \leq \overline{P}^{\text{ch}} \cdot x_t, \quad \forall t \in T \tag{1}$$

$$0 \leq p_t^{\text{dis}} \leq \overline{P}^{\text{dis}} \cdot (1 - x_t), \quad \forall t \in T \tag{2}$$

$$e_t \leq \overline{E}, \quad \forall t \in T \tag{3}$$

$$e_t \geq \underline{E}, \quad \forall t \in T \tag{4}$$

$$e_T \geq E_0, \tag{5}$$

$$e_t = e_{t-1} + \Delta T p_t^{\text{ch}} \eta^{\text{ch}} - \Delta T p_t^{\text{dis}} / \eta^{\text{dis}} - \Delta T p_t^{\text{loss}}, \quad \forall t \in T \tag{6}$$

Equation (1) constrains the ESS charging power below its charging power rating $\overline{P}^{\text{ch}}$ and Equation (2) does the same for the discharging power rating at $\overline{P}^{\text{dis}}$. In the ESS siting and sizing models, the right-hand-side (RHS) coefficients can be variables instead of parameters. Binary variable x_t ensures that the ESS is never charged and discharged at the same time. Generally, binary variables turn a model into a mixed-integer program, which complicates the solution procedure, which is the main reason for neglecting them in the models. Binary variables can be omitted without consequences if the considered market conditions are such that it would not be profitable for the ESS to be both charged and discharged at the same time. However, simultaneous charging and discharging is profitable, assuming imperfect efficiency of the charging/discharging cycle, during negative market prices.

In addition to constraints (1)–(6), Tejada-Arango et al. [12] constrained the ESS power by a ramping constraint for transition between charging and discharging mode as follows:

$$\left(e_t^{\text{dis}} - e_{t-1}^{\text{dis}}\right) - \left(e_t^{\text{ch}} - e_{t-1}^{\text{ch}}\right) + r_t^{+} \leq \tau RU \quad \forall t \in T \tag{7}$$

$$\left(e_t^{\text{ch}} - e_{t-1}^{\text{ch}}\right) - \left(e_t^{\text{dis}} - e_{t-1}^{\text{dis}}\right) + r_t^{-} \leq \tau RD \quad \forall t \in T \tag{8}$$

In Equations (7) and (8) e_t^{dis} and e_t^{ch} are discharged and charged energy during one time period, r_t^{+} and r_t^{-} are up and down ramping capacity reserves of the ESS and RU and RD are ramping limits. Ramping constraints of the ESS are often ignored because of the assumed instantaneous change in the power input or output levels. It was shown by Poncelet et al. [13] that generators' flexibility constraints play an important role in the ESS investment models. The same should be true for the ESS flexibility constraints, especially those ESS offering flexibility services to the system. Therefore, it is expected that more models with ESS ramping constraints will appear in the future papers.

Equation (3) limits the state of energy from above to its energy rating $\overline{E}$, and Equation (4) from below to $\underline{E}$. It is important to impose this lower limit to state of energy to decrease the rate of degradation for batteries or to preserve the minimum water levels in pumped-hydro storage units. Equation (5) ensures that the final state of energy is not lower than the initial one (E_0). This way it is certain that the ESS does not make profit by merely selling the leftover energy from the previous optimisation period and the model is simpler to incorporate in the long-term optimisation problems.

The last Equation (6) represents the state of energy calculation for all time periods of the considered time horizon. While Equations (1) and (2) constrain ESS power and Equations (3)–(5) constrain its energy levels, Equation (6) connects the two. Instead of state of energy (SOE), an absolute value that has a physical meaning in the field of power system economics, some authors calculate state of charge (SOC), which represents percentage of SOE relative to its energy rating. The first term in this equation represents the state of energy at the previous time step. This term is usually replaced with E_0 for the initial time period. Hemmati et al. [14], who included in their model initial state of energy as a decision variable, showed that the initial state of energy has an impact on planning decisions and should be selected with care. The second term is the amount of energy charged into the ESS during period ΔT, and the third is the amount of energy discharged from it during the same period, with the assumption that the said powers are constant over the time period. The parameters η^{ch} and η^{dis} are charging and discharging efficiencies of the ESS. The efficiencies used in this equation are not known for electrochemical ESS so one round-trip efficiency is commonly used instead of the separate charging/discharging efficiencies [15]. The last term represents the lost energy. This term can also be represented as a percentage of the previous state of energy and written as a coefficient of e_{t-1}. Losses materialise because of the self-discharge of the ESS over time. When modelling the ESS short-term operation, such as participation in the day-ahead or intraday markets, this term is often ignored [16,17], but it is taken into account in the long-term models [18].

Simple as it is, the presented generic model is not the best representation for every ESS technology. It works well for the energy storage technologies where the state of energy depends linearly on the charging power. However, batteries are usually charged with the the constant-current–constant-voltage (CC–CV) characteristic. In the CV charging phase, with the constant voltage and decreasing charging current, this model is not accurate. Several different battery models have been developed to address this issue. It was shown that more accurate models perform more realistically in the market environment. The first such model comes from [19], and reads:

$$p_t^{\text{ch}} \leq \overline{P}^{\text{ch}} \cdot x_t, \quad \forall t \in T \tag{9}$$

$$p_t^{\text{ch}} \leq \overline{P}^{\text{ch}} \cdot x_t \cdot \frac{\overline{E} - e_t}{\overline{E} - E^{\text{cc-cv}}}, \quad \forall t \in T \tag{10}$$

In these equations E^{cc-cv} is the state state of battery's energy at the CC–CV breakpoint. The authors here accounted for the decreasing charging current in the CV phase by assuming that the current is linearly decreasing and the maximum charging power is limited accordingly. Both Equations (9) and (10) limit the charging power, but if the state of energy is above the CC–CV breakpoint, only the (10) is the binding constraint.

The second approach, which uses the $\Delta e - e$ characteristic to represent how much energy is left to be charged to the ESS depending on the current state of energy, was presented in [15]. This approach uses a piece-wise linearisation of the $\Delta e - e$ function to limit the charging power at all time periods as follows:

$$e_t = \sum_{i=1}^{I-1} e_{ti}, \quad \forall t \in T \tag{11}$$

$$e_{ti} \leq R_{i+1} - R_i, \quad \forall t \in T, i \in I \tag{12}$$

$$p_t^{ch} = \frac{F_1}{\eta \Delta T} + \sum_{i=1}^{I-1} \frac{F_{i+1} - F_i}{R_{i+1} - R_i} \cdot \frac{\Delta e_t}{\eta \Delta T}, \quad \forall t \in T \tag{13}$$

In Equations (11)–(13) F_i are breakpoints on the Δe axis, and R_i are breakpoints on the e axis.

Gonzalez-Castellanos et al. [20] defined non-linear charging and discharging characteristics and approximated them by a convex combination of the sampling points. This way, the structure of Equations (1) and (2) does not change and the non-linearity is addressed through the RHS coefficients alone, specifically maximum charging and discharging power and state of charge. The authors in [20] also recognised the non-linear connection of the battery efficiency to the state of charge and the charging/discharging power. They used the input and output power as a third dimension for the convex approximation of the charging/discharging characteristic and defined them as follows:

$$P^{in} = P^{ch} \cdot \eta^{ch} \tag{14}$$

$$P^{out} = P^{dis} / \eta^{dis} \tag{15}$$

Discharging power is sampled by $[P^{dis}, SOC, P^{out}]$ and charging power by $[P^{ch}, E, P^{in}]$.

In the short-term models, battery cycling is not considered, but for the mid- to long-term optimisation, the cycle life constraints should be included in the model as well. There are several ways to include the cycle life in models. Duggal and Venkatesh [21], Kazemi and Zareipour [22], He et al. [23], Xu et al. [24] and Padmanabhan et al. [25] included battery life duration in the short-term and long-term scheduling objective functions. Duggal and Venkatesh [21] and He et al. [23] defined battery lifetime as exponential function of its depth-of-discharge (DOD) and showed that the battery lifetime decreases as the number of daily cycles increases. Padmanabhan et al. [25] linearised this function and showed that the linear expression is a good enough approximation of the original. The lifetime variable was used to analytically calculate the annual operating and maintenance cost that is, in turn, used to calculate daily operating cost, which is included in the objective function. Similarly, Kazemi and Zareipour [22] maximised the ESS profit function, calculated as a difference between the short-term revenues and the annual investment cost. Lifetime used to calculate the annual investment cost was approximated by the rainflow counting algorithm. This algorithm is commonly used to determine stress caused by cycling in any system. For batteries, the stress is proportional to the number of cycles and DOD [26]. Vejdan and Grijalva [27] included cycle life in the objective function as a coefficient calculated by dividing the capital cost per unit of capacity by double the number of life cycles. Qiu et al. [28] included battery lifetime in their model as a capacity coefficient in the state of energy calculation. This coefficient is calculated for each year as a combination of calendar and cycling ageing and decreases as years go by, reaching zero when the ESS is at the end of its life. Hajia et al. [29] included a nonlinear lifetime characteristic for battery ESS depending on the number of cycles and DOD in an expansion planning model. A different approach was presented by Gantz et al. [30] who limited the number of cycles in a month through a constraint. The limit in this constraint is calculated from data found in [31] by dividing the expected number of cycles by design life (in months). Mohsenian-Rad [32] used a constraint to limit the number of daily cycles as well and showed that choosing the number of daily cycles is important for finding a trade-off between the yearly profit and the battery lifetime.

Notice that all constraints mentioned until now refer to the real power modelling. However, papers that also consider reactive power injections by ESS have started to appear. These are mostly models focused on distribution grid applications of ESS, where AC power flow is more common than in the transmission-level studies. Some papers incorporate AC network constraints without considering possibility of ESS to offer reactive and real power [33]. However, there are also those that implement AC constraints for ESS [34–38]. Models with AC power flow are more rare because the non-linearity of the constraints renders them harder to solve. While DC power flow models are good enough for economic analysis, AC power flow is necessary for investigation of technical aspects of ancillary services such as voltage regulation. In addition to Equations (1)–(6), such models contain the following constraints on reactive power flows [39,40]:

$$p_t = p_t^{\text{dis}} - p_t^{\text{ch}} \tag{16}$$

$$p_t^2 + q_t^2 \leq s_t^2 \tag{17}$$

where p_t is net real power flow, q_t is reactive power flow and s_t is maximum apparent power flow of the inverter.

3. Market Participation of the ESS

In this section, an overview of the current state of the research on market-participating ESS is given. Figure 4 shows markets in which the ESS participation was investigated in the reviewed literature. It is evident from Figure 4 that most of the literature considers ESS participation in energy markets and fewer number of papers consider ancillary service markets. A big gap between these numbers is understandable because most papers in which ancillary service markets are considered also include energy market participation. The reasoning behind this is that it is cheaper for the ESS to buy electricity in the energy markets to prepare for offering ancillary services to the system than to participate only in an ancillary service market. Figure 4 also shows that there is much less literature that covers the market participation of distribution-level ESS. This is not a surprise, as the wholesale markets often have rules that limit the participation of smaller distributed resources. The number of papers addressing market-participating ESS at the distribution system level is expected to grow with the development of flexibility markets at this level.

3.1. Different Applications of ESS in Power Systems

The reviewed literature can be divided into three large groups based on the application of ESS: market arbitrage, supporting RES integration and long-term self scheduling. In recent years, both the scientific community and the industry are considering the ESS for other applications, such as voltage regulation and black starts. This section gives an overview of the key findings on these applications. A detailed overview of different ESS applications can be found in [41].

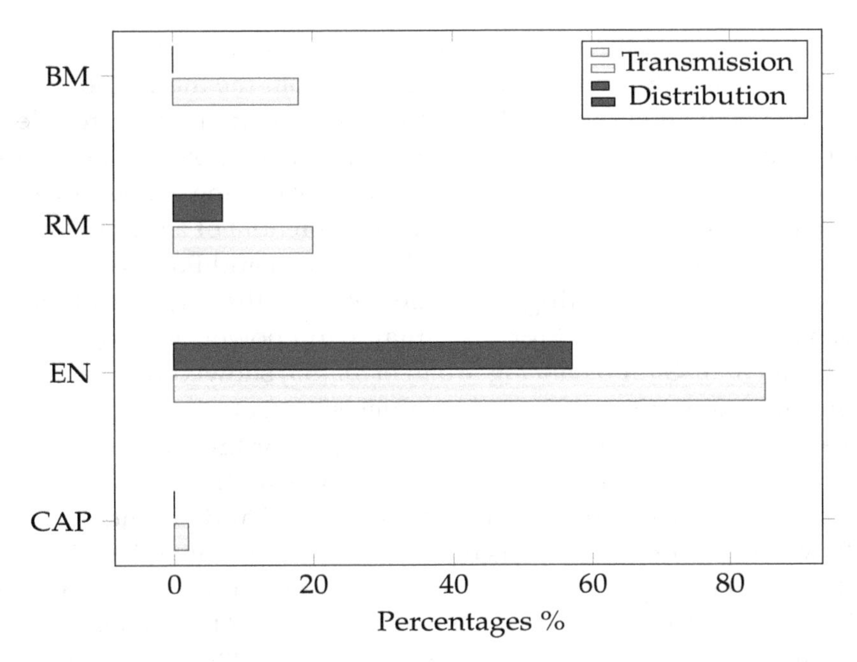

Figure 4. Share of the reviewed literature that considered ESS operation in different markets. Abbreviations: CAP, capacity; EN, energy market; RM, reserve market; BM, balancing services market.

3.1.1. Market Arbitrage

The most common market operation of an ESS is price arbitrage, i.e., buying energy when the prices are low and selling it when the prices are high. In electricity markets, this generally coincides with the low and high power consumption so the price arbitrage is also the energy arbitrage. However, the term energy arbitrage usually denotes operation outside of wholesale markets, e.g., balancing energy fluctuations from RES or variable loads in microgrids. Arbitrage was explicitly considered by many researchers. Thatte et al. [42] used robust optimisation to determine the optimal bidding strategy for ESS performing arbitrage in the day-ahead market. Xia et al. [43] showed by simultaneous perturbation method that an ESS in unit commitment can be used for energy and emissions arbitrage. Mohsenian-Rad [44] considered using distributed ESS for price arbitrage in a coordinated way so some ESS can be buying while others are selling electricity. The presented case study indicates that congestions are beneficial for this behaviour of ESS. Shafiee et al. [45] had an ESS performing arbitrage while taking into account uncertainty of price forecasting. Wang and Zhang [46] modelled ESS performing arbitrage in a real-time market as an arbitrage maximisation problem. Ciftci et al. [47] placed an ESS within a microgrid to be used for energy arbitrage and load following. They concluded that the possibility to offer multiple services to the microgrid might encourage investments in ESS.

A special type of arbitrage is the inter-temporal arbitrage between two time-scales, e.g., the day-ahead and the real-time markets, which was investigated in [48–50]. Braun [48] modelled a pumped-hydro storage and showed that inter-temporal arbitrage allows storage operators to exploit price differences between the two markets to optimise their short-term positions. Krishnamurthy et al. [50] showed that ESS performing inter-temporal arbitrage between day-ahead and intraday markets result in higher profits than those performing arbitrage only within the two markets. Zakeri and Syri [49] modelled a battery storage performing arbitrage between day-ahead and intraday markets in Nordic countries and concluded that a high share of hydropower plants in these markets reduces profitability potential for batteries.

3.1.2. RES Balancing

Another business case for an ESS is balancing the production from intermittent resources. Many papers investigated the possibility of applying an ESS for balancing variable wind production through coordinated optimisation or market mechanisms such as unit commitment.

Castronuovo et al. [51] recognised three major ways in which an ESS and an RES can be coordinated: (1) as one facility in which the ESS cannot purchase power from the market so it is only used for wind power balancing, (2) ESS is independent of the wind power plant but can provide reserve to balance its production and (3) both systems are a part of the same virtual power plant. This categorisation is on a trace of the general categorisation of ESS used for RES balancing. They can either be a part of the same or separate facilities. Usaola [52] analysed the profit potential of a thermal solar plant with liquid salt storage participating in the day-ahead market. Rahimiyan and Baringo [53] modelled combined ESS-wind power plant facility performing arbitrage between the day-ahead and intraday markets. Shu and Jirutitijaroen [54] concluded that for very small wind power plants (a few kilowatts) it would not be profitable to install an ESS for balancing. Yuan et al. [55] showed that capacity of an ESS in such combined facility influences profit and should be a decision factor along with the bidding strategy. Garcia-Gonzalez et al. [56] compared performances of the uncoordinated and coordinated operation of an ESS and a wind power plant and concluded that the coordination of the two increases profits for both facilities. The same conclusion was reached by Khodayar and Shahidehpour [57] and Daneshi and Srivastava [33], who compared the profits of a wind power plant and an ESS owned by the same company with and without coordination. Coordinated operation of an ESS and a wind power plant was also analysed by Sánchez de la Nieta et al. [58] who developed bidding strategies and tested them on Iberian market data. The same analysis was performed by Thatte et al. [59] on West Denmark market data. Jiang et al. [60] considered using ESS to mitigate risks related to wind forecast errors and to minimise system operation cost in centralised unit commitment. Li et al. [61] showed that ESS in unit commitment models with wind power plants decreases wind curtailment, load and reserve shortfalls and total system operation cost.

3.1.3. Self-Scheduling

Generalisation of the models that include coordination with intermittent resources are the self-scheduling models. The need for self-scheduling comes from the balancing responsibility and the goal of profit maximisation of market participants in decentralised markets. Varkani et al. [62] proposed a self-scheduling strategy for coordination of a pumped hydro plant and a wind power plant participating in the day-ahead market, while the pumped hydropower plant participates in the regulation reserve market as well. Parvania et al. [63] compared results of self-scheduling models for decentralised and centralised ESS, and the results indicate that the centralised ESS is more beneficial for the system. Self-scheduling models are often used for modelling the long-term operation of ESS. Such models are less common in the literature, which confirms that the long-term operation of ESS is still an open issue [2,3]. While Kazempour et al. [64] modelled a pumped-hydropower plant as part of a cascade to maximise profit of the whole system, Kazempour et al. [16] considered a self-scheduling pumped-hydropower plant aiming to optimise amount of energy to be store for one week ahead. Baslis and Bakirtzis [65] presented a pumped-hydropower plant in a long-term self scheduling model aware of its influence on prices and showed that such facility utilises strategy led by long-term objectives, independent of the short-term water inflows. Specifically, long-term contracts for large volumes can motivate ESS to behave counter-intuitively in the day-ahead market, dropping the price during the discharging hours because the energy is actually sold at the forward-contracted price. Thatte et al. [42] optimised the bidding strategy of an ESS performing arbitrage for one day and for period of 90 days. Kazemi and Zareipour [22] focused on the long-term scheduling of a battery considering the impact of the short-term operation on the battery lifetime in the long run. Pandžić et al. [66] analysed the influence of the optimisation period on profits of an ESS operator and concluded that it is best for a virtual storage plant to optimise operation over the scheduling horizon of at least two days. Alvarez et al. [18] used the future cost function to model the influence of ESS' long-term strategy on its short-term operation. They showed that fixing the ESS' state of energy to the previously scheduled long-term amounts increases operational costs and therefore decreases profit of the virtual power plant (VPP).

3.1.4. Ancillary Services

Besides energy arbitrage and RES balancing, an ESS can offer various other ancillary services to the network operators. Unlike the ESS performing arbitrage, those ESS that offer predominantly ancillary services make smaller numbers of cycles and therefore last longer. This was confirmed by Fleer et al. [67] who assessed profitability of ESS offering primary frequency reserve in German markets considering the uncertainties of reserve prices and investment costs. Thien et al. [68] considered ESS participation in German reserve markets but here the emphasis was placed on the market rules, showing that the decrease of duration of the traded products from 30 to 15 min is beneficial for the ESS. German balancing markets also allow ESS participation, the profitability of which was investigated by Olk et al. [69]. ESS considered for frequency reserve was already incorporated in many market environments, as is evident from Tables 1–3.

Table 1. Literature survey on the competitive ESS operating at the transmission level. (Abbreviations: CAP, capacity; EN, energy only; VR, voltage regulation; RM, reserve market; BM, balancing market).

Modelling Technique	Network Unconstrained	Network Constrained	Trading
Deterministic	[70]		CAP
	[48,49,52,64,71–75]	[20,25,33,43,63,76–78]	EN
	[71]		VR
	[49,64,71,75]	[25]	RM
	[64,71]		BM
Stochastic	[1,23,50,54,56,58,62,79,80]	[32,57,61,81]	EN
	[23,62]		RM
	[1,23,62]		BM
Interval		[82]	EN
Robust	[16,22,60,83]		EN
	[16,22,83]		RM
	[16,83]		BM
Chance-constrained	[58]	[51]	EN
Risk-contrained	[16]		EN RM BM

Other ancillary services are usually not procured through centralised markets, so the literature is less thorough in the areas of the market-based profitability of black start, voltage regulation and other services. There are, however, many examples of researchers investigating the possibilities of using ESS for such purposes. Black start is an interesting option for the ESS but even more so for hybrid facilities comprised of an ESS and a generator. Li et al. [84] presented a method of configuring an ESS combined with a wind power plant for the purpose of establishing the voltage and frequency for starting-up thermal power plants. Large shares of RES installed in distribution grids can cause nodal voltages to increase or flicker. The idea of using ESS to regulate voltages has been gaining popularity in recent years.

Most of the research in this area deals with voltage control in distribution networks, where ESS can impact voltages by changing its charging/discharging level. Sugihara et al. [85] proposed a subsidy programme offered by the DSO to the ESS owners in order to control their ESS when there is a need for voltage control and proved that it is possible to use customer-owned ESS for voltage regulation. Opathella et al. [71] showed that an ESS of any size can collect revenue by selling various ancillary services. Their results indicate that the largest sources of revenue for ESS are voltage and frequency regulation, while energy arbitrage, reserve and black start bring much smaller shares of the profits.

3.2. Market Opportunities for ESS at the Transmission and Distribution System Level

Based on the voltage level a facility is connected to, ESS are categorised as transmission- or distribution-level ESS. ESS connected to the high-voltage transmission networks are large-scale facilities that can participate in wholesale markets or offer various services to the power system and its users. For most of transmission-level ESS applications, DC load flow is an acceptable approximation and reactive power flows are not modelled. However, this does not stand for the distribution-level ESS that are connected to the medium and low voltage levels. They are generally smaller in size and thus unable to participate in wholesale markets directly. However, there are various solutions that allow their indirect participation, e.g., via aggregators. Directly, facilities from this category can offer energy and ancillary services in retail or local energy markets.

Retail markets were traditionally set up only for electricity consumers to procure necessary energy, and the adoption of the prosumer paradigm has started only lately. Markets at the distribution system level, both energy and ancillary services, are still not as developed like those at the transmission level; thus there are far fewer profit opportunities for the ESS there. Furthermore, while the transmission-level markets can have price-setting mechanisms based on network constraints, current distribution-level market design is much more complicated and specific to the location. In other words, the influence of the location on ESS profits is easy to calculate at the transmission system level, as opposed to the distribution system level where each DSO can have its own congestion-management mechanism and incentive scheme set in place.

In this work ESS are categorised as transmission or distribution-level based on the distinction in size and possible applications as the asset.

3.2.1. Transmission System-Level ESS

As shown by many researchers who modelled ESS operation at the transmission level, placement of an ESS within the grid influences its profitability. Li and Hedman [79] showed that transmission contingencies can limit the amount of power that an ESS can deliver or store, increasing the wind spillage. Wang et al. [76] showed that congestion is beneficial to the ESS profit in the markets where locational marginal pricing (LMP) is used. The same conclusion is supported by Mohsenian-Rad [44]. The LMPs are used in all papers covered by this study but one, where zonal markets are considered. Weibelzahl and Märtz [77] investigated how ESS influence optimal zonal decomposition of a transmission system. They compared optimal zone allocation for cases with and without ESS and concluded that storage facilities change absolute value and direction of transmission flows and zonal prices and can therefore greatly influence the optimal number of zones and their boundaries. Mohsenian-Rad [32] investigated the impact of many other parameters, such as seasonality, storage efficiency, charging and discharging rates and battery life on the ESS profit. Nasrolahpour et al. [86] considered the impact of ramping constraints of conventional generators on ESS operation and concluded that profit of an ESS is higher in systems with less flexible resources. Similar conclusions were reached by Poncelet et al. [13], who analysed relevance of flexibility constraints in unit commitment for planning models.

Most of the papers on the transmission-level ESS consider participation in energy-only markets (see Figure 4), both on the day-ahead and the real-time scale. However, the body of literature that considered ancillary services and capacity markets is expanding due to a general conclusion that an ESS can hardly be profitable by trading only energy and the investors should consider participation in different markets to maximise their profits [8,16]. Table 1 contains information on markets that are considered for ESS participation in the papers reviewed for this study.

ESS participation in capacity markets is generally constrained by its energy rating. Therefore, most of the existing capacity markets which allow for ESS participation have mechanisms to prevent ESS bidding more capacity than they can deploy. Opathella et al. [70] defined the capacity market for the ESS participation by assigning a capacity factor between zero and one to each market participant. The factor is zero for the ESS if at the considered moment the expected energy supply is greater than the demand, and if the demand is greater than the supply, the factor is calculated by dividing the necessary capacity by the available capacity of the ESS, capped off at 1.

3.2.2. Distribution System-Level ESS

Although electricity markets that allow participation of distribution-level resources are uncommon, an ESS at this level has many opportunities to ensure profit, both at retail and wholesale levels. Moreira et al. [35] analysed services ESS can provide to the distribution grid and the possible conflicts between them. Babacan et al. [87] modelled distribution-level ESS participating in the time-of-use pricing retail market, and Jiang et al. [88] investigated the possibility of load shaping through dynamic pricing. Gantz et al. [30] and Tushar et al. [89] considered a shared ESS, also called cloud energy storage, and optimal ways to divide its capacity among the users. Atzeni et al. [90] considered using behind-the-meter ESS and distributed generators to facilitate demand side management, objective being cost minimisation for each individual consumer. The result of this consumer behaviour is a flatter load curve, which shows that even selfish actors can be beneficial to the system if the goal is congestion relief or peak shaving. Gil-González et al. [91] showed that using an ESS in a distribution grid with high penetration of RES reduces its operating costs. Nazir et al. [40] explored the possibility of using ESS for loss minimisation in unbalanced distribution grids with high penetration of PV.

An ESS placed within the distribution grid can participate in the wholesale markets for energy and ancillary services through aggregators. Parvania et al. [63] investigated how decentralised ESSs scheduled by an aggregator influence the transmission grid and concluded that this way of scheduling might be prone to rescheduling after the power flows through the network are realised. Contreras-Ocaña et al. [78] modelled interactions between an aggregator and ESS units under its control and between an aggregator and a wholesale market. The interaction between the aggregator and the ESS was modelled as a Nash bargaining game. Their results showed that rational aggregator is always beneficial to the system but the same is not true for a strategic aggregator. Therefore, they developed pricing schemes to prevent the aggregator from exercising market power. Mortaz [92] demonstrated the impact of geographical diversity on an ESS aggregator's profit through a risk-measure: more diverse portfolios are almost always more efficient in handling the RES production. Wang and Kirschen [93] presented a two-stage model of an aggregator enabling trade between commercial consumer-owned ESS and day-ahead and real-time markets.

An ESS in the distribution system can participate in markets as a part of a microgrid or a virtual power plant. In such models, the aim is to optimise the operation of the system by utilising the ESS potential for energy arbitrage. Pandžić et al. [94] considered an ESS as a part of a virtual power plant offering in the day-ahead and balancing markets. Giuntoli and Poli [95] modelled a virtual power plant consisting of distributed generators, ESS and loads. They took into account grid locations of said resources but used a DC network model, an approximation which works well for transmission-level models, but is not very accurate for distribution-level ones. Ju et al. [96] investigated how participation of a virtual power plant containing ESS, distributed generators and loads in different types of demand response programmes, can benefit the grid. Their results showed that, while the incentive-based demand response has greater influence on the demand curve, it is most beneficial for the grid to introduce both the price and incentive-based programmes simultaneously. Ciftci et al. [47] modelled ESS used for load following and energy arbitrage within a microgrid with the objective of energy cost minimisation. Alvarez et al. [18] compared behaviour of a virtual power plant with and without and ESS and showed that ESS does increase the VPP's profit.Liu et al. [97] optimised a cloud ESS in a microgrid and define service pricing mechanisms for it. The microgrid in this paper contains households with rooftop PVs and a cloud energy storage that trades with them.

Table 2 presents the literature review of the distribution-level ESS. It shows that distribution-level ESS mostly participate in energy-only markets.

Table 2. Literature survey on the ESS operating at the distribution level. (Abbreviations: EN, energy only; VR, voltage regulation; RM, reserve market; BM, balancing market).

Modelling Technique	Network Unconstrained	Network Constrained	Trading
Deterministic	[87,88]		retail
	[30,78,89,90,95]	[35,71]	EN
		[71,85]	VR
		[71]	RM
		[35,71]	BM
Stochastic	[18,94]		EN
Robust	[53,96]		EN
Chance-constrained	[47]		EN

3.3. Strategic Market Participation

Two most known types of market competition are Cournot's and Bertrand's models. Cournot competition is signified by quantity bids and horizontal shifts of supply functions. On the contrary, Bertrand competition is signified by price bids and vertical shifts of supply functions. While Bertrand competition guarantees social welfare maximisation because no-one is motivated to bid more than their marginal cost, Cournot competition can lead to an equilibrium in which social welfare is not at the optimum. However, it was proven that for markets with a large number of competing players Cournot and Bertrand optima are equal [98].

Supply functions are the middle ground between the two extreme cases of competition. Supply function bids are the most common way the electricity markets are organised. Depending on the steepness of the supply function, the behaviour of the market participants can be said to follow the Cournot model (steep) or the Bertrand model (flat). Most electricity market participants follow a mixed strategy, biding their marginal cost to achieve Bertrand optimum and withholding capacity to exercise market power in Cournot context. This was confirmed by Lundin and Tangerås [99], who categorised the behaviour of suppliers in the Nordic power market as Cournot competition because of the horizontal shifts of supply functions between time-periods. Similar behaviour can be observed in the Alberta market by Shafiee et al. [100]. However, conclusions in [99] were based on the the changes in participants' behaviour between the day-ahead and intraday markets, thereby concluding that the increase in prices between the markets is a result of capacity withholding when it can easily be a consequence of capacity shortage.

ESS can arbitrarily behave as producers or consumers. Therefore, they are not simple market participants and can change market conditions in unexpected ways. Nasrolahpour et al. [72] and Sioshansi [73] compared the behaviour of an ESS as a strategic and non-strategic agent. They showed that adding a new ESS to an imperfectly competitive market can decrease social welfare, which is the opposite of what is expected in Cournot's model of competition, where an increase of the social welfare is expected when new participant joins the market.

As ESS do not have fuel costs which mostly comprise marginal costs of generators, the optimal Bertrand strategy for an ESS is to bid zero price for charging and market cap for discharging [66]. Nonetheless, it was shown that an ESS which strategically chooses prices and quantities can influence market price and increase its profit. Nasrolahpour et al. [86] proposed a bi-level model of an ESS that strategically sets prices and volumes in the upper level. They showed that the ESS utilising this strategy increased the market price while discharging and decreased it while charging, thereby maximising its own profit. A similar model was used by Wang et al. [76] to show that strategic ESS influence LMPs and therefore increase profit. A supply competition curve was modelled by Krishnamurthy et al. [50] as a price–quantity strategy for an ESS performing inter-temporal arbitrage. They showed that profits of an ESS utilising this strategy are higher on average than of an ESS bidding only quantities. Fang et al. [101] showed that an ESS strategically choosing when to charge or discharge chooses quantities in a way that does not alter LMPs. Shafiee et al. [100] noted the importance of assessing the impacts of ESS on prices during an economic analysis to avoid behaviour which causes ESS to be less profitable when behaving as a price-maker than as a price-taker. Zou et al. [102] found Nash–Cournot strategy for an ESS supporting large-scale RES. With the assumption that RES tend to increase the difference between peak and valley prices and strategic ESS tend to exploit this difference, they concluded that strategically behaving ESS can provide flexibility for the RES if they are driven by selfish objectives.

All these findings are true if only one strategic participant is assumed to exist in the market. In most of the literature only one strategic player is modelled and the other market participants are assumed to be behaving competitively. Pandžić et al. [66] modelled competition between three independent merchant-owned storage facilities via the diagonalisation algorithm. They showed that profitability of a strategic player significantly decreases when strategic behaviour of other market participants is neglected.

Table 3 presents a concise overview of the strategic ESS models. While strategic operation is mostly investigated for energy-only markets, there are several examples of ESS behaving strategically in ancillary service markets as well.

Table 3. Literature survey on the strategic ESS operating at the transmission level. (Abbreviations: EN, energy only; RM, reserve market; BM, balancing market).

Modelling Technique	Network Unconstrained	Network Constrained	Trading
Deterministic	[72,73,100,102–105]	[17,66,76]	EN
	[102]		RM
	[102,105]		BM
Stochastic	[27,55,65,106]	[18,44,86,101]	EN
	[106]		RM BM
Robust	[42,59,107]	[44]	EN
Chance-constrained	[55]	[92]	EN
Risk-contrained	[45]		EN

3.3.1. Bi-Level Models

The most common way of modelling strategic market participation is through a bi-level structure shown in Figure 5. This is a practical way of modelling a Stackelberg game. Stackelberg game is a strategic game with a leader and followers where the leader takes the first move to which the followers respond by optimising their position within the given circumstances. Therefore, multiple lower-level problems can appear, as in [86], where there is a lower-level problem for the day-ahead market and an additional one for each stochastic scenario in the real-time market. There can also be more than two levels in a model, as in [92] where the three levels optimise the positions of: (1) an aggregator, (2) an ESS owner and (3) a day-ahead electricity market operator.

A strategic ESS is represented by the upper-level profit maximisation model, while the lower level represents the market clearing process. The lower-level problem is included in the upper level as a set of constraints. Decisions from the upper level can be included in the lower-level problem as parameters. The following is a general mathematical representation of a bi-level problem, where the upper level minimises function $F(x,y)$ of the upper-level (x) and lower-level (y) decision variables, subject to the set of constraints represented by function $G(x,y)$; and the lower level minimises function $H(y)$ of lower-level decision variables, subject to set of constraints represented by the function $I(y)$:

$$\min_{x,y} F(x,y) \quad (18)$$
$$\text{subject to: } G(x,y) \geq 0$$
$$\{ \min_{y} H(y)$$
$$\text{subject to: } I(y) \geq 0 \}$$

To be able to solve this problem by a linear programming method, e.g., the simplex algorithm, the bi-level problem is transformed into a mathematical problem with equilibrium constraints (MPEC) using the primal-dual transformation and the strong duality constraint or Karush-Kuhn-Tucker (KKT) conditions. The resulting non-linear program is linearised using the KKT conditions and other linearisation techniques, such as Fortuny–Amat substitution [108]. The final program is solved by linear programming or mixed integer programming solvers, depending on the structure of the upper-level problem. The prices in these models are the dual variables of the lower-level power balance equations. To be able to apply KKT conditions, the lower-level problem must be convex. Having an ESS modelled using binary variables in the lower level is an issue because these types of models are non-convex. This issue is avoided either by modelling the ESS without binary variables [74] or by using some decomposition technique to solve the problem.

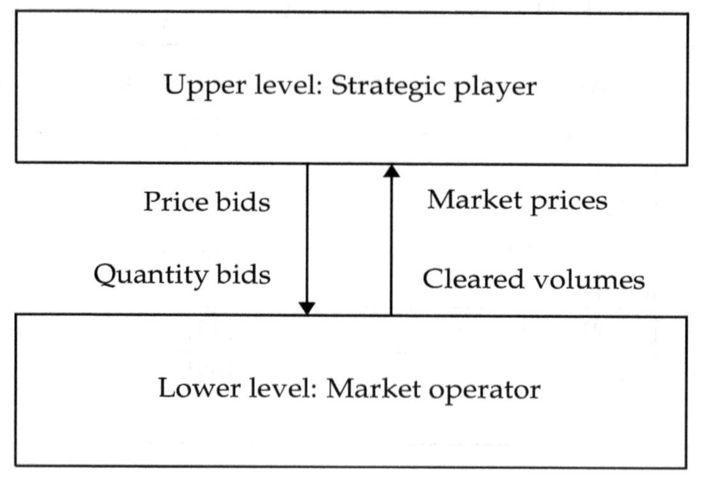

Figure 5. General structure of a bi-level model. The arrows represent the exchanges of decisions between the upper and the lower level.

One way to model a price-maker ESS is to place the full ESS model defined in Equations (1)–(5) in the upper level. Some papers in which this approach is used are [17,27,72,73,76,86,101,104,106]. A different approach is to include in the upper level only the objective function and an auxiliary variable that represents a strategic decision on bidding. This is the approach taken by Ye et al. [104], who concluded that while the strategic bidding increases the ESS profit, it decreases the social welfare. Another conclusion drawn in this paper is that the higher power rating of an ESS causes more strategic behaviour, and the higher energy rating causes more competitive behaviour. Ye et al. [74] used the same formulation to model a strategic generator in a bi-level setting and to investigate the impact of a price-taker ESS on the generator's market power. The conclusion was drawn that the ESS reduce the market power during peak and increase it during off-peak hours.

3.3.2. Price Quota Curves

A second approach to the strategic market participation modelling is the definition of the price quota curves (PQC), used by Shafiee et al. [100] and Shafiee et al. [107]. For the generators, the PQC is a step-wise decreasing function that shows dependence of the market price on the generated power, and for the demand it is a step-wise increasing function of the market price dependence on the power consumption. A PQC shows how the market price increases with the growing demand and decreases with the growing generation. In these models, the prices are variables, functions of the charging and discharging quantities. Baslis and Bakirtzis [65] used PQCs to model a strategic pumped-hydro plant that withholds capacity in peak hours to induce price spikes and discharges in off-peak hours, thereby increasing its profit.

3.3.3. Strategic Price-Setting

A third approach to modelling the strategic market participation is to predict prices from the historical data and place bids in a way that would maximise the profit. This results in an unusual strategy for electricity markets, bidding maximum available capacity at all times and choosing prices strategically. This is the approach taken by Thatte et al. [59], Sousa et al. [103] and Thatte et al. [42]. The major deficiency of this way of modelling the ESS behaviour in the markets is the limited accuracy of predictions.

4. ESS Investment Modelling

Capacity expansion planning is used to ensure that the system has enough resources to supply the demand at all times. While in the past, capacity planning mostly considered investments in transmission lines and generators, recent research in this area has taken ESS, demand response programmes, and other new technologies into account as well.

When planning investments in ESS, capital costs are generally separated into costs for energy and costs for power rating. Intuitively, this can be explained by pumped hydropower station investments where dam construction is considered energy cost while costs incurred by turbine and generator installation are power rating costs. On the other hand, when it comes to stationary battery storage, the costs of batteries themselves are energy costs, while the cost of a bidirectional AC/DC converter is reflected in power costs.

The objective of the model depends on the intended usage of the ESS. If the goal of the investor is performing price arbitrage between different time-scales or markets, the objective is profit maximisation and there are no budget constraints [72]. If, on the other hand, it is supposed to be used as a system asset, e.g., for congestion relief or load shifting, the objective is usually cost minimisation and one of the constraints is the total budget [109]. Dvorkin et al. [110] used a slightly different approach, constraining profit from below. They showed that, to ensure profitability of the ESS, this lower boundary should be set to the value of the investment cost.

There are two basic types of energy storage investment decisions: siting and sizing. Siting refers to the decisions on the optimal ESS placement within a grid, while sizing refers to the decisions on its power and energy ratings. These decisions are modelled as continuous variables for the continuous decisions or as binary variables for the yes/no decisions. Examples of continuous storage investment decisions include [14,111–115]. Storage investment decisions modelled as binary variables can be found in [109,116,117].

Siting decisions were made flexible by Kim and Dvorkin [36], where the influence of the mobile energy storage on a distribution grid was investigated. Models of mobile ESS include models of the transport systems used to move the storage around, such as the railway in Sun et al. [118].

Sizing-only decisions were considered by Nasrolahpour et al. [112] and Qiu et al. [28]. This is in accordance with the conclusion made by Lorente et al. [6], stating that siting is more critical than sizing due to complexity of the siting models. On the other hand, Zhao et al. [119], Chakraborty et al. [120] and Pandžić [121] considered only sizing decisions at a premise of a consumer as the placement of the ESS in these models is behind the meter.

4.1. Strategic ESS Investments

Until now, only a few papers have considered strategic ESS investment decisions. As in operational models, these decisions are modelled as bi-level mathematical programs.

Nasrolahpour et al. [112] compared results of strategic and perfectly competitive ESS sizing decisions to show how said decisions depend on the chosen strategy. They came to conclusions similar to those of Ye et al. [104] about energy and power ratings of installed ESS. A strategic investor will choose a facility with higher power, and a competitive one will choose one with a higher energy rating.

Dvorkin et al. [113] investigated the impact of transmission line investments on profitability of a strategic ESS investor, noting that investments in transmission lines reduce a number of opportunities for the ESS. Pandžić et al. [122] similarly investigated coordinated investments in transmission lines and energy storage but they gave the system operator a choice between investments in lines or storage. The system operator was placed in the upper level, where it can anticipate decisions of the strategic ESS investor who was placed in the middle level of the model. The lower level presented market clearing process. Their results showed that the system operator favours transmission line over ESS investments even for low ESS investment costs. A strategic investor invests more in storage than the system operator because it can ensure profits more securely by actively participating in the markets, which is forbidden for the system operator-owned ESS. Huang et al. [123] analysed the same situation if the system operator and ESS investor switch places so that the ESS investor is in the upper level and the system operator is in the middle level. The strategic investor takes an even bigger share of the market in this configuration. They showed that system operator-owned ESS has no profit for any investment cost scenario, which is in accordance with the assumption that the system operator uses its storage in the same way as transmission lines, for social welfare maximisation, and not turning profit.

The impact of optimal allocation of both strategically and non-strategically behaving ESS on price volatility in energy-only markets was investigated by Masoumzadeh et al. [124]. They showed that, although an ESS is able to decrease price volatility, it does not remove it completely, and the positive impact stops when the ESS reaches the profitability limit. Next, they showed that a non-strategic ESS has larger influence on the price volatility than a strategic one, which is explained by the former's social welfare maximisation objective, opposed to the latter's profit maximisation.

4.2. *Transmission System-Level ESS*

Investments in ESS can be planned independently or in coordination with other technologies. The technologies used for coordinated approaches depend largely on the applications of the ESS. At the transmission system level, besides the stand-alone approach, ESS planning has been investigated in coordination with transmission networks, generators and renewable energy sources. Table 4 presents various modelling approaches for ESS at the transmission system level. There are not many models without network constraints, which can be explained by the fact that the ESS exploits network congestions and LMPs to achieve profit. However, it must be noted that not all of these models include siting decisions.

Table 4. Literature survey on the ESS investment planning models at the transmission level.

Modelling Technique	Network Unconstrained	Network Constrained	Stages
Deterministic	[129]	[125]	multiple
		[12,126–128]	two
		[110,111,113,115,122,123,130–133]	single
Stochastic	[112]	[28]	multiple
		[14,114,134,135]	two
		[124,136,137]	single
Robust		[116]	multiple
		[117,134,138]	two
Chance-constrained		[139]	two

4.2.1. Investments in Standalone ESS

Pandžić et al. [125] considered ESS for minimisation of operating costs within a unit commitment model to determine the necessity for ESS at each bus. The model in this paper is divided into three stages: first the siting decisions are made; then sizing; and last is the operation stage. Dvorkin et al. [110] also considered ESS for congestion management and other transmission services, but they modelled the investments as a bi-level problem, just like Pandžić et al. [111]. Here, a bi-level formulation was used to model the relations of the siting and sizing decisions and LMPs from the perspective of a merchant trading off between energy and reserve markets. Hemmati et al. [14] modelled three time-scales of ESS operation: daily, weekly and seasonal, as three levels within the planning model.

Several private investors in distributed ESS were considered by Saber et al. [139]. The investors were differentiated by their treatment of risk and their zone within the transmission grid. The model structured this way can also represent one investor that treats risk differently in different bidding areas of a zonally structured market.

Zheng et al. [128] studied optimal ESS allocation within transmission system. Unlike most studies on the transmission system that consider only DC power flows, in this paper AC power flows were adopted. The model was of a bi-level structure where the upper level represented siting and sizing decisions and the lower level operational phase.

4.2.2. Coordinated Investments in ESS and Transmission System Assets

While the transmission lines have longer lifetimes, they are more costly and take longer to build than most ESS technologies. For this reason, an ESS is often considered as a substitute or support for the transmission lines. Hu et al. [131] showed that deploying ESS can reduce transmission grid investment costs and MacRae et al. [133] demonstrated that an ESS can be used to postpone investments in transmission system. Aguado et al. [137] showed that the net social welfare increases when ESS are included in transmission expansion planning. The relationship between the transmission lines and ESS expansion was further investigated by Bustos et al. [127], who concluded that the complementarity of the two depends on many system parameters, such as nodal demand, generation capacity, congestion and prices. A three-level model was proposed by Zhang and Conejo [116] for coordinated investments in transmission lines and ESS. The first level determined the investment decisions, while the second and the third modelled long-term and short-term operations. Nikoobakht and Aghaei [134] proposed a continuous-time model for coordinated planning of ESS and transmission network in order to better capture intermittent RES variability, and concluded that the proposed model utilises ESS in the operational phase better. The lowermost level constraints of the three-level problem used by García-Cerezo et al. [117] to model coordinated transmission and ESS expansion planning contained binary variables. Such a problem cannot be solved by standard methods, i.e., KKT optimality conditions, so the authors proposed a nested column-and-constraint generation algorithm. They solved the model with and without binary variables and showed that simultaneous charging and discharging occurs if binary variables are not used.

4.2.3. Coordinated Investments in ESS and Generators

Joint optimisation of generators, transmission lines and ESS was considered by Carrión et al. [135] where the ESS provision of frequency response is included in the planning stage. Wu et al. [129] considered investment planning for generators and pumped-hydro storage constrained by the low-carbon requirements of the system. Opathella et al. [126] analysed the influence of generator contingencies in a long-term planning model. Tejada-Arango et al. [12] proposed a change of approach for unit commitment used in planning models from energy-based to power-based. This new approach takes into account more granular data inputs in form of power demanded at each moment instead of hourly energy demand. The results show that this approach results in lower total investment and operating costs.

4.2.4. Coordinated Investments in ESS and RES

Energy storage was considered useful for incorporating intermittent wind production in the power system by Xiong and Singh [114] because it reduces daily operating costs by reducing wind spillage for high wind production scenarios and prevents load curtailment for low wind production scenarios. Fernández-Blanco et al. [136] also considered using ESS to reduce renewable energy spillage and tested the sensitivity of siting and sizing decisions on various model parameters, such as penalties for renewable spillage, marginal costs of conventional generators and maximum energy rating. They showed that ESS does not always reduce renewable spillage if this reduction is not a part of the objective function.

4.3. Distribution Level ESS

We consider distribution-level ESS all those connected to the low or medium voltage network. Just like with transmission-level ESS investments, the approach to investment planning depends on the intended usage of the ESS. Even when specific ESS purposes are considered, general decisions can be drawn. Hajia et al. [29] considered joint expansion planning of distributed generators and ESS. The nonlinear model was solved by various heuristic algorithms. They showed that the energy arbitrage opportunity is the most important factor in ESS sizing. Table 5 presents different modelling approaches to distribution-level ESS investment planning.

Table 5. Literature survey on the ESS investment planning models at the distribution level.

Modelling Technique	Network Unconstrained	Network Constrained	Stages
Deterministic	[140]	[141]	multiple
	[119,120]		two
	[121,142–144]	[145–147]	single
Stochastic		[109,148–152]	multiple
	[153]	[36]	two
	[121,154]	[34,37]	single
Robust	[121]		single
Chance-constrained	[155]		multiple
		[156,157]	single

4.3.1. Coordinated Investments in ESS and Distribution System Assets

Xing et al. [151] researched expansion planning of a distribution grid already containing distributed ESS, but did not consider ESS investments. Similarly, Saboori et al. [141] used a multi-stage planning model to investigate the impact of ESS on distribution grid expansion. Installation of ESS is shown to decrease the number of new lines. It was also shown that ESS have positive impact on grid voltages and reduce congestions. These benefits are increased with the size of ESS. Quevedo et al. [109] considered the impact of electric vehicles on distribution grid expansion planning, showing that additional demand from electric vehicles can incur high costs for system operators by causing need for network expansion, which can be avoided by installing stationary ESS. Besides ESS, the authors in this paper considered installation of distributed generators, substations, transformers and electric vehicle charging stations. Hassan and Dvorkin [146] investigated how ESS capital costs, distribution grid line ratings, penetration of photovoltaics within the distribution grid and placement of wind power plants within the transmission grid influence ESS siting and sizing decisions in a distribution grid operated in coordination with transmission grid. The coordination was represented as a bi-level model wherein the transmission system operation was in the upper level and distribution system operation in the lower level. They showed that, if only one-way power flow is allowed between the transmission and distribution systems, placement of RES within the transmission grid does not influence investments in distributed ESS.

Joint expansion planning of energy storage and the distribution grid was modelled by Shen et al. [150], Akhavan-Hejazi and Mohsenian-Rad [156] and Iria et al. [38]. Shen et al. [150] showed that a distribution grid relies on ESS for peak shaving and reliability enhancement. Akhavan-Hejazi and Mohsenian-Rad [156] took the research a step further by modelling both real and reactive power flows and considering ESS for voltage compensation within an active distribution network. Voltage regulation was also considered by Das et al. [147] who took into account both real and reactive power injection by ESS. Iria et al. [38] considered investments in ESS and on-load tap changer transformers in a distribution network with high penetration of renewable generation. The installation of the new network assets was performed for the purposes of congestion and voltage problem mitigation.

Recently, idea of active distribution networks has gotten very popular. These networks include controllable distributed resources such as generators and ESS. Compared to traditional distribution networks, active ones require changes in the planning approach. Nick et al. [34] investigated optimal allocation of distributed ESS in active distribution networks for various purposes: voltage control, congestion management, network loss and load curtailment minimisation. This work was taken a step further by Nick et al. [39] by incorporating grid reconfiguration possibility in the model. Active distribution network planning was further investigated by Kim and Dvorkin [36] and Abdeltawab and Mohamed [37], who researched the possibility of using mobile ESS for enhancing distribution grid stability, especially through voltage control. Li et al. [152] used a three-level structure to model coordinated investments in active distribution grid, RES and ESS. The upper level presented network structure planning, the middle level was the allocation of RES and ESS and the lower level was system operation model. Sekhavatmanesh and Cherkaoui [158] developed a method for using the ESS to ensure fast grid restoration of active distribution networks, a service similar to black start in transmission networks.

4.3.2. Investments in ESS for RES Integration

In order to accommodate large shares of RES, investment models for coordinated or stand-alone RES and ESS are considered. Santos et al. [148] and Santos et al. [149] is a two-part paper dealing with different RES-enabling technologies placed at the distribution grid level. Zhang et al. [157] investigated optimal ESS allocation in distribution grids with high wind power penetration. Their model contained wind curtailment cost in the objective function and minimum wind utilisation constraint. Their results showed that if more wind utilisation is required, a larger ESS needs to be installed. Xiao et al. [145] built a siting and sizing model for distributed energy storage systems in distribution grid to accommodate distributed RES. The grid was represented by an AC model which requires a solving approach able to handle non-linear models. The genetic algorithm was used in this case.

4.3.3. ESS in Aggregators' Investment Models

In most planning models that consider ESS interactions with an aggregator, ESS siting and sizing is not considered. Exceptions are the shared storage models. Zhao et al. [119] modelled an aggregator operating a shared storage aiming to maximise its profit. This model improves the utilisation of energy storage which means that the aggregator can invest in a smaller facility and still serve same number of customers. Shared storage optimisation was the objective in Chakraborty et al. [120] as well, where the cost sharing was modelled as a coalition game.

4.3.4. Investments in ESS in Microgrids and Vpps

Considering organisation, microgrids and VPPs are similar structures. Both structures are optimally run by a central operator with the aim of minimising operating costs or maximising profits. The difference between the two is structural—while a microgrid is a small portion of a distribution grid connected to the main grid through one point of common coupling, a VPP is spatially distributed set of resources. Yang and Nehorai [140] considered investments in a hybrid generator-storage facility within an islanded micro-grid at different geographical locations. Their results showed that the type of climate at the location influences selection of installed technologies. The planning problem in Khodaei et al. [159] focused on microgrids with distributed resources and energy storage and assessed the possibility of islanded operation. It was shown that, for shorter expected durations of the islanded operation, installation of an ESS is not justified. Cao et al. [155] presented a model for multi-stage microgrid expansion planning for stand-alone microgrids. Jacob et al. [144] proposed a sizing method for ESS within a microgrid with photovoltaic (PV) production based on design space approach. The method considered time-scale classification of the ESS and variability of PV output. Different time-scale storage requirements were met with different technologies: fuel cells and flow batteries for long-term, lithium and lead acid batteries for mid-term and flywheels and supercapacitors for short-term storage.

4.3.5. Behind-The-Meter ESS Investments

In behind-the-meter applications of ESS, the objective is usually minimisation of electricity cost. These costs can be a result of wholesale market participation, incentives paid by the system operators or participation in demand response programmes. These models can include RES and controllable loads as well. Sharma et al. [143] modelled a nearly zero-energy residential home with a PV system. Necessary power and energy ratings of the ESS were calculated for each time interval by subtracting the load from the produced power. Optimal ESS capacity was then determined by minimising overall costs heuristically by using the genetic algorithm and two other minimisation functions provided in MATLAB. Although this approach might lead to a local optimum, all three techniques converged to the same solution, so the authors concluded that they reached the global optimum. They showed that after the installation of ESS, energy exchanged with the grid decreased significantly. Zhu et al. [142] presented a method for sizing ESS within distribution grid with high PV penetration and tested the method on three types of ESS: behind-the-meter, utility owned and merchant owned. They showed that it is more economical for the DSO to procure services from the latter two types of ESS. They also showed that existence of demand side management reduces the size of installed ESS. Pandžić [121] optimised investments in ESS for a hotel participating in a two-tariff retail market. Bayram et al. [154] developed an analytical method for sizing a shared ESS used by consumers to achieve various benefits, such as improved power quality and cost reduction, by participating in demand response programs and avoiding peak power charges. Wang and He [153] presented a model for making optimal decisions on behind-the-meter ESS installation and demand response programme participation. The model was suitable for commercial consumers and the paper presented two case studies for a smaller and a larger consumer. Depending on the distributed generator's capacity, the model makes different decisions on ESS size and participation in a demand response scheme. They showed that for low electricity prices no ESS is installed.

5. Computational Complexity of ESS Models

The computational complexity of a model depends on its various properties, such as the number and types of variables, the number of time periods and whether or not uncertainties are considered. Table 6 shows three main types of models encountered during this literature review. Linear programming (LP) is the simplest type of models, as it contains only continuous variables and therefore can usually be solved by algorithms like Simplex. In using these types of models for ESS modelling there is a risk of simultaneous charging and discharging in certain cases, as discussed in Section 4.3.1. Mixed-integer linear programming (MILP) models contain integer variables. In the case of ESS modelling, these are usually binary variables used to prohibit simultaneous charging and discharging of the ESS or investment indicators. In the next level of complexity, second order cone programs (SOCP), are used for modelling AC power flows. They are being used more and more for modelling ESS operation within distribution grids. These three types of models are not the only ones used in the literature. For example, Shafiee et al. [45] used a mixed-integer non-linear model, Huang et al. [123] used a mixed-integer quadratic programming model and Abdeltawab and Mohamed [37] used a mixed-integer convex programming model.

Table 6. Types of mathematical models used in the considered literature. (Abbreviations: LP, linear programming; MILP, mixed-integer linear programming; SOCP, second order cone programming).

Operation	Planning	Solving Technique
[20,22,23,30,43,46,49–51,53,54,57,59, 73,74,78,88–90,105]	[119,125,130,156]	LP
[1,16–18,20,25,32,33,35,44,47,50,52, 56–58,60,61,63,65,66,70–72,76,79–83, 86,92,94–96,100–102,104,106,107]	[14,28,109–115,117,121,122,126,127, 131,133–137,139,148,149,155,156, 159]	MILP
[40,91]	[34,36,38,39,116,146,151,158]	SOCP

The main ways to deal with intractability issues in ESS models involve choosing the right modelling and solving techniques. The intractability of models is caused by large numbers of input parameters. Although many clustering techniques were developed to deal with large numbers of input parameters in power system modelling, they are not appropriate for ESS modelling. Intertemporal constraints, especially for long-term energy storage, cause large errors in models where clustering techniques are applied. For this reason there are not as many papers dealing with long-term storage operation as there are with short-term operation in the intraday and day-ahead markets (Figure 6). However, ESS-friendly clustering techniques are being developed. To reduce computational burden of planning models, Pineda and Morales [130] proposed a chronological time-period clustering different from the standard hours-days-weeks clustering. The proposed technique separates longer time periods into clusters in which smallest time-period is one hour and only consecutive and similar clusters can be merged. Tejada-Arango et al. [132] also presented clustering techniques for decreasing computational burden of the planning models. They proposed improved system states and representative periods clustering techniques and showed that they do indeed shorten the computation time.

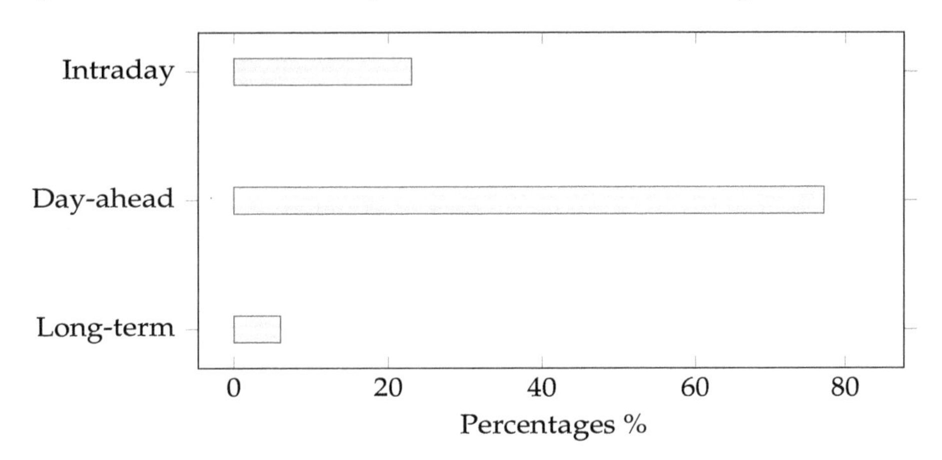

Figure 6. Share of the reviewed literature that considered ESS operational models at three timescales: intraday, day-ahead and long-term.

5.1. Modelling Techniques

Power system planning solves investment problems by taking into account technical and economic constraints, and it has always been concerned with uncertainties. For a long time, the main sources of uncertainty in the planning models were load growth and production of hydro-power plants. Nowadays, the largest source of uncertainties is the production of intermittent resources, i.e., wind and solar power plants. This is evident from Figure 7, which shows that, while load is the dominant source of uncertainties in planning models, price is considered to be uncertain more often in operational than planning models. RES as a source of uncertainty is commonly considered in both the operational and the planning models. Label "none" in Figure 7 signifies the share of deterministic models where no uncertainties were considered.

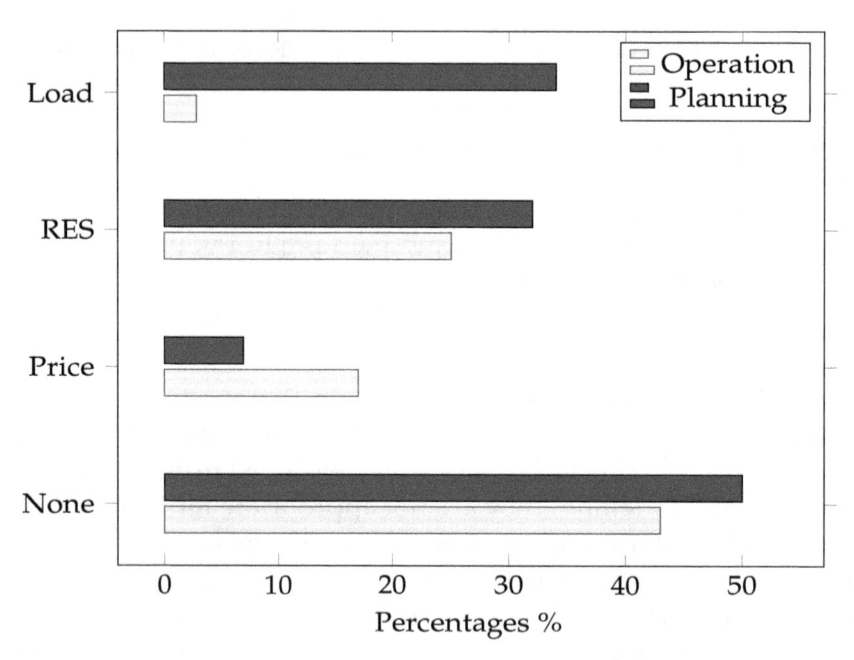

Figure 7. Sources of uncertainties in the reviewed operational and planning models (the sample contains 70 operational and 55 planning models).

There are several ways to include uncertainties in a mathematical model. The three main types of mathematical programming methods for dealing with uncertainties are robust, stochastic and chance-constrained optimisation. Comparisons of the methods were performed with general conclusions that stochastic optimisation yields larger problems and takes longer to solve than the other methods. It was also concluded that robust optimisation gives good results for border-line cases, but for most cases it is better to use stochastic optimisation. It was shown by Khodayar and Shahidehpour [57], who compared the results from deterministic and stochastic problems, that neglecting uncertainties generally results in higher expected profits. The uncertainties in the models stem from the stochastic nature of RES output, load levels and market prices.

Besides mathematical programming methods, the approach of machine learning for dealing with uncertainties in computationally tractable way is gaining momentum in recent years. Until now, it was applied only to a few ESS modelling problems. Machine learning techniques are mostly used to forecast values of the uncertain parameters such as wind power by Varkani et al. [62]. Another useful application of machine learning algorithms is as a heuristic used to solve complex stochastic programming models. For example, Yuan et al. [55] used a neural network with genetic algorithm to solve a large stochastic problem of coordinating wind power plant and ESS.

However, reinforcement learning can be used to model market-participating ESS. This approach takes on the uncertainties by "teaching" the model to respond to them, instead of modelling them mathematically. The technique that has gotten the most attention from the power systems modelling community is Q-learning. Q-learning is a model-less reinforcement learning algorithm that describes the behaviour of a model by a Markov decision process with actions to which rewards are assigned and the goal is to maximise cumulative reward. This algorithm was used by Wang and Zhang [46] to model market participation of a strategic ESS. Although machine learning algorithms are widely used for data fitting and forecasting, Wang and Zhang [46] were the first ones to use this approach to model the ESS market participation. Q-learning was also used by Ye et al. [160] to model the strategic market participation of a generator.

The biggest flaw of this approach is that the model trained on specific data set learns to behave only in the circumstances described by the data. This can be problematic in a changing environment, e.g., a market in which new players are appearing.

5.2. Solving Techniques

Based on the reviewed literature, the general conclusion is that solving mathematical models with uncertainties is quite complex. This is especially the case with multi-stage models, which can be intractable even for small number of stages. The complexity of the models with ESS is even greater because of the intertemporal constraints and the necessity for the inclusion of binary variables. Many different solving techniques were developed with an aim of simplifying this process. Some of them are:

- Heuristic algorithms [55,128,143,147];
- Decomposition techniques [43,105,122], especially Benders decomposition [22,57,60,106,112,134] and column-and-constraint generation [92,113];
- Dynamic programming [16,18,27,54,161].

Some of these techniques are based on duality theory, and are therefore unsuitable for solving models with binary variables, common in ESS modelling. For a detailed analysis of solving techniques used for complex planning problems with ESS, an interested reader should refer to Lorente et al. [6] and Zidar et al. [8].

6. Conclusions

The scope of the literature dealing with ESS is vast, which indicates that a significant number of issues have already been solved in both the operation and the investment planning models including ESS. However, the potential for improvements in model development exists in both areas. New use cases for ESS need to be created, especially for distribution-level ESS to ensure reliable grid operation in the context of active network design, distributed generation and demand response programs. Allowing distributed resources to participate in the wholesale energy and ancillary service markets, and the deployment of innovative local energy markets, should open even more possibilities for ESS. At the transmission level, different ownership structures of ESS (privately or TSO-owned) necessitate different operating strategies so there is room for inventiveness as well.

As a conclusion, some of the specific areas where study of ESS can be expanded are:

- ESS operating in zonal market structure, especially considering uncertainties. While many electricity markets are structured zonally (Europe, Australia), not so many studies on ESS participating in zonal markets were conducted.
- Studies on strategic ESS will have us believe that it is very simple to ensure large profits for investors by employing an aggressive strategy during operational phase. More studies on competition between more than one strategic ESS are therefore necessary to reconcile the results of ESS with general economic theory.
- Strategic investment models, on the other hand, do not deal with bidding decisions and only siting and sizing decisions are strategic in these models. Nonetheless, competition between more than one strategic investor was not yet investigated.
- Market participation of the distribution-level ESS was not widely investigated in the past. The design of distribution-level markets will have a large role in enabling profitability of distributed ESS. Aggregation, local energy markets and peer-to-peer trading between distribution-level ESS must be tested thoroughly before they are implemented in real-world situations.
- Although load, RES and prices have been considered as the main sources of uncertainties so far, other sources are appearing in the models, such as contingencies and EV loads. Although ESS has proven to be of most use in situations where flexibility of the system is jeopardised, their application in mitigating various uncertainty risks should be analysed.

- Tractability of models is a challenge even without ESS. However, long-term operation of ESS does not play well into most clustering techniques used to simplify models with time-spans of a year or longer. Therefore, it is necessary to develop ESS-oriented clustering techniques which would enable including long-term operation in operational and investment models with ESS.

Author Contributions: Conceptualisation, M.M. and H.P.; methodology, M.M. and H.P.; formal analysis, M.M.; investigation, M.M. and H.P.; resources, M.M. and H.P.; data curation, M.M. and H.P.; writing—original draft preparation, M.M.; writing—review and editing, H.P. and D.Y.; visualization, M.M.; supervision, H.P.; project administration, H.P.; funding acquisition, H.P. and D.Y. All authors have read and agreed to the published version of the manuscript.

Abbreviations

The following abbreviations are used in this manuscript:

ESS	Energy storage system
RHS	Right-hand side
SOC	State of charge (percentage)
SOE	State of energy (absolute value)
DA	Day-ahead
CAP	Capacity market
EN	Energy market
BM	Balancing services market
RM	Reserve market
VPP	Virtual power plant
RES	Renewable energy source
LP	Linear programming
MILP	Mixed-integer linear programming
SOCP	Second order cone programming

Nomenclature

Sets

T	set of time points indexed by t from 1 to N_T

Parameters

ΔT	duration of a time-step t, h
η	round trip energy efficiency
η^{ch}	charging efficiency
η^{dis}	discharging efficiency
$\overline{E}$	maximum state of energy, Wh
$\overline{P}^{\text{ch}}$	maximum charging power, W
$\overline{P}^{\text{dis}}$	maximum discharging power, W
$\underline{E}$	minimum state of energy, Wh
$E^{\text{cc}-\text{cv}}$	state of energy at the boundary between CC and CV charging phases, Wh
E_0	initial state of energy, Wh
RU	ramping-up reserve, W

Variables

e_t	state of energy at time t, Wh
e_t^{ch}	charged energy during period t, Wh
e_t^{dis}	discharged energy during period t, Wh
p_t^{loss}	lost power at time t, W
p_t^{ch}	charging power at time t, W
p_t^{dis}	discharging power at time t, W
r_t^{+}	ramp up at time t, W
r_t^{-}	ramp down at time t, W
x_t	binary variable: 1 when the ESS is charging, 0 when discharging

References

1. Suazo-Martínez, C.; Pereira-Bonvallet, E.; Palma-Behnke, R.; Zhang, X. Impacts of Energy Storage on Short Term Operation Planning Under Centralized Spot Markets. *IEEE Trans. Smart Grid* **2014**, *5*, 1110–1118. [CrossRef]
2. Luo, X.; Wang, J.; Dooner, M.; Clarke, J. Overview of current development in electrical energy storage technologies and the application potential in power system operation. *Appl. Energy* **2015**, *137*, 511–536. [CrossRef]
3. Koohi-Fayegh, S.; Rosen, M.A. A review of energy storage types, applications and recent developments. *J. Energy Storage* **2020**, *27*, 101047. [CrossRef]
4. Khan, N.; Dilshad, S.; Khalid, R.; Kalair, A.R.; Abas, N. Review of energy storage and transportation of energy. *Energy Storage* **2019**, *1*, e49. [CrossRef]
5. Awadallah, M.A.; Venkatesh, B. Energy Storage in Distribution System Planning and Operation: Current Status and Outstanding Challenges. *Can. J. Electr. Comput. Eng.* **2019**, *42*, 10–19. [CrossRef]
6. Lorente, J.L.; Liu, X.A.; Best, R.; Morrow, D.J. Energy storage allocation in power networks—A state-of-the-art review. In Proceedings of the 2018 53rd International Universities Power Engineering Conference (UPEC), Glasgow, Scotland, 4–7 September 2018; pp. 1–6.
7. Sheibani, M.R.; Yousefi, G.R.; Latify, M.A.; Hacopian Dolatabadi, S. Energy storage system expansion planning in power systems: A review. *IET Renew. Power Gener.* **2018**, *12*, 1203–1221. [CrossRef]
8. Zidar, M.; Georgilakis, P.S.; Hatziargyriou, N.D.; Capuder, T.; Škrlec, D. Review of energy storage allocation in power distribution networks: applications, methods and future research. *IET Gener. Transm. Distrib.* **2016**, *10*, 645–652. [CrossRef]
9. Miller, L.; Carriveau, R. A review of energy storage financing—Learning from and partnering with the renewable energy industry. *J. Energy Storage* **2018**, *19*, 311–319. [CrossRef]
10. Directive (EU) 2019/944 of the European Parliament and of the Council of 5 June 2019 on Common Rules for the Internal Market for Electricity and Amending Directive 2012/27/EU. 2019. Available online: https://www.legislation.gov.uk/eudr/2019/944/contents (accessed on 2 September 2020).
11. Mejia, C.; Kajikawa, Y. Emerging topics in energy storage based on a large-scale analysis of academic articles and patents. *Appl. Energy* **2020**, *263*, 114625. [CrossRef]
12. Tejada-Arango, D.A.; Morales-España, G.; Wogrin, S.; Centeno, E. Power-Based Generation Expansion Planning for Flexibility Requirements. *IEEE Trans. Power Syst.* **2019**. [CrossRef]
13. Poncelet, K.; Delarue, E.; D'haeseleer, W. Unit commitment constraints in long-term planning models: Relevance, pitfalls and the role of assumptions on flexibility. *Appl. Energy* **2020**, *258*, 113843. [CrossRef]
14. Hemmati, R.; Shafie-Khah, M.; Catalão, J.P.S. Three-Level Hybrid Energy Storage Planning Under Uncertainty. *IEEE Trans. Ind. Electron.* **2019**, *66*, 2174–2184. [CrossRef]
15. Pandžić, H.; Bobanac, V. An Accurate Charging Model of Battery Energy Storage. *IEEE Trans. Power Syst.* **2019**, *34*, 1416–1426. [CrossRef]
16. Kazempour, S.J.; Moghaddam, M.P.; Haghifam, M.; Yousefi, G. Risk-constrained dynamic self-scheduling of a pumped-storage plant in the energy and ancillary service markets. *Energy Convers. Manag.* **2009**, *50*, 1368–1375. [CrossRef]

17. Pandžić, H.; Kuzle, I. Energy storage operation in the day-ahead electricity market. In Proceedings of the 2015 12th International Conference on the European Energy Market (EEM), Lisbon, Portugal, 19–22 May 2015; pp. 1–6.
18. Alvarez, M.; Rönnberg, S.K.; Bermúdez, J.; Zhong, J.; Bollen, M.H.J. A Generic Storage Model Based on a Future Cost Piecewise-Linear Approximation. *IEEE Trans. Smart Grid* **2019**, *10*, 878–888. [CrossRef]
19. Vagropoulos, S.I.; Bakirtzis, A.G. Optimal Bidding Strategy for Electric Vehicle Aggregators in Electricity Markets. *IEEE Trans. Power Syst.* **2013**, *28*, 4031–4041. [CrossRef]
20. Gonzalez-Castellanos, A.J.; Pozo, D.; Bischi, A. Non-ideal Linear Operation Model for Li-ion Batteries. *IEEE Trans. Power Syst.* **2019**. [CrossRef]
21. Duggal, I.; Venkatesh, B. Short-Term Scheduling of Thermal Generators and Battery Storage With Depth of Discharge-Based Cost Model. *IEEE Trans. Power Syst.* **2015**, *30*, 2110–2118. [CrossRef]
22. Kazemi, M.; Zareipour, H. Long-Term Scheduling of Battery Storage Systems in Energy and Regulation Markets Considering Battery's Lifespan. *IEEE Trans. Smart Grid* **2018**, *9*, 6840–6849. [CrossRef]
23. He, G.; Chen, Q.; Kang, C.; Pinson, P.; Xia, Q. Optimal Bidding Strategy of Battery Storage in Power Markets Considering Performance-Based Regulation and Battery Cycle Life. *IEEE Trans. Smart Grid* **2016**, *7*, 2359–2367. [CrossRef]
24. Xu, B.; Zhao, J.; Zheng, T.; Litvinov, E.; Kirschen, D.S. Factoring the Cycle Aging Cost of Batteries Participating in Electricity Markets. *IEEE Trans. Power Syst.* **2018**, *33*, 2248–2259. [CrossRef]
25. Padmanabhan, N.; Ahmed, M.; Bhattacharya, K. Battery Energy Storage Systems in Energy and Reserve Markets. *IEEE Trans. Power Syst.* **2020**, *35*, 215–226. [CrossRef]
26. Alam, M.J.E.; Saha, T.K. Cycle-life degradation assessment of Battery Energy Storage Systems caused by solar PV variability. In Proceedings of the 2016 IEEE Power and Energy Society General Meeting (PESGM), Boston, MA, USA, 17–21 July 2016; pp. 1–5.
27. Vejdan, S.; Grijalva, S. Maximizing the Revenue of Energy Storage Participants in Day-Ahead and Real-Time Markets. In Proceedings of the 2018 Clemson University Power Systems Conference (PSC), Charleston, SC, USA, 4–7 September 2018; pp. 1–6.
28. Qiu, T.; Xu, B.; Wang, Y.; Dvorkin, Y.; Kirschen, D.S. Stochastic Multistage Coplanning of Transmission Expansion and Energy Storage. *IEEE Trans. Power Syst.* **2017**, *32*, 643–651. [CrossRef]
29. Hajia, N.; Venkatesh, B.; Awadallah, M.A. Optimal Asset Expansion in Distribution Networks Considering Battery Nonlinear Characteristics Expansion optimale des actifs dans les réseaux de distribution en tenant compte des caractéristiques non linéaires des batteries. *Can. J. Electr. Comput. Eng.* **2018**, *41*, 191–199. [CrossRef]
30. Gantz, J.M.; Amin, S.M.; Giacomoni, A.M. Optimal Capacity Partitioning of Multi-Use Customer-Premise Energy Storage Systems. *IEEE Trans. Smart Grid* **2014**, *5*, 1292–1299. [CrossRef]
31. EPRI. *Electric Energy Storage Technology Options: A White Paper Primer on Applications, Costs, and Benefits*; Technical Report; Palo Alto: Santa Clara, CA, USA, 2010.
32. Mohsenian-Rad, H. Optimal Bidding, Scheduling, and Deployment of Battery Systems in California Day-Ahead Energy Market. *IEEE Trans. Power Syst.* **2016**, *31*, 442–453. [CrossRef]
33. Daneshi, H.; Srivastava, A.K. Security-constrained unit commitment with wind generation and compressed air energy storage. *IET Gener. Transm. Distrib.* **2012**, *6*, 167–175. [CrossRef]
34. Nick, M.; Cherkaoui, R.; Paolone, M. Optimal Allocation of Dispersed Energy Storage Systems in Active Distribution Networks for Energy Balance and Grid Support. *IEEE Trans. Power Syst.* **2014**, *29*, 2300–2310. [CrossRef]
35. Moreira, R.; Moreno, R.; Strbac, G. Synergies and conflicts among energy storage services. In Proceedings of the 2016 IEEE International Energy Conference (ENERGYCON), Leuven, Belgium, 4–8 April 2016; pp. 1–6. [CrossRef]
36. Kim, J.; Dvorkin, Y. Enhancing Distribution System Resilience With Mobile Energy Storage and Microgrids. *IEEE Trans. Smart Grid* **2019**, *10*, 4996–5006. [CrossRef]
37. Abdeltawab, H.; Mohamed, Y.A.I. Mobile Energy Storage Sizing and Allocation for Multi-Services in Power Distribution Systems. *IEEE Access* **2019**. [CrossRef]
38. Iria, J.; Heleno, M.; Cardoso, G. Optimal sizing and placement of energy storage systems and on-load tap changer transformers in distribution networks. *Appl. Energy* **2019**, *250*, 1147–1157. [CrossRef]

39. Nick, M.; Cherkaoui, R.; Paolone, M. Optimal Planning of Distributed Energy Storage Systems in Active Distribution Networks Embedding Grid Reconfiguration. *IEEE Trans. Power Syst.* **2018**, *33*, 1577–1590. [CrossRef]
40. Nazir, N.; Racherla, P.; Almassalkhi, M. Optimal Multi-Period Dispatch of Distributed Energy Resources in Unbalanced Distribution Feeders. *IEEE Trans. Power Syst.* **2020**, *35*, 2683–2692. [CrossRef]
41. Miletić, M.; Luburić, Z.; Pavić, I.; Capuder, T.; Pandžić, H.; Andročec, I.; Marušić, A. A review of energy storage systems applications. In Proceedings of the Mediterranean Conference on Power Generation, Transmission, Distribution and Energy Conversion (MEDPOWER 2018), Dubrovnik, Croatia, 12–15 November 2018; pp. 1–6.
42. Thatte, A.A.; Xie, L.; Viassolo, D.E.; Singh, S. Risk Measure Based Robust Bidding Strategy for Arbitrage Using a Wind Farm and Energy Storage. *IEEE Trans. Smart Grid* **2013**, *4*, 2191–2199. [CrossRef]
43. Xia, Y.; Ghiocel, S.G.; Dotta, D.; Shawhan, D.; Kindle, A.; Chow, J.H. A Simultaneous Perturbation Approach for Solving Economic Dispatch Problems With Emission, Storage, and Network Constraints. *IEEE Trans. Smart Grid* **2013**, *4*, 2356–2363. [CrossRef]
44. Mohsenian-Rad, H. Coordinated Price-Maker Operation of Large Energy Storage Units in Nodal Energy Markets. *IEEE Trans. Power Syst.* **2016**, *31*, 786–797. [CrossRef]
45. Shafiee, S.; Zareipour, H.; Knight, A.M.; Amjady, N.; Mohammadi-Ivatloo, B. Risk-Constrained Bidding and Offering Strategy for a Merchant Compressed Air Energy Storage Plant. *IEEE Trans. Power Syst.* **2017**, *32*, 946–957. [CrossRef]
46. Wang, H.; Zhang, B. Energy Storage Arbitrage in Real-Time Markets via Reinforcement Learning. In Proceedings of the 2018 IEEE Power Energy Society General Meeting (PESGM), Portland, OR, USA, 5–9 August 2018; pp. 1–5. [CrossRef]
47. Ciftci, O.; Mehrtash, M.; Safdarian, F.; Kargarian, A. Chance-Constrained Microgrid Energy Management with Flexibility Constraints Provided by Battery Storage. In Proceedings of the 2019 IEEE Texas Power and Energy Conference (TPEC), College Station, TX, USA, 7–8 February 2019; pp. 1–6.
48. Braun, S. Hydropower Storage Optimization Considering Spot and Intraday Auction Market. *Energy Procedia* **2016**, *87*, 36–44. [CrossRef]
49. Zakeri, B.; Syri, S. Value of energy storage in the Nordic Power market - benefits from price arbitrage and ancillary services. In Proceedings of the 2016 13th International Conference on the European Energy Market (EEM), Porto, Portugal, 6–9 June 2016; pp. 1–5. [CrossRef]
50. Krishnamurthy, D.; Uckun, C.; Zhou, Z.; Thimmapuram, P.R.; Botterud, A. Energy Storage Arbitrage Under Day-Ahead and Real-Time Price Uncertainty. *IEEE Trans. Power Syst.* **2018**, *33*, 84–93. [CrossRef]
51. Castronuovo, E.D.; Usaola, J.; Bessa, R.; Matos, M.; Costa, I.; Bremermann, L.; Lugaro, J.; Kariniotakis, G. An integrated approach for optimal coordination of wind power and hydro pumping storage. *Wind Energy* **2014**, *17*, 829–852. [CrossRef]
52. Usaola, J. Operation of concentrating solar power plants with storage in spot electricity markets. *IET Renew. Power Gener.* **2012**, *6*, 59–66. [CrossRef]
53. Rahimiyan, M.; Baringo, L. Strategic Bidding for a Virtual Power Plant in the Day-Ahead and Real-Time Markets: A Price-Taker Robust Optimization Approach. *IEEE Trans. Power Syst.* **2016**, *31*, 2676–2687. [CrossRef]
54. Shu, Z.; Jirutitijaroen, P. Optimal Operation Strategy of Energy Storage System for Grid-Connected Wind Power Plants. *IEEE Trans. Sustain. Energy* **2014**, *5*, 190–199. [CrossRef]
55. Yuan, Y.; Li, Q.; Wang, W. Optimal operation strategy of energy storage unit in wind power integration based on stochastic programming. *IET Renew. Power Gener.* **2011**, *5*, 194–201. [CrossRef]
56. Garcia-Gonzalez, J.; Muela, R.M.R.d.l.; Santos, L.M.; Gonzalez, A.M. Stochastic Joint Optimization of Wind Generation and Pumped-Storage Units in an Electricity Market. *IEEE Trans. Power Syst.* **2008**, *23*, 460–468. [CrossRef]
57. Khodayar, M.E.; Shahidehpour, M. Stochastic Price-Based Coordination of Intrahour Wind Energy and Storage in a Generation Company. *IEEE Trans. Sustain. Energy* **2013**, *4*, 554–562. [CrossRef]
58. Sánchez de la Nieta, A.A.; Contreras, J.; Muñoz, J.I. Optimal coordinated wind-hydro bidding strategies in day-ahead markets. *IEEE Trans. Power Syst.* **2013**, *28*, 798–809. [CrossRef]

59. Thatte, A.A.; Viassolo, D.E.; Xie, L. Robust bidding strategy for wind power plants and energy storage in electricity markets. In Proceedings of the 2012 IEEE Power and Energy Society General Meeting, Grand Hyatt, San Diego, CA, 22–26 July 2012; pp. 1–7.
60. Jiang, R.; Wang, J.; Guan, Y. Robust Unit Commitment With Wind Power and Pumped Storage Hydro. *IEEE Trans. Power Syst.* **2012**, *27*, 800–810. [CrossRef]
61. Li, N.; Uçkun, C.; Constantinescu, E.M.; Birge, J.R.; Hedman, K.W.; Botterud, A. Flexible Operation of Batteries in Power System Scheduling With Renewable Energy. *IEEE Trans. Sustain. Energy* **2016**, *7*, 685–696. [CrossRef]
62. Varkani, A.K.; Daraeepour, A.; Monsef, H. A new self-scheduling strategy for integrated operation of wind and pumped-storage power plants in power markets. *Appl. Energy* **2011**, *88*, 5002–5012. [CrossRef]
63. Parvania, M.; Fotuhi-Firuzabad, M.; Shahidehpour, M. Comparative Hourly Scheduling of Centralized and Distributed Storage in Day-Ahead Markets. *IEEE Trans. Sustain. Energy* **2014**, *5*, 729–737. [CrossRef]
64. Kazempour, S.J.; Hosseinpour, M.; Moghaddam, M.P. Self-scheduling of a joint hydro and pumped-storage plants in energy, spinning reserve and regulation markets. In Proceedings of the 2009 IEEE Power Energy Society General Meeting, Calgary, AB, Canada, 26–30 July 2009; pp. 1–8.
65. Baslis, C.G.; Bakirtzis, A.G. Mid-Term Stochastic Scheduling of a Price-Maker Hydro Producer With Pumped Storage. *IEEE Trans. Power Syst.* **2011**, *26*, 1856–1865. [CrossRef]
66. Pandžić, K.; Pandžić, H.; Kuzle, I. Virtual storage plant offering strategy in the day-ahead electricity market. *Int. J. Electr. Power Energy Syst.* **2019**, *104*, 401–413. [CrossRef]
67. Fleer, J.; Zurmühlen, S.; Meyer, J.; Badeda, J.; Stenzel, P.; Hake, J.F.; Sauer, D.U. Techno-economic evaluation of battery energy storage systems on the primary control reserve market under consideration of price trends and bidding strategies. *J. Energy Storage* **2018**, *17*, 345–356. [CrossRef]
68. Thien, T.; Schweer, D.; Stein, D.v.; Moser, A.; Sauer, D.U. Real-world operating strategy and sensitivity analysis of frequency containment reserve provision with battery energy storage systems in the german market. *J. Energy Storage* **2017**, *13*, 143–163. [CrossRef]
69. Olk, C.; Sauer, D.U.; Merten, M. Bidding strategy for a battery storage in the German secondary balancing power market. *J. Energy Storage* **2019**, *21*, 787–800. [CrossRef]
70. Opathella, C.; Elkasrawy, A.; Mohamed, A.A.; Venkatesh, B. A Novel Capacity Market Model With Energy Storage. *IEEE Trans. Smart Grid* **2019**, *10*, 5283–5293. [CrossRef]
71. Opathella, C.; Elkasrawy, A.; Mohamed, A.A.; Venkatesh, B. Optimal Scheduling of Merchant-Owned Energy Storage Systems With Multiple Ancillary Services. *IEEE Open Access J. Power Energy* **2020**, *7*, 31–40. [CrossRef]
72. Nasrolahpour, E.; Zareipour, H.; Rosehart, W.D.; Kazempour, S.J. Bidding strategy for an energy storage facility. In Proceedings of the 2016 Power Systems Computation Conference (PSCC), Genoa, Italy, 20–24 June 2016; pp. 1–7.
73. Sioshansi, R. When energy storage reduces social welfare. *Energy Econ.* **2014**, *41*, 106–116. [CrossRef]
74. Ye, Y.; Papadaskalopoulos, D.; Strbac, G. An MPEC approach for analysing the impact of energy storage in imperfect electricity markets. In Proceedings of the 2016 13th International Conference on the European Energy Market (EEM), Porto, Portugal, 6–9 June 2016; pp. 1–5.
75. Nguyen, T.A.; Byrne, R.H.; Chalamala, B.R.; Gyuk, I. Maximizing The Revenue of Energy Storage Systems in Market Areas Considering Nonlinear Storage Efficiencies. In Proceedings of the 2018 International Symposium on Power Electronics, Electrical Drives, Automation and Motion (SPEEDAM), Amalfi, Italy, 20–22 June 2018; pp. 55–62.
76. Wang, Y.; Dvorkin, Y.; Fernández-Blanco, R.; Xu, B.; Kirschen, D.S. Impact of local transmission congestion on energy storage arbitrage opportunities. In Proceedings of the 2017 IEEE Power Energy Society General Meeting, Chicago, IL, USA, 16–20 July 2017; pp. 1–5.
77. Weibelzahl, M.; Märtz, A. On the effects of storage facilities on optimal zonal pricing in electricity markets. *Energy Policy* **2018**, *113*, 778–794. [CrossRef]
78. Contreras-Ocaña, J.E.; Ortega-Vazquez, M.A.; Zhang, B. Participation of an Energy Storage Aggregator in Electricity Markets. *IEEE Trans. Smart Grid* **2019**, *10*, 1171–1183. [CrossRef]
79. Li, N.; Hedman, K.W. Economic Assessment of Energy Storage in Systems With High Levels of Renewable Resources. *IEEE Trans. Sustain. Energy* **2015**, *6*, 1103–1111. [CrossRef]

80. Ding, H.; Hu, Z.; Song, Y. Rolling Optimization of Wind Farm and Energy Storage System in Electricity Markets. *IEEE Trans. Power Syst.* **2015**, *30*, 2676–2684. [CrossRef]
81. Pozo, D.; Contreras, J.; Sauma, E.E. Unit Commitment With Ideal and Generic Energy Storage Units. *IEEE Trans. Power Syst.* **2014**, *29*, 2974–2984. [CrossRef]
82. Bruninx, K.; Dvorkin, Y.; Delarue, E.; Pandžić, H.; D'haeseleer, W.; Kirschen, D.S. Coupling Pumped Hydro Energy Storage With Unit Commitment. *IEEE Trans. Sustain. Energy* **2016**, *7*, 786–796. [CrossRef]
83. Kazemi, M.; Zareipour, H.; Amjady, N.; Rosehart, W.D.; Ehsan, M. Operation Scheduling of Battery Storage Systems in Joint Energy and Ancillary Services Markets. *IEEE Trans. Sustain. Energy* **2017**, *8*, 1726–1735. [CrossRef]
84. Li, C.; Zhang, S.; Zhang, J.; Qi, J.; Li, J.; Guo, Q.; You, H. Method for the Energy Storage Configuration of Wind Power Plants with Energy Storage Systems used for Black-Start. *Energies* **2018**, *11*, 3394. [CrossRef]
85. Sugihara, H.; Yokoyama, K.; Saeki, O.; Tsuji, K.; Funaki, T. Economic and Efficient Voltage Management Using Customer-Owned Energy Storage Systems in a Distribution Network With High Penetration of Photovoltaic Systems. *IEEE Trans. Power Syst.* **2013**, *28*, 102–111. [CrossRef]
86. Nasrolahpour, E.; Kazempour, J.; Zareipour, H.; Rosehart, W.D. Impacts of Ramping Inflexibility of Conventional Generators on Strategic Operation of Energy Storage Facilities. *IEEE Trans. Smart Grid* **2018**, *9*, 1334–1344. [CrossRef]
87. Babacan, O.; Ratnam, E.L.; Disfani, V.R.; Kleissl, J. Distributed energy storage system scheduling considering tariff structure, energy arbitrage and solar PV penetration. *Appl. Energy* **2017**, *205*, 1384–1393. [CrossRef]
88. Jiang, T.; Cao, Y.; Yu, L.; Wang, Z. Load Shaping Strategy Based on Energy Storage and Dynamic Pricing in Smart Grid. *IEEE Trans. Smart Grid* **2014**, *5*, 2868–2876. [CrossRef]
89. Tushar, W.; Chai, B.; Yuen, C.; Huang, S.; Smith, D.B.; Poor, H.V.; Yang, Z. Energy Storage Sharing in Smart Grid: A Modified Auction-Based Approach. *IEEE Trans. Smart Grid* **2016**, *7*, 1462–1475. [CrossRef]
90. Atzeni, I.; Ordóñez, L.G.; Scutari, G.; Palomar, D.P.; Fonollosa, J.R. Demand-Side Management via Distributed Energy Generation and Storage Optimization. *IEEE Trans. Smart Grid* **2013**, *4*, 866–876. [CrossRef]
91. Gil-González, W.; Montoya, O.D.; Grisales-Noreña, L.F.; Cruz-Peragón, F.; Alcalá, G. Economic Dispatch of Renewable Generators and BESS in DC Microgrids Using Second-Order Cone Optimization. *Energies* **2020**, *13*, 1703. [CrossRef]
92. Mortaz, E. Portfolio Diversification for an Intermediary Energy Storage Merchant. *IEEE Trans. Sustain. Energy* **2019**. [CrossRef]
93. Wang, Z.; Kirschen, D.S. Two-stage optimal scheduling for aggregators of batteries owned by commercial consumers. *Transm. Distrib. IET Gener.* **2019**, *13*, 4880–4887. [CrossRef]
94. Pandžić, H.; Morales, J.M.; Conejo, A.J.; Kuzle, I. Offering model for a virtual power plant based on stochastic programming. *Appl. Energy* **2013**, *105*, 282–292. [CrossRef]
95. Giuntoli, M.; Poli, D. Optimized Thermal and Electrical Scheduling of a Large Scale Virtual Power Plant in the Presence of Energy Storages. *IEEE Trans. Smart Grid* **2013**, *4*, 942–955. [CrossRef]
96. Ju, L.; Tan, Z.; Yuan, J.; Tan, Q.; Li, H.; Dong, F. A bi-level stochastic scheduling optimization model for a virtual power plant connected to a wind–photovoltaic–energy storage system considering the uncertainty and demand response. *Appl. Energy* **2016**, *171*, 184–199. [CrossRef]
97. Liu, Z.; Yang, J.; Song, W.; Xue, N.; Li, S.; Fang, M. Research on cloud energy storage service in residential microgrids. *IET Renew. Power Gener.* **2019**, *13*, 3097–3105. [CrossRef]
98. Daughety, A.F. (Ed.) *Cournot Oligopoly: Characterization and Applications*; Cambridge University Press: Cambridge, UK, 1989.
99. Lundin, E.; Tangerås, T.P. Cournot competition in wholesale electricity markets: The Nordic power exchange, Nord Pool. *Int. J. Ind. Organ.* **2020**, *68*, 102536. [CrossRef]
100. Shafiee, S.; Zamani-Dehkordi, P.; Zareipour, H.; Knight, A.M. Economic assessment of a price-maker energy storage facility in the Alberta electricity market. *Energy* **2016**, *111*, 537–547. [CrossRef]
101. Fang, X.; Li, F.; Wei, Y.; Cui, H. Strategic scheduling of energy storage for load serving entities in locational marginal pricing market. *IET Gener. Transm. Distrib.* **2016**, *10*, 1258–1267. [CrossRef]
102. Zou, P.; Chen, Q.; Xia, Q.; He, G.; Kang, C. Evaluating the Contribution of Energy Storages to Support Large-Scale Renewable Generation in Joint Energy and Ancillary Service Markets. *IEEE Trans. Sustain. Energy* **2016**, *7*, 808–818. [CrossRef]

103. Sousa, J.A.; Teixeira, F.; Faias, S. Impact of a price-maker pumped storage hydro unit on the integration of wind energy in power systems. *Energy* **2014**, *69*, 3–11. [CrossRef]
104. Ye, Y.; Papadaskalopoulos, D.; Moreira, R.; Strbac, G. Strategic capacity withholding by energy storage in electricity markets. In Proceedings of the 2017 IEEE Manchester PowerTech, Manchester, UK, 18–22 June 2017; pp. 1–6.
105. Ning Lu.; Chow, J.H.; Desrochers, A.A. Pumped-storage hydro-turbine bidding strategies in a competitive electricity market. *IEEE Trans. Power Syst.* **2004**, *19*, 834–841. [CrossRef]
106. Nasrolahpour, E.; Kazempour, J.; Zareipour, H.; Rosehart, W.D. A Bilevel Model for Participation of a Storage System in Energy and Reserve Markets. *IEEE Trans. Sustain. Energy* **2018**, *9*, 582–598. [CrossRef]
107. Shafiee, S.; Zareipour, H.; Knight, A.M. Developing Bidding and Offering Curves of a Price-Maker Energy Storage Facility Based on Robust Optimization. *IEEE Trans. Smart Grid* **2019**, *10*, 650–660. [CrossRef]
108. Fortuny-Amat, J.; McCarl, B. A Representation and Economic Interpretation of a Two-Level Programming Problem. *J. Oper. Res. Soc.* **1981**, *32*, 783–792. [CrossRef]
109. Quevedo, P.M.d.; Muñoz-Delgado, G.; Contreras, J. Impact of Electric Vehicles on the Expansion Planning of Distribution Systems Considering Renewable Energy, Storage, and Charging Stations. *IEEE Trans. Smart Grid* **2019**, *10*, 794–804. [CrossRef]
110. Dvorkin, Y.; Fernández-Blanco, R.; Kirschen, D.S.; Pandžić, H.; Watson, J.; Silva-Monroy, C.A. Ensuring Profitability of Energy Storage. *IEEE Trans. Power Syst.* **2017**, *32*, 611–623. [CrossRef]
111. Pandžić, H.; Dvorkin, Y.; Carrión, M. Investments in merchant energy storage: Trading-off between energy and reserve markets. *Appl. Energy* **2018**, *230*, 277–286. [CrossRef]
112. Nasrolahpour, E.; Kazempour, S.J.; Zareipour, H.; Rosehart, W.D. Strategic Sizing of Energy Storage Facilities in Electricity Markets. *IEEE Trans. Sustain. Energy* **2016**, *7*, 1462–1472. [CrossRef]
113. Dvorkin, Y.; Fernández-Blanco, R.; Wang, Y.; Xu, B.; Kirschen, D.S.; Pandžić, H.; Watson, J.; Silva-Monroy, C.A. Co-Planning of Investments in Transmission and Merchant Energy Storage. *IEEE Trans. Power Syst.* **2018**, *33*, 245–256. [CrossRef]
114. Xiong, P.; Singh, C. Optimal Planning of Storage in Power Systems Integrated With Wind Power Generation. *IEEE Trans. Sustain. Energy* **2016**, *7*, 232–240. [CrossRef]
115. Xu, B.; Wang, Y.; Dvorkin, Y.; Fernández-Blanco, R.; Silva-Monroy, C.A.; Watson, J.; Kirschen, D.S. Scalable Planning for Energy Storage in Energy and Reserve Markets. *IEEE Trans. Power Syst.* **2017**, *32*, 4515–4527. [CrossRef]
116. Zhang, X.; Conejo, A.J. Coordinated Investment in Transmission and Storage Systems Representing Long- and Short-Term Uncertainty. *IEEE Trans. Power Syst.* **2018**, *33*, 7143–7151. [CrossRef]
117. García-Cerezo, A.; Baringo, L.; García-Bertrand, R. Robust Transmission Network Expansion Planning Problem Considering Storage Units. *arXiv* **2019**, arXiv:1907.04775.
118. Sun, Y.; Li, Z.; Shahidehpour, M.; Ai, B. Battery-Based Energy Storage Transportation for Enhancing Power System Economics and Security. *IEEE Trans. Smart Grid* **2015**, *6*, 2395–2402. [CrossRef]
119. Zhao, D.; Wang, H.; Huang, J.; Lin, X. Virtual Energy Storage Sharing and Capacity Allocation. *IEEE Trans. Smart Grid* **2019**. [CrossRef]
120. Chakraborty, P.; Baeyens, E.; Poolla, K.; Khargonekar, P.P.; Varaiya, P. Sharing Storage in a Smart Grid: A Coalitional Game Approach. *IEEE Trans. Smart Grid* **2019**, *10*, 4379–4390. [CrossRef]
121. Pandžić, H. Optimal battery energy storage investment in buildings. *Energy Build.* **2018**, *175*, 189–198. [CrossRef]
122. Pandžić, K.; Pandžić, H.; Kuzle, I. Coordination of Regulated and Merchant Energy Storage Investments. *IEEE Trans. Sustain. Energy* **2018**, *9*, 1244–1254. [CrossRef]
123. Huang, Q.; Xu, Y.; Courcoubetis, C. Stackelberg Competition Between Merchant and Regulated Storage Investment under Locational Marginal Pricing. In Proceedings of the 2019 IEEE 15th International Conference on Control and Automation (ICCA), Hokkaido, Japan, 6–9 July 2019; pp. 687–692.
124. Masoumzadeh, A.; Nekouei, E.; Alpcan, T.; Chattopadhyay, D. Impact of Optimal Storage Allocation on Price Volatility in Energy-Only Electricity Markets. *IEEE Trans. Power Syst.* **2018**, *33*, 1903–1914. [CrossRef]
125. Pandžić, H.; Wang, Y.; Qiu, T.; Dvorkin, Y.; Kirschen, D.S. Near-Optimal Method for Siting and Sizing of Distributed Storage in a Transmission Network. *IEEE Trans. Power Syst.* **2015**, *30*, 2288–2300. [CrossRef]

126. Opathella, C.; Elkasrawy, A.; Adel Mohamed, A.; Venkatesh, B. MILP formulation for generation and storage asset sizing and sitting for reliability constrained system planning. *Int. J. Electr. Power Energy Syst.* **2020**, *116*, 105529. [CrossRef]
127. Bustos, C.; Sauma, E.; Torre, S.d.l.; Aguado, J.A.; Contreras, J.; Pozo, D. Energy storage and transmission expansion planning: Substitutes or complements? *Transm. Distrib. IET Gener.* **2018**, *12*, 1738–1746. [CrossRef]
128. Zheng, L.; Hu, W.; Lu, Q.; Min, Y. Optimal energy storage system allocation and operation for improving wind power penetration. *Transm. Distrib. IET Gener.* **2015**, *9*, 2672–2678. [CrossRef]
129. Wu, J.; Qiu, J.; Wang, X.; Ni, Y.; Han, X.; Dai, J.; Du, Z.; Xie, X. Study on Medium and Long-Term Generation Expansion Planning Method Considering the Requirements of Green Low-Carbon Development. In Proceedings of the 2018 IEEE PES Asia-Pacific Power and Energy Engineering Conference (APPEEC), Sabah, Malaysia, 7–10 October 2018; pp. 689–694.
130. Pineda, S.; Morales, J.M. Chronological Time-Period Clustering for Optimal Capacity Expansion Planning With Storage. *IEEE Trans. Power Syst.* **2018**, *33*, 7162–7170. [CrossRef]
131. Hu, Z.; Zhang, F.; Li, B. Transmission expansion planning considering the deployment of energy storage systems. In Proceedings of the 2012 IEEE Power and Energy Society General Meeting, Grand Hyatt, San Diego, USA, 22–26 July 2012; pp. 1–6.
132. Tejada-Arango, D.A.; Domeshek, M.; Wogrin, S.; Centeno, E. Enhanced Representative Days and System States Modeling for Energy Storage Investment Analysis. *IEEE Trans. Power Syst.* **2018**, *33*, 6534–6544. [CrossRef]
133. MacRae, C.A.G.; Ernst, A.T.; Ozlen, M. A Benders decomposition approach to transmission expansion planning considering energy storage. *Energy* **2016**, *112*, 795–803. [CrossRef]
134. Nikoobakht, A.; Aghaei, J. Integrated transmission and storage systems investment planning hosting wind power generation: Continuous-time hybrid stochastic/robust optimisation. *IET Gener. Transm. Distrib.* **2019**, *13*, 4870–4879. [CrossRef]
135. Carrión, M.; Dvorkin, Y.; Pandžić, H. Primary Frequency Response in Capacity Expansion With Energy Storage. *IEEE Trans. Power Syst.* **2018**, *33*, 1824–1835. [CrossRef]
136. Fernández-Blanco, R.; Dvorkin, Y.; Xu, B.; Wang, Y.; Kirschen, D.S. Optimal Energy Storage Siting and Sizing: A WECC Case Study. *IEEE Trans. Sustain. Energy* **2017**, *8*, 733–743. [CrossRef]
137. Aguado, J.; de la Torre, S.; Triviño, A. Battery energy storage systems in transmission network expansion planning. *Electr. Power Syst. Res.* **2017**, *145*, 63–72. [CrossRef]
138. Dehghan, S.; Amjady, N. Robust Transmission and Energy Storage Expansion Planning in Wind Farm-Integrated Power Systems Considering Transmission Switching. *IEEE Trans. Sustain. Energy* **2016**, *7*, 765–774. [CrossRef]
139. Saber, H.; Heidarabadi, H.; Moeini-Aghtaie, M.; Farzin, H.; Karimi, M.R. Expansion Planning Studies of Independent-Locally Operated Battery Energy Storage Systems (BESSs): A CVaR-Based Study. *IEEE Trans. Sustain. Energy* **2019**. [CrossRef]
140. Yang, P.; Nehorai, A. Joint Optimization of Hybrid Energy Storage and Generation Capacity With Renewable Energy. *IEEE Trans. Smart Grid* **2014**, *5*, 1566–1574. [CrossRef]
141. Saboori, H.; Hemmati, R.; Abbasi, V. Multistage distribution network expansion planning considering the emerging energy storage systems. *Energy Convers. Manag.* **2015**, *105*, 938–945. [CrossRef]
142. Zhu, X.; Yan, J.; Lu, N. A Graphical Performance-Based Energy Storage Capacity Sizing Method for High Solar Penetration Residential Feeders. *IEEE Trans. Smart Grid* **2017**, *8*, 3–12. [CrossRef]
143. Sharma, V.; Haque, M.H.; Aziz, S.M. Energy cost minimization for net zero energy homes through optimal sizing of battery storage system. *Renew. Energy* **2019**, *141*, 278–286. [CrossRef]
144. Jacob, A.S.; Banerjee, R.; Ghosh, P.C. Sizing of hybrid energy storage system for a PV based microgrid through design space approach. *Appl. Energy* **2018**, *212*, 640–653. [CrossRef]
145. Xiao, J.; Zhang, Z.; Bai, L.; Liang, H. Determination of the optimal installation site and capacity of battery energy storage system in distribution network integrated with distributed generation. *Transm. Distrib. IET Gener.* **2016**, *10*, 601–607. [CrossRef]
146. Hassan, A.; Dvorkin, Y. Energy Storage Siting and Sizing in Coordinated Distribution and Transmission Systems. *IEEE Trans. Sustain. Energy* **2018**, *9*, 1692–1701. [CrossRef]

147. Das, C.K.; Bass, O.; Mahmoud, T.S.; Kothapalli, G.; Mousavi, N.; Habibi, D.; Masoum, M.A.S. Optimal allocation of distributed energy storage systems to improve performance and power quality of distribution networks. *Appl. Energy* **2019**, *252*, 113468. [CrossRef]
148. Santos, S.F.; Fitiwi, D.Z.; Shafie-khah, M.; Bizuayehu, A.W.; Cabrita, C.M.P.; Catalão, J.P.S. New Multi-Stage and Stochastic Mathematical Model for Maximizing RES Hosting Capacity—Part II: Numerical Results. *IEEE Trans. Sustain. Energy* **2017**, *8*, 320–330. [CrossRef]
149. Santos, S.F.; Fitiwi, D.Z.; Shafie-Khah, M.; Bizuayehu, A.W.; Cabrita, C.M.P.; Catalão, J.P.S. New Multistage and Stochastic Mathematical Model for Maximizing RES Hosting Capacity—Part I: Problem Formulation. *IEEE Trans. Sustain. Energy* **2017**, *8*, 304–319. [CrossRef]
150. Shen, X.; Shahidehpour, M.; Han, Y.; Zhu, S.; Zheng, J. Expansion Planning of Active Distribution Networks With Centralized and Distributed Energy Storage Systems. *IEEE Trans. Sustain. Energy* **2017**, *8*, 126–134. [CrossRef]
151. Xing, H.; Cheng, H.; Zhang, Y.; Zeng, P. Active distribution network expansion planning integrating dispersed energy storage systems. *Transm. Distrib. IET Gener.* **2016**, *10*, 638–644. [CrossRef]
152. Li, R.; Wang, W.; Xia, M. Cooperative Planning of Active Distribution System With Renewable Energy Sources and Energy Storage Systems. *IEEE Access* **2018**, *6*, 5916–5926. [CrossRef]
153. Wang, Z.; He, Y. Two-stage optimal demand response with battery energy storage systems. *Transm. Distrib. IET Gener.* **2016**, *10*, 1286–1293. [CrossRef]
154. Bayram, I.S.; Abdallah, M.; Tajer, A.; Qaraqe, K.A. A Stochastic Sizing Approach for Sharing-Based Energy Storage Applications. *IEEE Trans. Smart Grid* **2017**, *8*, 1075–1084. [CrossRef]
155. Cao, X.; Wang, J.; Zeng, B. A Chance Constrained Information-Gap Decision Model for Multi-Period Microgrid Planning. *IEEE Trans. Power Syst.* **2018**, *33*, 2684–2695. [CrossRef]
156. Akhavan-Hejazi, H.; Mohsenian-Rad, H. Energy Storage Planning in Active Distribution Grids: A Chance-Constrained Optimization With Non-Parametric Probability Functions. *IEEE Trans. Smart Grid* **2018**, *9*, 1972–1985. [CrossRef]
157. Zhang, Y.; Dong, Z.Y.; Luo, F.; Zheng, Y.; Meng, K.; Wong, K.P. Optimal allocation of battery energy storage systems in distribution networks with high wind power penetration. *IET Renew. Power Gener.* **2016**, *10*, 1105–1113. [CrossRef]
158. Sekhavatmanesh, H.; Cherkaoui, R. Optimal Infrastructure Planning of Active Distribution Networks Complying With Service Restoration Requirements. *IEEE Trans. Smart Grid* **2018**, *9*, 6566–6577. [CrossRef]
159. Khodaei, A.; Bahramirad, S.; Shahidehpour, M. Microgrid Planning Under Uncertainty. *IEEE Trans. Power Syst.* **2015**, *30*, 2417–2425. [CrossRef]
160. Ye, Y.; Qiu, D.; Sun, M.; Papadaskalopoulos, D.; Strbac, G. Deep Reinforcement Learning for Strategic Bidding in Electricity Markets. *IEEE Trans. Smart Grid* **2019**. [CrossRef]
161. Jiang, D.R.; Powell, W.B. Optimal Hour-Ahead Bidding in the Real-Time Electricity Market with Battery Storage using Approximate Dynamic Programming. *arXiv* **2014**, arXiv:1402.3575.

6

Fuzzy Logic Weight based Charging Scheme for Optimal Distribution of Charging Power among Electric Vehicles in a Parking Lot

Shahid Hussain [1], Mohamed A. Ahmed [2,3], Ki-Beom Lee [1] and Young-Chon Kim [1,*]

[1] Division of Electronic and Information, Department of Computer Science and Engineering, Jeonbuk National University, Jeonju 54896, Korea; shahiduop@jbnu.ac.kr (S.H.); keywii@jbnu.ac.kr (K.-B.L.)
[2] Department of Electronic Engineering, Universidad Técnica Federico Santa María, Valparaíso 2390123, Chile; mohamed.abdelhamid@usm.cl
[3] Department of Communications and Electronics, Higher Institute of Engineering & Technology–King Marriott, Alexandria 23713, Egypt
* Correspondence: yckim@jbnu.ac.kr.

Abstract: Electric vehicles (EVs) parking lots are representing significant charging loads for relatively a long period of time. Therefore, the aggregated charging load of EVs may coincide with the peak demand of the distribution power system and can greatly stress the power grid. The stress on the power grid can be characterized by the additional electricity demand and the introduction of a new peak load that may overwhelm both the substations and transmission systems. In order to avoid the stress on the power grid, the parking lot operators are required to limit the penetration level of EVs and optimally distribute the available power among them. This affects the EV owner's quality of experience (QoE) and thereby reducing the quality of performance (QoP) for the parking lot operators. The QoE is represents the satisfaction level of EV owners; whereas, the QoP is a measurement representing the ratio of EVs with QoE to the total number of EVs. This study proposes a fuzzy logic weight-based charging scheme (FLWCS) to optimally distribute the charging power among the most appropriate EVs in such a way that maximizes the QoP for the parking lot operators under the operational constraints of the power grid. The developed fuzzy inference mechanism resolves the uncertainties and correlates the independent inputs such as state-of-charge, the remaining parking duration and the available power into weighted values for the EVs in each time slot. Once the weight values for all EVs are known, their charging operations are controlled such that the operational constraints of the power grid are respected in each time slot. The proposed FLWCS is applied to a parking lot with different capacities. The simulation results reveal an improved QoP comparing to the conventional first-come-first-served (FCFS) based scheme.

Keywords: charging scheduling; electric vehicles; fuzzy logic weight; optimal distribution of power; parking lot

---

1. Introduction

The growing concerns of carbon dioxide emissions, the effect of global warming and the reliance on fossil fuel motivated the use of electric vehicles (EVs) in the transportation sector. As a result, the transportation sector is rapidly moving towards the use of EVs including both the plug-in hybrid electric vehicles (PHEVs) and battery electric vehicles (BEVs). A PHEV has the option to use energy either from the electric battery or from the on-board engine–generator and has the flexibility to be recharged from the external power socket as well as from an on-board engine–generator [1]. A BEV uses an electric battery to run, which can be recharged only from external electrical sources [2]. The charging of both PHEVs and BEVs is solely dependent on the power grid; therefore, a considerable energy demand of vehicles will shift from the fossil fuel to the electric power grid [3].

The advancement in the Internet of things (IoT) technology is playing an important role in the intelligent transport systems (ITS) including smart mobility, vehicle-to-vehicle (V2V) communication, vehicle-to-infrastructure (V2I) communication and autonomous vehicles. Smart mobility is a modern, efficient and sustainable system that offers a revolution in all modes of transportation with respect to vehicles, infrastructures and people. The V2V communication enables nearby vehicles to exchange information together in order to improve driver safety and avoid accidents. The V2I communication enables moving vehicles to exchange valuable information with roadside units (RSU) in order to improve road efficiency and optimize travel time. In autonomous vehicles, the automated driving system relies completely on the vehicle's onboard computer, hardware and software in order to monitor both the environment condition as well as the road status without any human intervention [4–6]. As a result, the EVs including autonomous and connected EVs are rapidly growing in the transportation market and could potentially influence the electricity distribution infrastructure [7–9]. This is because EVs are moving across the city and are representing a spatial and temporal based varying charging load. The future public parking lots will represent a huge load for relatively a long period of time that may coincide with the residential peak load and will overload the power grid [10]. A relation between the vehicles on the street and the residential load profile has been identified in [11], which demonstrated an overlap between vehicles on the street and the residential peak load from 2:00 PM to 6:00 PM. During such a time period, a high penetration level of charging EVs can stress the power grid [12]. The stress on the power grid can be characterized as an additional electricity demand that may introduces a new peak load and will overwhelms the substations in the low-voltage distribution network. To avoid the stress on the power grid, the parking lot operators are required to limit the penetration of charging EVs and distribute the power within a limited number of EVs. As a result, the requirements of the power grid can be satisfied and several cost factors (i.e., the upgrade of the low-voltage distribution transformer, the upgrade of the transmission infrastructure, generation of more power for mitigating the new peak load) can be saved. However, on the other side, this can have a significant effect on the desired quality of experience (QoE) level for the EV owners during their parking duration.

The QoE defines the EV owner's satisfaction level and is a function of the battery capacity, current and required state-of-charge (SoC) of an EV as illustrated in Figure 1. The figure demonstrates the different status of an EV battery highlighted with different colors. The battery SoC is 30% (green highlighted), the QoE is 40% (white highlighted) and the lower and upper limits are 20% and 80%, respectively (red highlighted) for maintaining the efficiency of the battery. The satisfaction of QoE is a base to measure the quality-of-performance (QoP) for the parking lot operators. The QoP can be defined as the ratio of EVs with satisfied QoE to the total number of requesting EVs during the operational hours of the parking lot. Considering 12 h as the parking operational hours, a higher value of QoP corresponds to better performance and vice versa. Therefore, at any time instant, the selection of the most appropriate EVs for charging among all the EVs candidates such that maximizing the QoP while respecting all the constraints from the power grid is a complex and challenging task for the parking lot operators. The complexity of this problem is due to the dependency of QoE satisfaction level on multiple and independent factors, including the battery capacity, the required SoC, the remaining parking duration, the current parking occupancy, the charging power of charging stations (CSs), the current baseload on the low-voltage substation, and the amount of available power from the power grid. These are spatial and temporal based varying parameters with a high degree of uncertainty which results in a more complex system. Considering the required SoC of an EV battery as an example, the drivers usually determine the required SoC in terms of the battery level such as low battery level (i.e., high the required SoC), medium battery level (i.e., medium the required SoC) and high battery level (i.e., low the required SoC). The complexity and nonlinearity of temporal and spatial-based varying real-time systems can be resolved into a simple weighted sum of linear subsystems through the fuzzy logic inference mechanism [13,14].

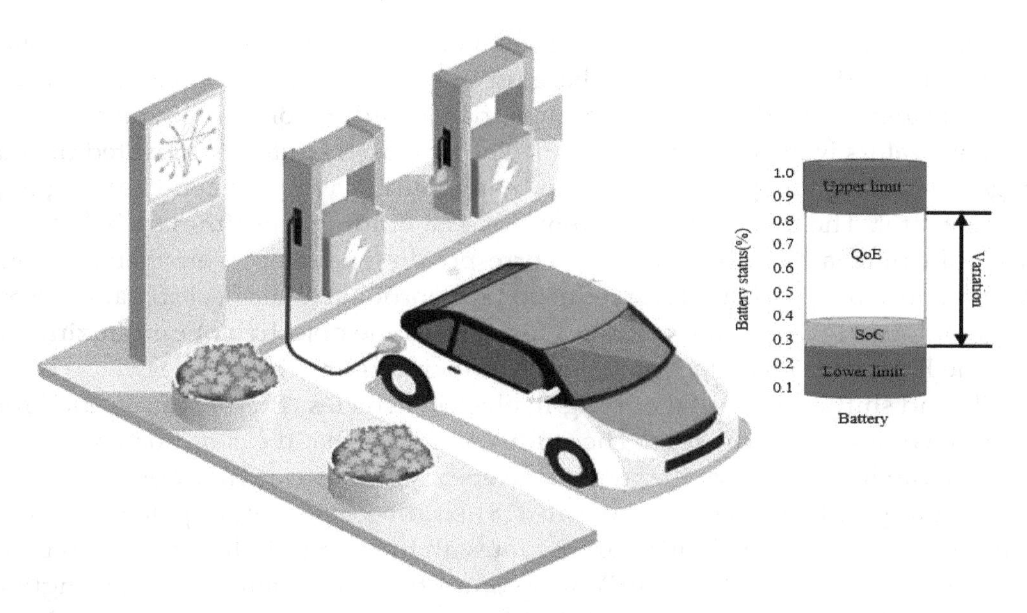

Figure 1. Illustration of the different statuses of an electric vehicle (EV) battery at the plug-in time.

This work aims to develop a charging strategy for EVs in a parking lot that maximizes the QoE and the QoP under the bounded constraints of the power grid, using a fuzzy logic inference mechanism. Moreover, the study aims to answer the research questions which include: what are the main entities involved in the charging system? What are the requirements of these entities? How to control the charging EVs such that it minimizes the PL load under the permissible limit of the power grid while maximizing the EV owner's satisfaction? The contributions of this research work are as follows.

- The requirements of EV owners and the power grid are identified, and a charging scheme based on a fuzzy inference mechanism for EVs in the parking lot is developed with the aim to maximize the QoP under the bounded constraints of the power grid;
- The problem is formulated with an objective function and solved through the fuzzy logic inference mechanism. Among the different parameters, three of the most relevant parameters (i.e., the required SoC, remaining parking duration and available power) that influence the QoP are selected to model the fuzzy logic inference mechanism;
- The developed fuzzy inference mechanism correlates the required SoC, remaining parking duration and available power in real time and compute weight values for each of the EVs requesting for the charging operations. Once the weight values for each of the competing EVs are known, their charging operations are controlled, and the available power is distributed among the optimal number of charging EVs;
- An algorithm for FLWCS is developed and applied to a parking lot with different parking capacities. The performance of the algorithm is validated against the FCFS-based scheme and the results are verified in terms of QoP.

2. Literature Review

With the growing penetration of EVs in the transportation market, it is indispensable for the fleet operators to effectively manage the charging load of EVs considering the requirements of both the power grid and the EV owners. In literature, the problem of managing the charging load of EVs in parking lots has been studied from different perspectives and objectives.

The authors in [15] studied the problem of charging cost minimization by considering three different types of public EV fleets attached with a photovoltaic (PV) system. The three parking lots included: (1) commercial customer's fleet, where the charging operation is mainly performed at night time, (2) commuter customer's fleet, where the charging is performed during day time and (3) opportunity customer's fleet, for commuters with short parking duration. Three different options for forecasting electricity generation from the PV system along with controlled and uncontrolled charging strategies were considered. In each fleet, the charging cost is optimized by utilizing electricity from the PV system. By considering the day-ahead energy market, coordination and payment mechanism for a group of sub-aggregators were introduced in [16]. The proposed strategy motivated the participation of sub-aggregators through incentives, where sub-aggregators reports their charging requirements to the main aggregator. The main aggregator employed a bidding algorithm on behalf of requesting aggregators and the purchase of energy and the corresponding payment were then distributed among them. The results showed a substantial cost reduction proportional to the fleet size and the participation of the sub-aggregators. However, these studies focused on the objective of minimizing the charging cost; whereas, the EV owner's satisfaction in terms of their required energies is yet to be explored.

In ref. [17], load shifting potential of plug-in electric vehicles (PEVs) was studied for domestic, work and public charging infrastructures. The study concluded that the coordinated charging through demand response can help to utilize the renewable energy sources and support to shift a significant amount of EVs charging load. The authors in [18] studied a rectangle placement algorithm for scheduling the charging load of EVs at a parking lot with the aim to reduce variation in load. In this algorithm, the energy requirement for a PEV was computed as a rectangle whose length is time and height is the power. The results verified that the rectangle placement algorithm combined with the charging level selection reduced the average load variation, improved the load factor and flatten the total load profile comparing to the traditional first-come-first-served based charging. An optimal charging scheduling strategy was studied in [19], which considered multiple factors, such as transport system information (road length), vehicle characteristics (velocity and wait time) and power grid information (load deviation and node voltage) for managing the EVs. The proposed optimal strategy showed reduced losses, small voltage drop in nodes and optimized the load curve. A peak load minimization strategy based on binary linear programming coupled with a bisection algorithm for parking lot was proposed in [20]. The proposed strategy was simulated with a fast CS and improved results were obtained compared to the uncontrolled charging strategy. All these studies mainly proposed solutions for optimizing the load of the power grid but lacking to address the requirements of the EV owners and the parking lot operators.

The authors in [21] presented a multi-objective optimization control strategy to minimize the charging cost of PEV owners and load variance in the low-voltage network. Several strategies such as uncontrolled charging, smart charging, smart charging with voltage unbalanced reduction (VUR) and smart charging with VUR and vehicle-to-grid (V2G) were simulated by considering a low-voltage distribution network in Denmark. The results concluded that the proposed multi-objective strategy can reduce both the energy losses, charging cost and can support a high penetration rate of PEVs. The scope of this work was limited to residential customers where the EVs are staying overnight and have enough time to be recharged using slow charging rate, but the proposed solution may not be effective for public parking lots with EVs having shorter stay time. Two-phase optimization method for optimizing the charging cost and smoothing the total load profile was presented in [22]. In the first phase of the model, the electricity price was defined according to the status of the historical daily power curve, for ensuring the maximum profit to both power grid and PEV owners. The second phase then reduces the load fluctuation by optimizing the charging and discharging power of EVs according to the power grid constraints. The results showed the effectiveness of their proposed strategy by smoothing the total load profile than the uncontrolled charging strategy. Inspired from the gray wolf optimizer (GWO), the authors in [23] proposed an improved binary gray wolf optimization (IBGWO) algorithm for parking lot coupled with an energy storage system (ESS) and a PV system. The work aimed to reduce the charging load and cost by utilizing the usages of PV and ESS in the parking lot. The simulation results showed that their proposed IBGWO has a superior performance over the other meta-heuristic algorithms. These studies were tested for a small number of EVs with limited battery capacities and yet their feasibility need to be explored for sizeable parking lots with larger battery capacities.

The authors in [24] proposed a CS selection algorithm based on the fuzzy logic controller with the aim to balance the charging load of EVs and reduce their waiting time. Taking into account the service time (charging time), the speed of EVs and the distance between the EVs current position and the CSs, the fuzzy logic controller was used to determine a weighted priority value for each pair of EVs and CSs. The simulation results showed superior performance by reducing the average waiting time than the random and maximum weight-based scheduling schemes. Re-routing of moving EVs towards an appropriate CS based on multi-agents system for distributing the charging load of EVs among multiple geographically dispersed CSs was proposed in [25]. In this scheme, the EV agent was developed through fuzzy logic controller, which was requesting the other high-level agents to provide reservation services for charging. A total of 21 EVs with a battery capacity of 100 kWh and 6 CSs dispersed within a defined virtual block were simulated. The results showed that the proposed multi-agents based scheme supported a cognitive distribution of the charging load of EVs among the CSs. However, these studies proposed solutions for minimizing the waiting time and balancing the charging load among CSs but lacking to address the charging level satisfactions requirements of EV owners.

It is worth mentioning that all the requirements, including the constraints of the power grid, the EV owners QoE and the QoP of parking lots are of utmost importance while scheduling the charging operations of EVs in parking lots. In order to achieve these requirements, multiple uncertain parameters such as battery SoC, parking duration, required SoC, and available energy [26] needs to be considered. Most of the work solved the charging optimization problem by using dynamic & stochastic programing and heuristic algorithms with the assumption of perfect knowledge on SoC and required SoC which may result in an imprecise decision. To the best of our knowledge, none of the above work focused on an arbitrage consideration of the aforementioned requirements while scheduling the charging of EVs in parking lots. The proposed FLWCS utilizes the services of the fuzzy logic inference mechanism and correlates the information from EVs (i.e., required SoC and remaining parking duration) and from the power grid (i.e., available power) into weighted values for the EVs competing each time slot. Based on the weight values, the charging operations of EVs are controlled in such a way that help to maximize the QoE for EV owners and thereby the QoP for the parking lot under the operational constraints of the power grid.

The ongoing part of this research work is to elaborate on the behavioral aspects of electric vehicle charging including driving (social behavior of owners travels) and charging behavior (suitable charging locations, market-economics, the impact of charging load, waiting and charging time) from the socioeconomic perspective.

3. Proposed Fuzzy Logic Weight Based Charging Scheme

The scheduling problem for a large scale of EVs in a parking lot involves various parameters from multiple domains which results in a more complex and system [27]. The selection of the most relevant parameters and their corresponding correlation can enhance the efficiency of the algorithm. This section gives a comprehensive presentation of different domains and the correlation of their parameters through the fuzzy inference mechanism for the proposed FLWCS. An overview of the conventional system and its associated deficiencies are exemplified in the following.

Let us assume that there are five EVs in a parking lot and the operators are expected to schedule their charging operations. For the sake of simplicity, all the EVs are considered to be of the same battery capacities (i.e., 40 kWh each), however; they have different arrival times, departure times, SoCs and parking durations. The arrival sequence of these EVs is such as EV1 arrived first, then EV2, etc. The SoCs at the arrival time are 25%, 25%, 37%, 50% and 62% for EV1, EV2, EV3, EV4 and EV5, respectively. Based on the battery capacities and SoCs the required QoEs are 75%, 75%, 63%, 50% and 38% for EV1, EV2, EV3, EV4 and EV5, respectively. The total operational time period of the parking lot (in this example) is assumed to be 3 h, which is normalized into a total of 12 equal time slots with a slot size of 15 min.

Given the arrival and departure time sequence of each of the EVs, the parking durations are computed as 10, 6, 7, 6 and 3 time slots for EV1, EV2, EV3, EV4 and EV5, respectively. Considering that each of the parking spots is equipped with fast CS, which can support a charging rate of 20 kW/h, such that each of the CSs is providing charging power of 5 kW/time slot. It is further assumed that at any time slot t, the power grid can support the charging of three EVs simultaneously. The parking lot operators are required to satisfy the QoE requirements of all the five EVs while respecting the power grid operational constraints. The charging scheduling of these EVs with respect to the conventional FCFS scheme and the proposed FLWCS and their corresponding output in terms of power consumption, their QoE and QoP, are visualized in Figure 2. In the case of FCFS-based scheme (Figure 2a), the EVs start charging immediately upon their arrivals; whereas, the proposed FLWCS scheduled them based on their weight values computed through multiple factors, including the updated SoC, the remaining parking duration (RPD) and the power grid operational constraints, etc. as depicted in Figure 2d. In this example, both schemes are able to follow the power grid constraints as shown in Figure 2b,e. However, in contrast to the FCFS-based scheme, the proposed FLWCS is able to fulfill the QoE requirements for all the EV owners and thereby improve the QoP for the parking lot. Considering the QoE as charging until the full battery capacity, the proposed FLWCS is able to improve the quality of performance by 40% comparing to the FCFS-based scheme, as can be observed from Figure 2c,f.

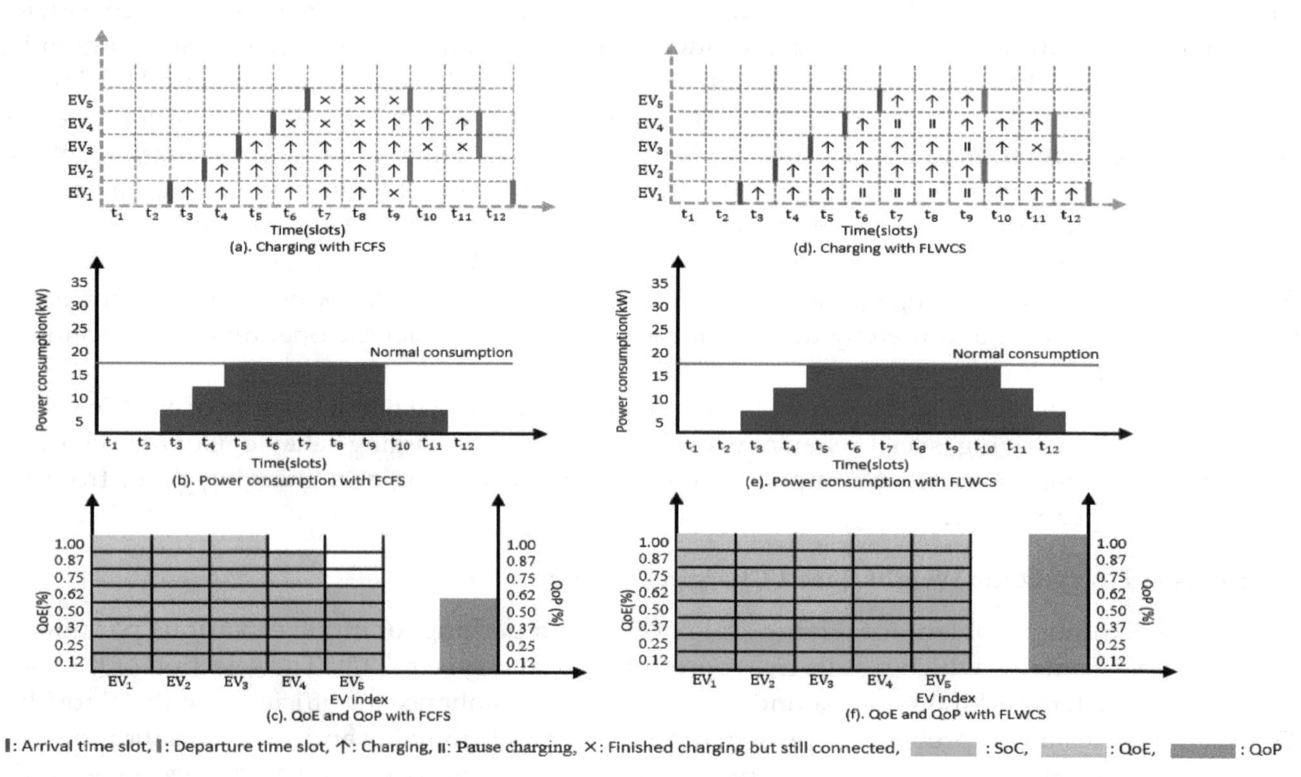

Figure 2. Example that illustrates the charging operations of EVs with first-come-first-serve (FCFS) and fuzzy logic weight based charging (FLWCS) (**a**) FCFS and (**d**) FLWCS; power consumption with (**b**) FCFS and (**e**) FLWCS; QoE of EV owners and the QoP of parking lot (**c**) FCFS and (**f**).

3.1. System Model of the Proposed FLWCS

The system model of the proposed FLWCS is presented in this section, as illustrated in Figure 3. It consists of several functional components including the power grid, the power distribution infrastructure (substations & power line), the distribution system operators (DSO), the baseload (BL) of electricity consumption for residential and commercial buildings, the EVs parking lot and the communication network. The power grid controls the electricity production from different energy sources such as fossil fuels, natural gas and nuclear. The generated electricity is transmitted to the

DSO through a high voltage (HV) power network covering a long distance and needs to be converted to medium voltage (MV) through the HV/MV substations. The functions of DSO include the collection of demands from residential and commercial buildings and allocating power to the low-voltage distribution network. The low-voltage distribution network is supporting two kinds of load: the BL and an EV charging load (parking lot load). The BL is the electrical demand for daily needs such as lighting, water/room heating, air condition, laundry machine, etc. This consumption of electricity is the basic requirement of daily life and depends on the occupancies of peoples, lifestyles and conveniences. Therefore, the BL is considered to be an average consumption of the residential and commercial buildings; whereas, the parking lot load represents the charging load of EVs in parking lot, connected to the low-voltage distribution network. Assuming the futuristic smart parking lot scenarios, this work considers a parking lot with installed electrical infrastructures such that each of the parking spots is equipped with a CS. Furthermore, each of the CSs has a J1772 connector that can be plugged into the inlet of EV and is coupled with a power supply of 208–240 Volt alternate-current (AC) for feeding about 19.2 kWh energy (i.e., level 2 charging option) [28]. The parking lot controller is a central entity, responsible for running the proposed FLWCS and the overall management of the parking lot. The proposed FLWCS is classified into three main components according to their functions.

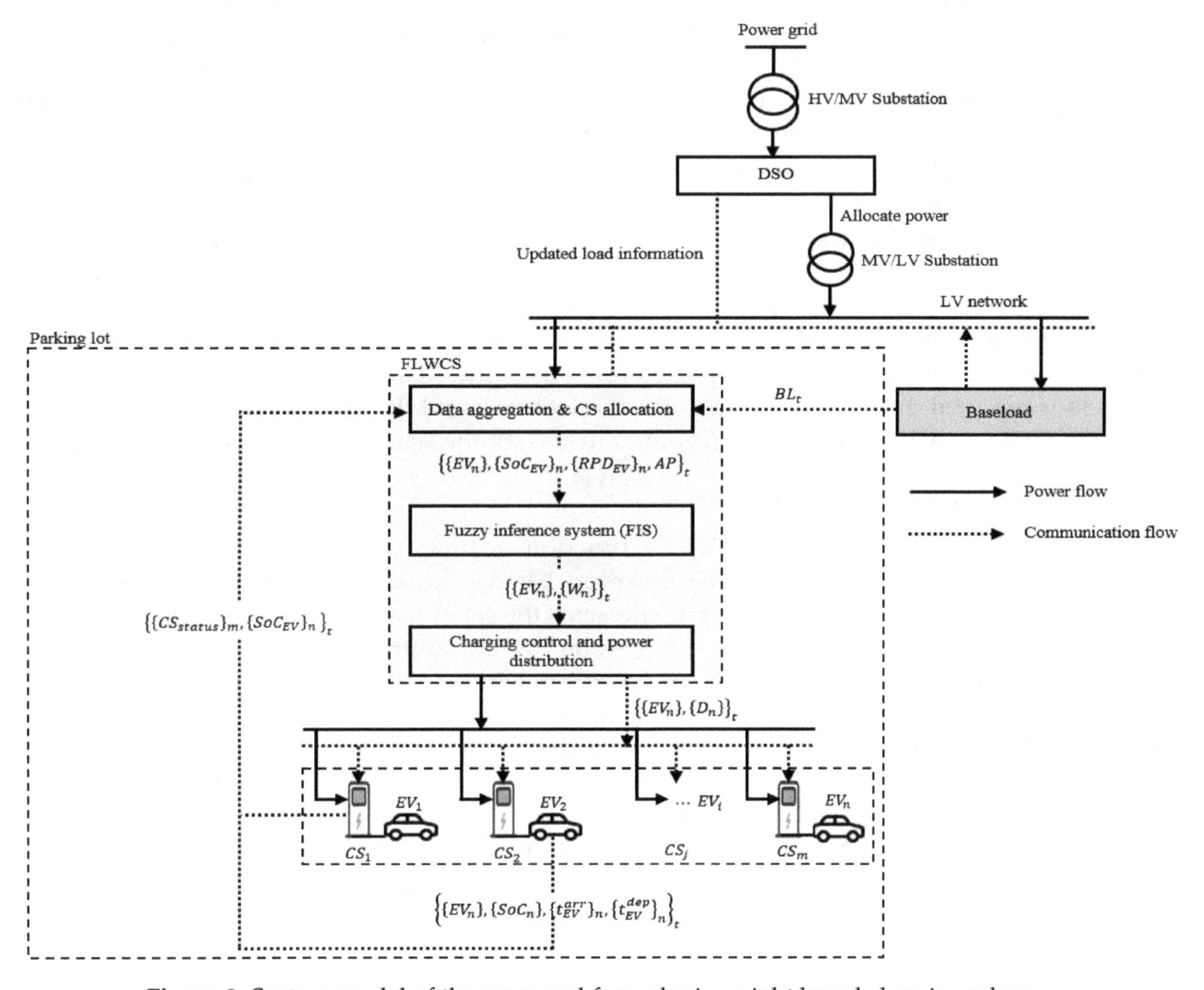

Figure 3. System model of the proposed fuzzy logic weight based charging scheme.

- **Data aggregation and CS allocation:** The EV owners are expected to provide their information such as arrival time, departure time and SoC to the parking lot controller upon their arrival. The information is initially processed and any of the available CSs are allocated to the newly arrived EVs. The FLWCS manages and controls the charging operations of all the connected EVs in each scheduling period and requires the status of the CSs and the BL information in real time. It is envisioned that a bidirectional communication network is established between each of the CSs and the parking lot controller, and smart meters installed at the CSs are used to detect the status (charging/idle) and measure the amount of energy consumption for the connected EVs [29]. The power consumption of the residential and commercial buildings connected to the low-voltage distribution system is measured through the advanced metering infrastructures (AMI) installed at the customer's premises and the BL is updated to the DSO and the parking lot controller through a wide area network [30].

- **Fuzzy logic controller:** The charging scheduling problem in this work is for a sizeable public parking lot which represents a significant charging load if all the EVs are charged simultaneously in the current time slot. Based on the EV owner's behaviors, EVs are classified into a routine and non-routine EVs [31]. The routine represents the EVs commuting on a daily basis between the home and workplaces and EVs are parking during the duty hours. The non-routine are the EVs which can be parked for a long or short duration depending on the type of their owners activities such as visiting a shopping mall, theaters, an appointment with a doctor or other social events [32]. Depending on the type of EVs in the parking lot, the operational data of EVs and the current status of the power grid play an important role in the fuzzy logic controller. The operational data of each of the EV in the set of EVs (N_{EV}) (including required SoC and RPD and the amount of available power (AP) computed through the BL obtained in real time (t) are the inputs to the fuzzy logic controller. The developed fuzzy inference mechanism evaluates the required SoC, the remaining parking duration and the available power and computes weight values normalized in [0, 1] range for the EVs in each time slot.
- **Charging control and power distribution:** Considering the weight values obtained through the fuzzy inference mechanism (according to the updated status of the power grid and the EVs information), the number of charging operations is controlled, and the power is distributed among the most appropriate EVs. The current status of the CSs and the updated SoC (power consumption) of each the EVs are measured and reported for consideration in the next scheduling period. The process is repeated during the parking operational hours and the optimized power consumption and the QoE for each of the EVs are recorded in each of the scheduling periods.

3.2. Problem Formulation and Objective Function

The arrival and departure of an EV is a function of time and therefore, at any time slot t, a new arrived EV has to be added while a served EV has to be removed from the set of EVs. Let $N_{EV}(t) = \{EV_1(t), EV_2(t), \cdots, EV_{l-1}(t)\}$ represents the set of parked EVs at time slot t, the arrival and departure of an EV can be handled by using union ($\cup$) and subtraction ($\backslash$) operations of set theory as given in Equation (1), where EV_l and EV_i represents newly arrived and served EVs. The parking lot operators record the current and future necessary information obtained from the EV owners for each of the new EVs. The required SoC of the newly (last) arrived EV_l is a function of the SoC and its battery capacity, and for any ith EV it can be computed according to Equation (2). The total load of the parking lot is the aggregated demand of all the existing EVs and the new arrived EV in the current time slot t and can be computed according to Equation (3). The total power consumption of the low-voltage distribution system at time slot t can be obtained through summing up the baseload of the residential and commercial building and the total energy demand of the parking lot, as given by Equation (4).

$$N_{EV}(t) = \begin{cases} N_{EV}(t) \cup EV_l(t), & \text{if } t^{arr}_{EV_l} \le t \\ N_{EV}(t) \backslash EV_i(t), & \text{if } t^{dep}_{EV_i} = t \end{cases} \quad (1)$$

$$SoC^{req}_{EV_l}(t) = \big(1 - SoC_{EV_l}(t)\big) * BC_{EV_l} \quad (2)$$

$$E^{PL}_{total\_demand}(t) = \sum_{i=1}^{l-1} SoC^{req}_{EV_i}(t) + SoC^{req}_{EV_l}(t) \quad (3)$$

$$TL_{dist\_grid}(t) = BL(t) + E^{PL}_{total\_demand}(t) \quad (4)$$

where t is the current time slot, $t^{arr}_{EV_l}$ is the arrival time of newly arrived EV, $t^{dep}_{EV_i}$ is the departure time of any ith departing EV, BC_{EV_l} and $SoC^{req}_{EV_l}$ are the battery capacity and the required SoC of the newly arrived EV, $E^{PL}_{total\_demand}$ is the total energy demand of parking lot and TL_{dist_grid} is the total load. To avoid the overloading of the power grid, the total load must be within the nominal capacity of the low-voltage distribution transformers. As mentioned earlier, the baseload represents the fundamental requirements of the customers and is assumed to be an uncontrolled load, whereas considering the flexibility of EV owner's behavior, the parking lot load is assumed to be a controllable load. In order to keep the total load within the normal operation, certain limits are required to be considered. The authors in [33] defined an upper reference power limit (URPL) based on the transformer capacity. However, for the sake of safe operation, we maintained some margin between the URPL and transformer capacity. This work defines the URPL by considering the transformer capacity and the previous day baseload profile (assuming that the current and the previous day have a similar pattern of power consumption) as given by Equation (5).

$$URPL(t) = Trans_{cap} - \left(\frac{1}{T} \sum_{t=1}^{T} BL(t) \times \omega \right) \quad (5)$$

where $Trans_{cap}$ is the transformer capacity, T is the total number of time slots, BL is the baseload and ω is a percentage stability factor define by the low-voltage distribution operators for voltage and frequency maintenance. The second part in first term in Equation (5) represents some margin between the $Trans_{cap}$ and $URPL$. The available power (AP) varies according to the varying BL profile and can be computed based on the URPL and the current value of BL profile, as given by Equation (6). The relationship between the total power demand $\left(E^{PL}_{total\_demand}\right)$ of parking lot and the AP influences the overloading of the distribution network. The $E^{PL}_{total\_demand}$ (Equation (3)) and AP (Equation (6)) can be correlated in any of the two possible cases [34]. In the first case, the AP is enough to support the charging load of all the requesting EVs in the current time slot t; whereas, in the second case, the power demand is higher than the AP, as expressed by Equations (7) and (8). Depending on the AP and charging power (P_C) of a CS, the latter case will allow charging a certain number $\left(N^{Cha}_{EV}\right)$ of EVs as calculated by Equation (9).

$$AP(t) = URPL(t) - BL(t) \quad (6)$$

$$E^{PL}_{total\_demand}(t) \leq AP(t) \quad (7)$$

$$E^{PL}_{total\_demand}(t) > AP(t) \quad (8)$$

$$N^{Cha}_{EV}(t) = \left| \frac{AP(t)}{P_C} \right| \quad (9)$$

At any time slot t, allowing to charge more than N^{Cha}_{EV} number of EVs will abruptly affect the peak-load and may worsen the performance of the distribution network. In this case, the parking lot operators have either to request more power allocation or to cut down their charging load. Depending on the power generation and infrastructure capacities, the allocation of more power is costly and time consuming; whereas, reducing the power demand is more a feasible solution, but the complexity presents challenges on how to choose the most appropriate EVs for charging while restricting/holding the others. This work defines the objective function of minimizing the parking lot power demand by controlling the charging of EVs through their weight values, as given in Equation (10).

$$\min_{E^{PL}_{total\_demand}} \sum_{t=1}^{P^T} C\left[\sum_{i=1}^{N_{EV}}\left\{\left(BC_{EV_i} \times SoC_{EV_i}\right) + D_{EV_i} n P_C\right\}\right](t) \tag{10}$$

where P^T is the total parking duration, C is a binary variable representing whether the parking lot is empty or not, D_{EV_i} is the decision variable used to control the charging of the ith EV and n is the charging efficiency. Depending on the weight value (W) of the ith EV, the accumulated load and the $URPL$, the value of D_{EV_i} can be defined as given by Equation (11).

$$\begin{cases} D_{EV_i}(t) = 1, & \text{if } W_{EV_i}(t) \text{ is highest \& } TL_{dist\_grid}(t) \leq URPL(t) \\ D_{EV_i}(t) = 0, & \text{Otherwise} \end{cases} \tag{11}$$

Each of the ith EV has a defined parking duration and a time period ($\mathcal{T}^C_{EV_i}$) for the charging operation, such that the charging time is the subset of parking duration. The parking duration is computed based on the arrival and the departure time sequence for each of the EVs. The charging time period of the ith EV can be defined according to its battery capacity, required SoC and the charging power per time slot, as expressed by Equation (12).

$$\mathcal{T}^C_{EV_i} = \left(\frac{BC_{EV_i} - \left(SoC^{req}_{EV_l} \times BC_{EV_i}\right)}{P_c}\right) \tag{12}$$

The optimization function defined in Equation (10) is subject to several technical and non-technical constraints. The parking lot has a known operational hour, defined by a pair of starting and ending time (t^P_{st}, P^T). The arrival and departure of the ith EV must be within the parking operational hours. The charging time period must be within the arrival and departure time slots of EV. These constraints are defined in Equations (13)–(15).

$$t^P_{st} \leq t^{arr}_{EV_i} \tag{13}$$

$$t^{dep}_{EV_i} \leq P^T \tag{14}$$

$$t^{arr}_{EV_i} < \mathcal{T}^C_{EV_i} \leq t^{dep}_{EV_i} \tag{15}$$

To maintain the battery efficiency, the SoC, charging cycles ($B^{cyc}_{EV_i}$) of the battery and the charging power of the ith EV must be within the defined maximum $SoC^{max}_{EV_i}$, maximum number of battery charging cycles $\left(B^{\max\_cyc}_{EV_i}\right)$ *and* maximum charging power P^{max}_C [35] as given in Equations (16)–(18). The total load at any time slot t must be within the URPL, as given in Equation (19).

$$SoC_{EV_i}(t) \leq SoC^{max}_{EV_i} \tag{16}$$

$$B^{cyc}_{EV_i} \leq B^{\max\_cyc}_{EV_i} \tag{17}$$

$$P^{EV_i}_C(t) \leq P^{max}_C \tag{18}$$

$$TL(t) \leq URPL(t) \tag{19}$$

The charging operation of EVs in each time slot influences the total load of the power grid. The charging impact on total load is measured in percentage and can be computed with respect to the highest peak load and the URPL, as given in Equation (20). The QoE for the ith EV is the function of the SoC_{EV_i}, $SoC^{req}_{EV_i}$ and BC_{EV_i} and can be computed according to Equation (21). Similarly, the parking lot QoP is function of the number of satisfied EVs $\left(\mathcal{N}_{EVsatisfied}\right)$, the number of unsatisfied EVs $\left(\mathcal{N}_{EVUnsatisfied}\right)$, the QoE and the total number of EVs ($\mathcal{N}_{EV}$) during the parking lot operational hours and can be computed according to Equation (22).

$$Load\ impact\ (\%) = \begin{cases} \left(\frac{peak_{load} - URPL}{peak_{load}}\right) \times 100, & \text{If } peak_{load_{bus}} > URPL_{bus} \\ 0, & \text{Otherwise} \end{cases} \tag{20}$$

$$QoE_{EV_i} = \begin{cases} 1, & \text{if } SoC_{EV_i}^{req} \geq 1 \\ SoC_{EV_i}^{req} - SoC_{EV_i}, & \text{if } SoC_{EV_i} < SoC_{EV_i}^{req} < 1 \end{cases} \tag{21}$$

$$QoP = \left(\frac{|\mathcal{N}_{EV}| - \sum_{i=1}^{N_{EV}} EV_{Unsatisfied\_QoE}(i)}{|\mathcal{N}_{EV}|}\right) \tag{22}$$

3.3. *Fuzzy Logic Inference Mechanism*

Definition 1. *The crisp sets are based on the theory of complete knowledge, for instance, an element is either a member of a set or not. Whereas in fuzzy sets the degree of membership function determines the belonging of an element to the set. An element x in fuzzy set $A \in X$ (universal set) can be represented through the degree of its membership function as expressed in Equation (23) [36].*

$$A = \{(x,\ \mu_A(x))\ :\ x\ \in\ X\} \tag{23}$$

where $\mu_A(x)$ is the degree of membership function which represents the belonging of x to the fuzzy set A in the range [0, 1]. The degree of membership function defines how closely the element x belongs to the set A. A higher degree represents a strong whereas a lower degree represents a weak belonging of x to the fuzzy set A. The concepts of membership functions used in this work are as follows:

- *Triangular membership function:* A triangular membership function reflects the shape of a triangle and can be defined by three parameters a, b and m such that $a < m < b$, as given in Equation (24) [37].

$$\mu_A(x) = \begin{cases} 0, \text{ if } x \leq a \\ \frac{x-a}{m-a}, \text{ if } a < x \leq m \\ \frac{b-x}{b-m}, \text{ if } m < x \leq b \\ 0, \text{ if } b \leq x \end{cases} \tag{24}$$

- *Left-Right open shoulder trapezoidal membership function:* The left–right open membership functions can be defined by two parameters a and b and graphically represented by ⏋& Γ symbols and the functions can be written as Equations (25) and (26).

$$\mu_A(x) = \begin{cases} 1, \text{ if } x \leq a \\ \frac{b-x}{b-a}, \text{ if } a < x \leq b \\ 0, \text{ if } x > b \end{cases} \tag{25}$$

$$\mu_A(x) = \begin{cases} 0, \text{ if } x \leq a \\ \frac{x-a}{b-a}, \text{ if } a < x \leq b \\ 1, \text{ if } x > b \end{cases} \tag{26}$$

- *Trapezoidal membership function:* The trapezoidal membership function resembles a trapezoidal shape and can be defined by four parameters a, b, c and d. The parameters a and d defines the abscissa of two vertices at the bottom while the parameters b and c denotes the abscissa of the two vertices at the top of the trapezoidal [37]. Mathematically, it can be expressed as Equation (27).

$$\mu_A(x) = \begin{cases} 0, \text{ if } (x \le a) \text{ } or \text{ } (x > d) \\ \frac{x-a}{b-a}, \text{ if } a < x \le b \\ 1, \text{ if } b < x \le c \\ \frac{d-x}{d-c}, \text{ if } c < x \le d \end{cases} \quad (27)$$

3.3.1. Fuzzification of Crisp Inputs and Their Fuzzy Membership Functions

The charging operation of the ith EV is controlled through the decision variable D_{EV_i} (Equation (10)), which is based on its weight value. There are multiple parameters such as arrival and departure time, the SoC^{req}, the RPD, the BL and the AP from both the EVs and the power grid domains which needs to be considered while computing the W value for the ith EV. The independent nature and the temporal-based variation of these parameters are introducing a higher degree of uncertainty, which presents complexity and challenges in the task of weight computation. It is believed that the SoC^{req}, the RPD and the AP are the most relevant inputs that influence the weight value in each time slot [38]. Therefore, to compute an adequate weight value for the ith EV, this work correlates the SoC^{req}, the RPD and the AP through the fuzzy inference mechanism. These crisp inputs should be linearly structured between the minimum and maximum boundaries with their corresponding units and should be defined through the set of linguistic variables for representing them through the membership functions. The RPD input is based on the operating hours (12 h) of the parking lot, which is normalized into 48 time slots such that each time slot represents 15 min. Considering the dynamic behavior of EV owners, the RPD is modeled with three membership functions and is represented with linguistic terms short duration (SD), average duration (AD) and long duration (LD) [25,32]. The linguistic terms SD and LD are implemented as left and right open shoulder membership functions, whereas the term AD is implemented as trapezoidal membership functions using Equations (25)–(27). The fuzzy set SD and AD contains the degree of membership functions for the set of EVs having RPD in the range of $0 \le \mu_{n_{EV}}(RPD) \le 8$ time slots and $4 \le \mu_{n_{EV}}(RPD) \le 20$ time slots, respectively. The n_{EV} is the number of EVs in the set. The fuzzy set LD holds the degree of membership functions for the set of EVs with RPD in the range of $16 \le \mu_{n_{EV}}(RPD) \le 48$ time slots. The implementation detail of RPD is given in Table 1 and is virtualized in Figure 4a. The SoC^{req} is a function of SoC and the battery capacity and is measured in the range of [0–1]. It is modeled with five membership functions which are represented with linguistic terms very low (VL), low (L), medium (M), high (H) and very high (VH), respectively. The fuzzy sets VL and VH contain the degree of membership functions for the set of EVs with SoC^{req} in the ranges of $0 \le \mu_{n_{EV}}(SoC^{req}) \le 0.3$ and $0.7 \le \mu_{n_{EV}}(SoC^{req}) \le 1$, respectively. Similarly, the fuzzy sets L, M and H contains the degree of membership functions for the EVs with SoC^{req} in the ranges of $0.1 \le \mu_{n_{EV}}\ (SoC^{req}) \le 0.5$, $0.3 \le \mu_{n_{EV}}\ (SoC^{req}) \le 0.7$ and $0.5 \le \mu_{n_{EV}}(SoC^{req}) \le 0.9$, respectively. The details of all parameters for the implementation of SoC^{req} is given in Table 2 and is shown in Figure 4b. The third input is the AP which is measured in kW and is normalized from low available power to high available power in the range [0–100]. The AP is modeled with five membership functions and is represented with linguistic terms very low AP (VLAP), low AP (LAP), medium AP (MAP), high AP (HAP) and very high AP (VHAP). These linguistic terms are implemented with two left–right open shoulders and three triangular membership functions using Equations (25)–(27).

Furthermore, the VLAP and VHAP contain the degree of membership functions for the time slots with AP in the ranges $0 \le \mu_t(AP) \le 30$ and $70 \le \mu_t(AP) \le 100$, respectively. By this way, the fuzzy sets LAP, MAP and HAP contain the degree of membership functions for the time slots in the ranges of $10 \le \mu_t(AP) \le 50$, $30 \le \mu_t(AP) \le 70$ and $50 \le \mu_t(AP) \le 90$, respectively. The implementation detail of AP is given in Table 3 and is shown in Figure 4c.

Table 1. Implementation detail of membership functions for RPD.

Fuzzy Sets	Type of MF	Arguments (Time Slots)
SD	Left open shoulder	a = 4, b = 8
AD	Trapezoidal	a = 4, b = 8, c = 16, d = 20
LD	Right open shoulder	a = 16, b = 20

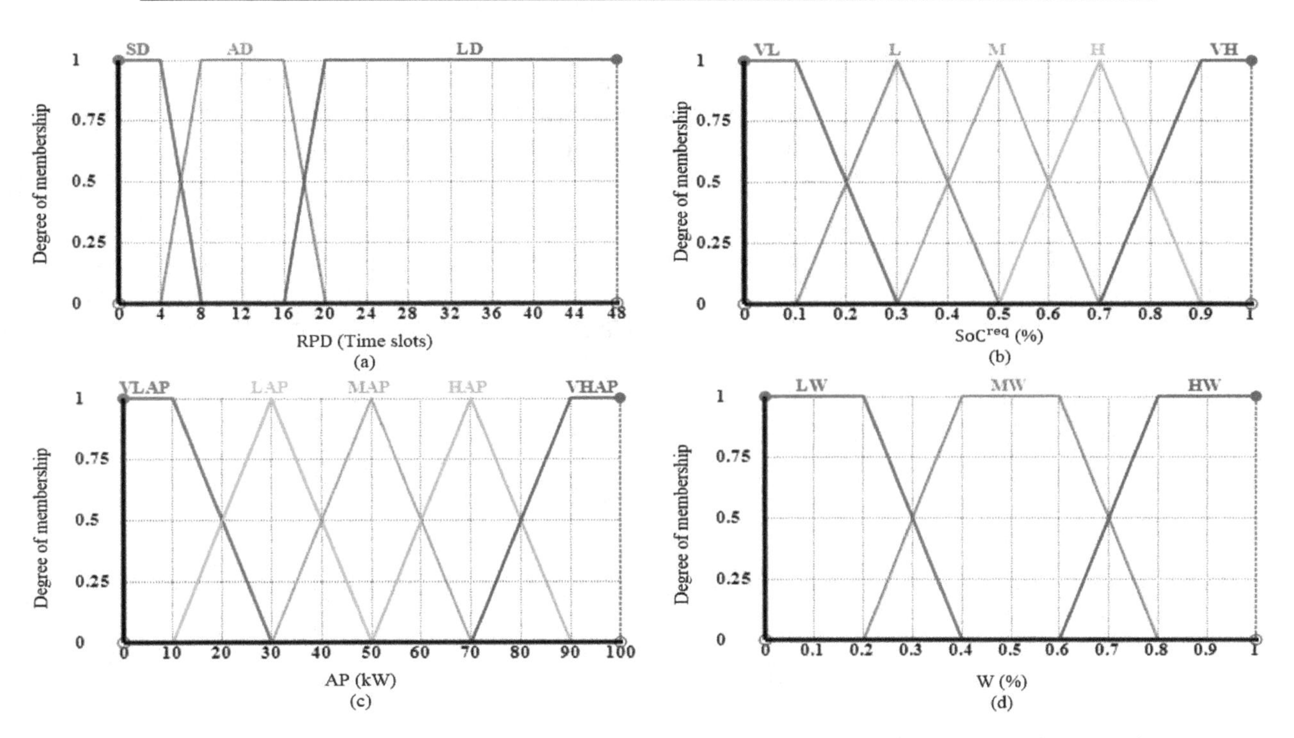

Figure 4. Membership functions of the fuzzified input and output variables. (**a**) Membership functions of remaining parking duration (RPD); (**b**) membership functions of required state of charge (SoC); (**c**) membership functions of available power (AP)and (**d**) membership functions of weight value (W) variable.

Table 2. Implementation detail of membership functions for required SoC.

Fuzzy Sets	Type of MF	Arguments (%)
VL	Left open shoulder	a = 0.1, b = 0.3
L	Triangular	a = 0.1, m = 0.3, b = 0.5
M	Triangular	a = 0.3, m = 0.5, b = 0.7
H	Triangular	a = 0.5, m = 0.7, b = 0.9
VH	Right open shoulder	a = 0.7, b = 0.9

Table 3. Implementation detail of membership functions for AP.

Fuzzy Sets	Type of MF	Arguments (kW)
VLAP	Left open shoulder	a = 10, b = 30
LAP	Triangular	a = 10, m =30, b = 50
MAP	Triangular	a = 30, m = 50, b = 70
HAP	Triangular	a = 50, m = 70, b = 90
VHAP	Right open shoulder	a = 70, b = 90

3.3.2. Fuzzy Inference Mechanism for Obtaining the Fuzzified Weight Variable

The set of input memberships and the set of expert's rules are evaluated through the fuzzy inference system (FIS) to generate the fuzzified output. Therefore, it is of utmost importance to define the output variable and the set of fuzzy expert's rules. In this work, the FIS computes the W_{EV_i}

for each of the ith requesting EVs. The value of W_{EV} variable for each of the EVs is measured in the range of [0–1] and is fuzzified with three membership functions using Equations (25)–(27). The membership functions of the output variable are represented with linguistic terms low weight (LW), medium weight (MW) and high weight (HW). The linguistic terms LW and HW contains the set of EVs with the degree of memberships in the ranges of $0 < \mu_{n_{EV}}(W) \leq 0.4$ and $0.6 < \mu_{n_{EV}}(W) \leq 1$ and are modeled with left and right open shoulder membership functions. Whereas the linguistic term MW holds the set of EVs having the degree of memberships in the range of $0.2 \leq \mu_{n_{EV}}(W) \leq 0.8$. The implementation detail of the output variable is given in Table 4 and shown in Figure 4d. The fuzzy rules represents a set of process that correlates the degree of memberships of a set of inputs to the degree of memberships of the output variable using IF–THEN logical statements [39]. The set of rules is usually designed according to the expert's knowledge of the problem domain [40]. The sequence of IF–THEN statements forms an algorithm which captures the currently known information and infers the output using fuzzy rules implication. In the logical IF–THEN statement, the IF part represents the antecedents (conditions) which capture the observed information and the THEN part shows the consequent (conclusion). The consequent is fuzzified knowledge and is represented in the form of linguistic variable and degree of membership. The antecedents relate multiple inputs through AND/OR logical operators, while the consequent infers the output by using the intersection, union and composition operations of the fuzzy set theory.

Table 4. Implementation detail of membership functions for W.

Fuzzy Sets	Type of MF	Arguments (%)
LPF	Left open shoulder	a = 0.2, b = 0.4
APF	Triangular	a =0.2, b = 0.4, c = 0.6, d = 0.8
HPF	Right open shoulder	a = 0.6, b = 0.8

Definition 2. *The relation of two fuzzy sets A and B is represented by R = A → B and can be defined as the Cartesian product in X* Y space, where X and Y are the universal sets such that A ⊆ X and B ⊆ Y. The mathematical representation of two fuzzy sets and multiple fuzzy sets is given in Equations (28) and (29) [37,41].*

$$R(x, y) = \left\{ \frac{\mu_R(x, y)}{(x, y)} \middle| (x, y) \in X \times Y \right\} \tag{28}$$

$$R = \begin{matrix} & y_1 & \cdots & y_n \\ x_1 & \mu_R(x_1, y_1) & \cdots & \mu_R(x_1, y_n) \\ \vdots & \vdots & \ddots & \vdots \\ x_m & \mu_R(x_m, y_1) & \cdots & \mu_R(x_m, y_n) \end{matrix} \tag{29}$$

Definition 3. *For two fuzzy relations R = A → B and Q = B → C, a new relation S can be computed using the fuzzy composition operation, such that S relates the elements of C in Q and elements of A in R, as given by Equation (30).*

$$S = R \circ Q \tag{30}$$

The symbol " " is the composition operator which connects the elements of R and Q based on their membership functions. The Mamdani *min–max* is a famous composition method which can be used to infer the degree of input membership functions to the fuzzy set S, as given in Equations (31) and (32).

$$\mu_S(x,\ z) = \left\{ \frac{\mu_S(x,\ z)}{(x,\ z)} \middle| (x,\ z) \in X \times Z \right\} \tag{31}$$

$$\mu_S(x,\ z) = \max\Big(min\big(\mu_R(x,\ y),\ \mu_Q(x,\ z)\big)\Big) \tag{32}$$

Definition 4. *The set of fuzzy rules* R = $\{R_1,\ R_2,\ \cdots,\ R_n\}$ *along with their corresponding antecedents and consequences using IF–THEN statement can be expressed as given in Equation (33) and can be generalized as given in Equation (34).*

$$\begin{cases} R_1 : \text{ if } x_1 \text{ } is \text{ } A^1 \text{ THEN } y_1 \text{ } is \text{ } B^1 \\ R_2 : \text{ if } x_2 \text{ } is \text{ } A^2 \text{ THEN } y_2 \text{ } is \text{ } B^2 \\ \vdots \\ R_n : \text{ if } x_n \text{ } is \text{ } A^n \text{ THEN } y_m \text{ } is \text{ } B^m \end{cases} \tag{33}$$

$$\text{IF } x_S \text{ } is \text{ } A^S \text{ THEN } y_S \text{ } is \text{ } B^S \tag{34}$$

The sets $x_S = \{x_1, x_2, \cdots, x_n\}$ and $y_S = \{y_1, y_2, \cdots, y_m\}$ are the n input fuzzy variables and the sets $A^S = \{A^1,\ A^2 \cdots A^n\}$ and $B^S = \{B^1,\ B^2 \cdots B^m\}$ are the linguistic representation of the antecedents and consequences in universes of discourses X and Y, respectively [42]. Considering the generalized form of rules defined in Equation (35), the *min* and *max* operation on the degree of membership functions of A^S and B^S for the x_S and y_S input variables are expressed in Equations (35) and (36), respectively.

$$\mu_{A^S B^S}(x_S, y_S) = \min[\mu_{A^n}(x_n), \mu_{B^m}(y_m)] \tag{35}$$

$$\mu_{A^S B^S}(x_S, y_S) = \max[\mu_{A^n}(x_n), \mu_{B^m}(y_m)] \tag{36}$$

The approximate reasoning feature of FIS is used to infer the most appropriate knowledge when multiple rules are applicable for the given inputs. The approximate reasoning is the process of matching the degree of input data to each of the applicable rules. The higher the matching degree of input data to the rules the closer is the inferred conclusion to those rules and vice versa. The approximate reasoning can be done by considering all the applicable IF–THEN rules and using any aggregation method such as Mamdani *min–max* operation. Considering all the combinations of three inputs and their corresponding output variable, this work defines the set of fuzzy rules for computing the weight values for the EVs, as given in Tables 5–7. In the case of multiple rules say r applicable rules such that $i = 1, 2, 3 \ldots r$, the aggregated inferred weight value for each of the ith EV can be obtained by *min–max* operation on r applicable rules as given by Equation (37).

$$\mu_{EV_i}(W) = max\left[\begin{array}{c} min\left\{\ \mu(RPD_t)^1, \mu\left(SoC_t^{req}\right)^1, \mu(AP_t)^1\right\}, \\ \cdots, min\left\{\ \mu(RPD_t)^r, \mu\left(SoC_t^{req}\right)^r, \mu(AP_t)^r\right\} \end{array}\right] \tag{37}$$

Table 5. Fuzzy mapping rules of the fuzzy inference system (FIS) when RPD is short duration (SD).

W		**AP**				
		VLAP	**LAP**	**MAP**	**HAP**	**VHAP**
SoC^{req}	**VL**	LW	LW	LW	LW	MW
	L	LW	LW	MW	MW	MW
	M	LW	MW	MW	MW	HW
	H	MW	AW	HW	HW	HW
	VH	HW	HW	HW	HW	HW

Table 6. Fuzzy mapping rules of FIS when RPD is average duration (SD).

W		AP				
		VLAP	**LAP**	**MAP**	**HAP**	**VHAP**
	VL	LW	LW	LW	MW	MW
	L	LW	LW	MW	MW	MW
SoC^{req}	**M**	LW	LW	HW	HW	HW
	H	MW	HW	HW	HW	HW
	VH	MW	HW	HW	HW	HW

Table 7. Fuzzy mapping rules of FIS when RPD is long duration (SD).

W		AP				
		VLAP	**LAP**	**MAP**	**HAP**	**VHAP**
	VL	LW	LW	LW	LW	MW
	L	LW	LW	LW	MW	MW
SoC^{req}	**M**	LW	LW	MW	MW	MW
	H	LW	LW	HW	HW	HW
	VH	MW	HW	HW	HW	HW

3.3.3. Defuzzification for Obtaining the Crisp Weight Variable

The fuzzy inference results in a fuzzified output, which must be converted into crisp weight value through the defuzzification process. There are several defuzzification methods, including center of gravity (COG), middle of maxima (MOM), first of maxima (FOM) and last of maxima (LOM) and random choice of maxima (RCOM). The use of a specific defuzzification method depends on the type of input membership functions such as overlapping or non-overlapping membership functions. For the non-overlapping membership functions, the MOM is a suitable choice while for overlapping membership functions, the COG is the most feasible solution. This is because in the case of non-overlapping membership functions a slight change in the input data reflects an abrupt change in the output, whereas in the case of overlapping membership functions any minor change does not influence the output significantly. This work uses overlapping membership functions for input data and consider the COG method to compute the crisp value for the weight variable. To compute the crisp weight value for the ith EV, the standard equations of the COG method can be utilized as given in Equations (38) and (39).

$$W_{EV_i} = \frac{\sum_{k=1}^{m} \mu_{W_{EV_i}}(x_k) \, * \, x_k}{\sum_{k=1}^{m} \mu_{W_{EV_i}}(x_k)}, \forall\, k = 1,\, 2, \cdots m \text{ and } x \in W_{EV} \tag{38}$$

$$W_{EV_i} = \frac{\int x * \mu_{W_{EV_i}}(x)dx}{\int \mu_{W_{EV_i}}(x)\, dx}, \text{ for } x \in \, W_{EV} \tag{39}$$

The input data can either be discrete or continuous values. For the case of discrete inputs Equation (38) can be used while for the case of continuous values Equation (39) can be used to compute the crisp value of weight variable.

3.4. Flowchart of the Proposed Algorithm

The flowchart of the proposed FLWCS is shown in Figure 5. The detailed procedure of the algorithm is explained in the following steps.

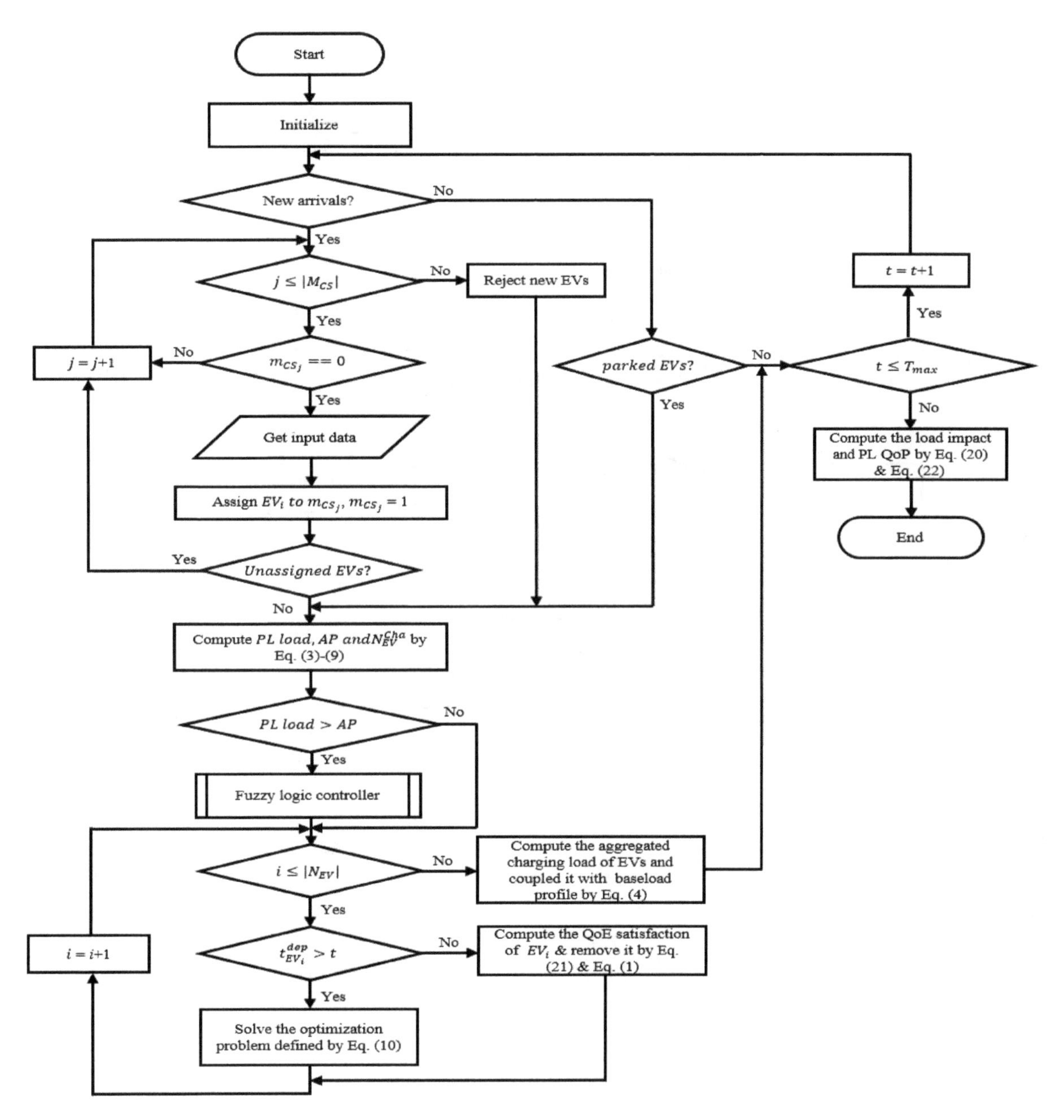

Figure 5. Flowchart of the algorithm in the proposed FLWCS.

Step 1. Initialize all the input parameters of the system, such as the initial and maximum simulation time, the number of CSs and their charging power and other control variables.

Step 2. Check for the new arrivals of EVs in the current time slot t. If there is new arrival of EVs the algorithm check for any available parking spot and CS by iterating through each of the CSs. Note that this work considered futuristic parking scenarios which assume that each of the parking spots is equipped with a CS. If there is an available spot and CS, the algorithm registers each of the new EVs into the system using Equations (1) and (2) and collects the inputs from the new EVs and assign them to the CSs. The status of the CSs is updated from idle to busy.

Step 3. Compute the total energy demand of parking lot, the AP and the number of EVs that can be supported by the AP according to Equations (3)–(9).

Step 4. Check whether the energy demand of parking lot is greater than the AP or not, as stated by Equations (7) and (8). If the condition is true, i.e., the energy demand is higher than the AP go to the next (Step 5) and call the fuzzy logic controller subroutine as shown in Figure 6. However, if the AP is enough to support the parking lot energy demand, then go to Step 6.

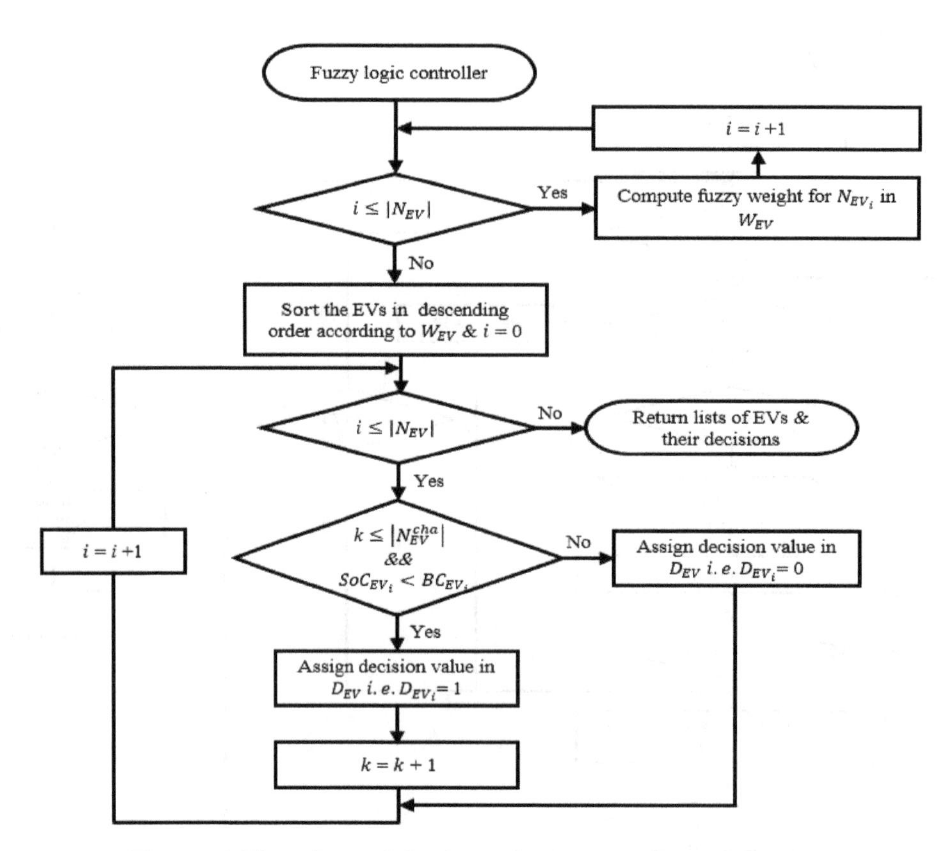

Figure 6. Flowchart of the fuzzy logic controller subroutine.

Step 5. Construct the list of weight values W_{EV} by computing the W_{EV_i} for each of the ith EV according to the input data using the developed fuzzy inference mechanism. Once the list of the weight values is computed, sort the list of EVs (N_{EV}) in descending order according to W_{EV} list. Furthermore, construct the list of decisions (D_{EV}) by checking the SoC against the battery capacity of each the EVs and the total number of allowed EVs for charging within the AP. Considering these conditions the list D_{EV} is updated with 0 and 1 values. Finally, the EVs and their corresponding decision lists are returned to the main calling algorithm.

Step 6. Check the departure time $t_{EV_i}^{dep}$ of each of the ith EV against the current time slot t. If in the current time slot, any of the ith EV is departing, then compute its QoE using Equation (21) and remove the departing EV from the set of EVs using Equation (1). However, if the EV has still to stay in the parking lot, the algorithm solves the optimization problem defined by Equation (10) for each of the EVs, according to their corresponding decision D_{EV_i} values. Once the optimization problem is solved for all of the EVs, their charging operations are performed in the current time slot. The algorithm then couples the aggregated charging load to the current baseload and computes the total load. If the current time slot is not reached to the maximum simulation time, increment the current time slot t and repeat the process from Step 2 to Step 6. However, if the simulation time reached to its maximum time limit, compute the load impact and the parking lot QoP using Equations (20) and (22).

4. Simulation Results and Discussion

This work assume a low-voltage distribution network, which feeds electricity to the residential houses and a parking lot. The transformer capacity of the distribution network is based on the lumped load of the node-820 in the IEEE 34 bus system [43]. The total baseload depends upon the number of houses in the distribution network and their electricity consumption. The average electricity consumption of a typical household is assumed to be about 2.78 kW and load factor of the houses is about 70% of the lumped load of node-820 in the IEEE 34 bus system [44,45]. As a result, a total of 34 houses was computed for the low-voltage distribution network and their aggregated baseload profile is visualized in Figure 7. The parking lot operational hours is considered to be from 7:00 AM to 7:00 PM [46]. These 12 h are normalized into 48 time slots with a 15 min resolution. The parking spots are equipped with fast CSs of 20 kW/h supporting a charging power of 5 kW/time slot. Furthermore, four different types of EVs with battery capacities of 40 kWh, 60 kWh, 80.5 kWh and 100 kWh are considered for the simulation [47–50].

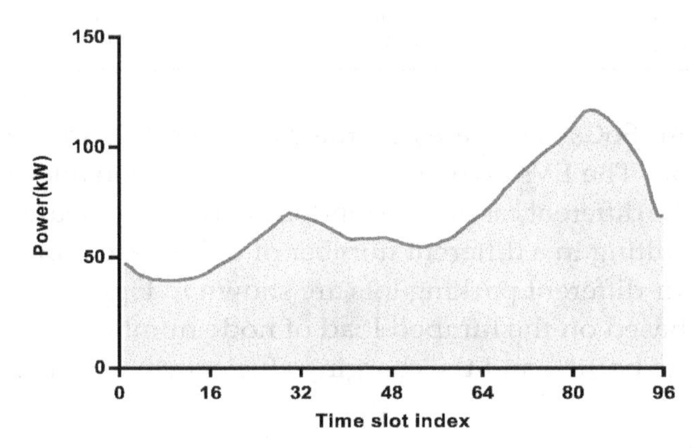

Figure 7. Aggregated baseload profile of household consumption in low-voltage distribution network.

The simulation is developed using java language, where the open source jFuzzyLogic libraries are utilized for implementing the fuzzy logic inference system [51]. The simulation is performed for four different parking capacities of 50 EVs (case-1), 100 EVs (case-2), 150 EVs (case-3) and 200 EVs (case-4). The four different types of EVs are distributed with a random penetration level as given in Table 8. The arrivals of EVs are randomly generated with $\mu = 42$ slot number and $\sigma = 6$ time slots, while their stay time are generated with $\mu = 20$ time slot number and $\sigma = 4$ time slots, using Gaussian distribution. Their corresponding departure times are then computed by summing up their arrivals and stay time distribution. The arrival and departure time distribution of EVs for the four different parking capacities is plotted in Figure 8.

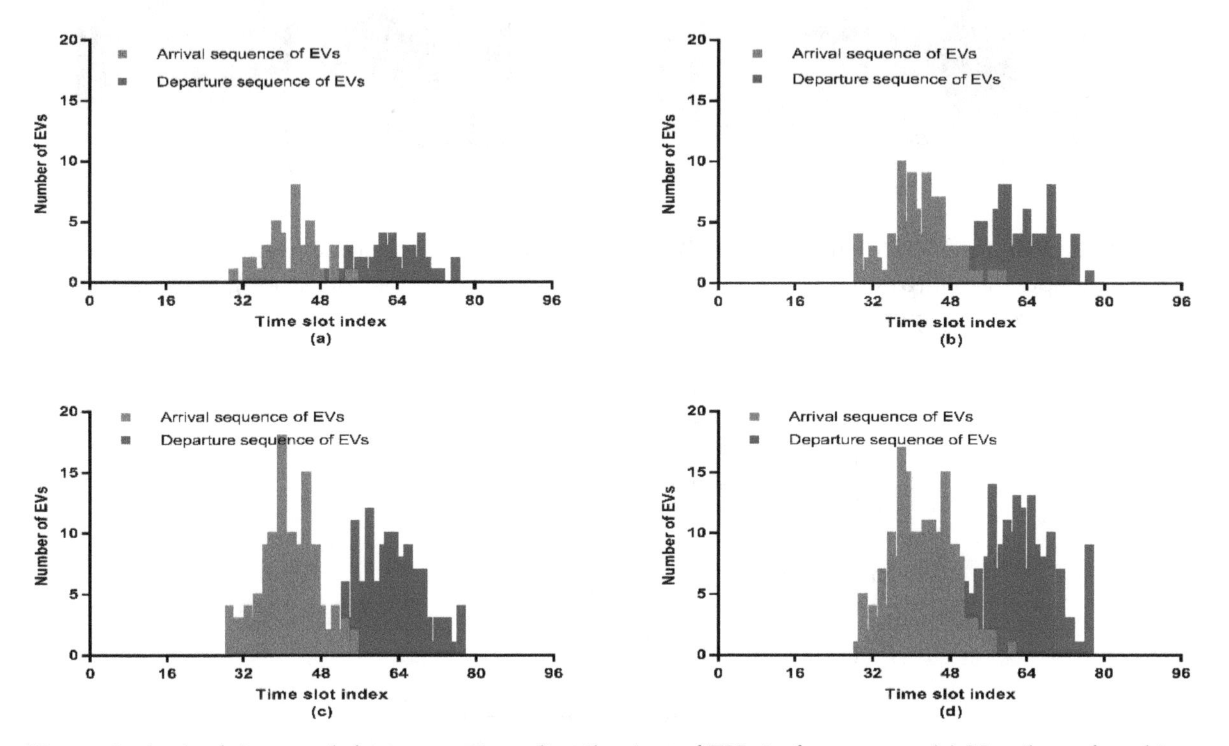

Figure 8. Arrival time and departure time distribution of EVs in four cases. (**a**) Number of parking spots is 50; (**b**) number of parking spots is 100; (**c**) number of parking spots is 150 and (**d**) number of parking spots is 200.

Table 8. Penetration levels of different type of EVs.

Cases	Nissan LEAF-40 kWh [47]	Tesla S-60 kWh [48]	Tesla Model-3 80.5 kWh [49]	Tesla Model X-100 kWh [50]
1	14%	24%	32%	30%
2	27%	25%	21%	27%
3	30%	18%	25%	27%
4	27%	21%	22%	30%

Similarly, the arrival time SoCs of EVs are generated between 20% and 50% of the battery capacities using a uniform distribution. The EVs arrival time SoCs distribution and their battery capacities are plotted in Figure 9 for four the different cases. The random arrival, departure sequences of EVs and their corresponding SoCs are resulting in a different number of EVs in each time slot. The temporal-based varying occupancies for four different parking lots are shown in Figure 10. The transformer capacity $Trans_{cap}$ is assumed to be based on the lumped load of node number 820 of the IEEE 34 bus system. The value of ω is assumed to be 10% and the charging efficiency η is considered to be about 0.90 [52].

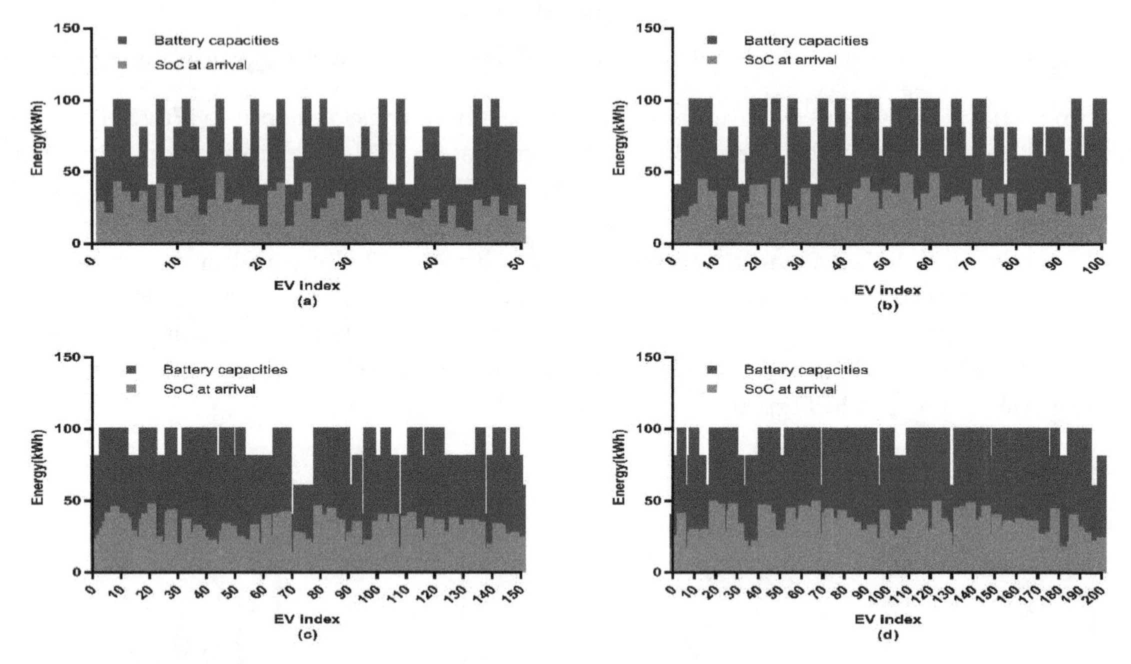

Figure 9. Arrival time SoC distribution of EVs and their battery capacities in four cases. (**a**) Number of parking spots is 50; (**b**) number of parking spots is 100; (**c**) number of parking spots is 150 and (**d**) number of parking spots is 200.

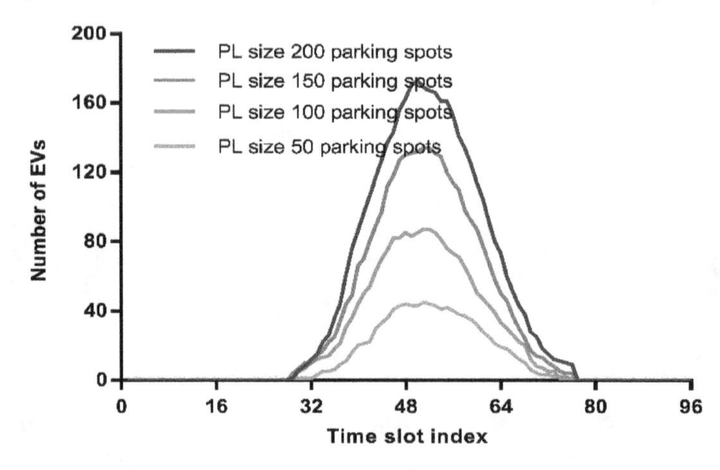

Figure 10. Temporal varying occupancies of EVs parking duration in four cases.

The simulations are performed in four different cases, where each case corresponds to different parking capacity. The performance of the proposed FLWCS is evaluated against the conventional FCFS-based scheme by considering the QoE and the QoP as the performance metrics. The FCFS-based scheme performs the charging operation of EVs according to their arrival sequence; therefore, an EV with the earliest arrival time has the highest priority to be charged. In contrast, the FLWCS computes weight values for EVs in each of the time slots using fuzzy inference mechanism. The weight values are dynamically computed in each time slot and are used to choose the most appropriate number of charging EVs that help to maximize the QoE and QoP while maintaining the grid constraints. The concept is almost the same as the authors in [53] dynamically controlled a threshold value between a normal and guard channels based on the people's mobility.

The results in Figure 11 show the number of EVs requesting for charging operations, the number of EVs that can be supported by the AP under the normal operational limit of the power grid and their scheduling with FCFS and the proposed FLWCS. In each case, it can be observed that with the increasing parking occupancies the number of EVs with charging requests are also increasing. However, the variation of the baseload profile and the operational constraints of the power grid limit the number of EVs to be charged in each time slot. Following the operational constraints of the power grid, the total charging demand and the different behaviors of EV owners, in each time slot the two schemes perform their scheduling in a different manner. The FCFS-based scheme prioritizes the early arriving EVs and thereby with the passage of time most of the later arriving EVs with shorter staying duration are unable to get the opportunities for charging operations.

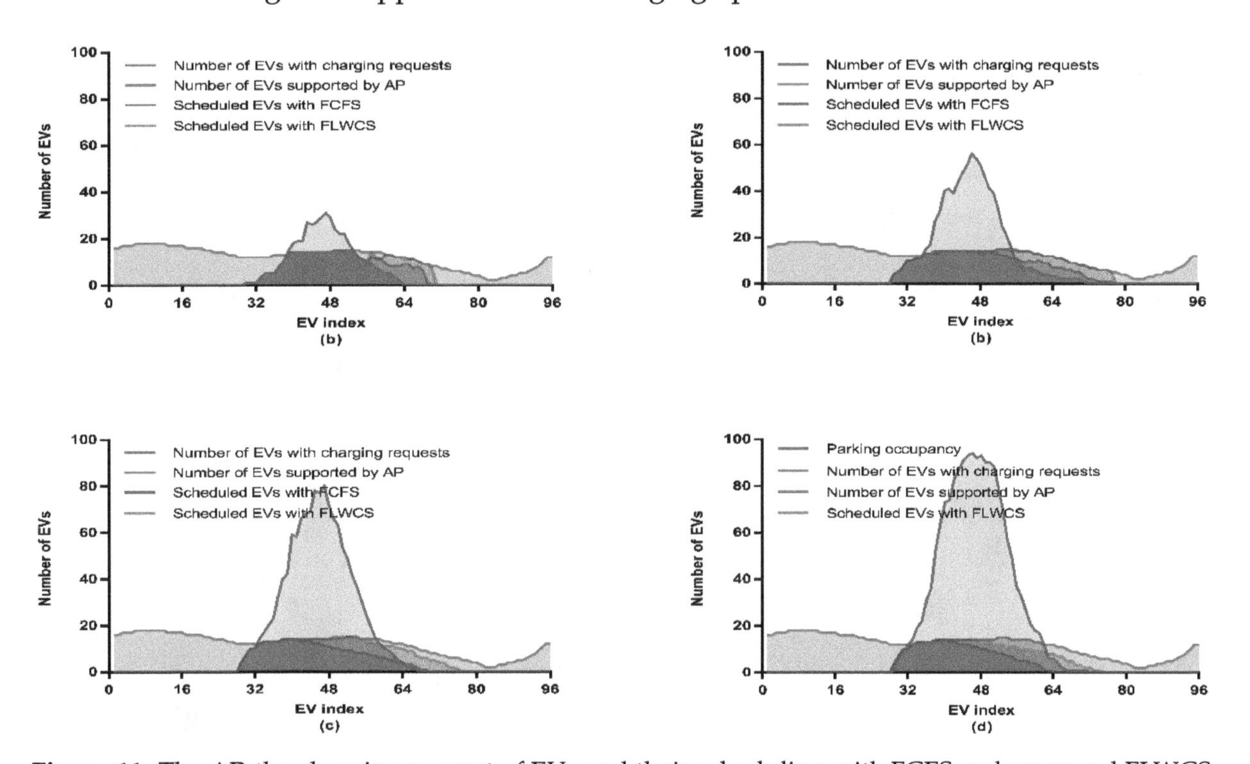

Figure 11. The AP, the charging request of EVs and their scheduling with FCFS and proposed FLWCS in four different cases. (**a**) Number of parking spots is 50; (**b**) number of parking spots is 100; (**c**) number of parking spots is 150 and (**d**) number of parking spots is 200.

Whereas, the proposed FLWCS maximize the charging operations by scheduling the most appropriate EVs for charging according to their weight values. In each time slot, the charging load of EVs with respect to the FCFS-based scheme and FLWCS is shown in Figure 12. It can be observed that the proposed FLWCS can utilize the AP in a more efficient manner compared to the FCFS-based scheme. The difference in AP utilization between the two charging schemes is more obvious from case-1 to case-4 with the increasing parking size and occupancy. The parking occupancies, the number of EVs with charging request and their scheduling for the charging operations affect the total load on the power grid. The aggregation of the baseload and the charging load of EVs results in the formation of the total load profile of the power grid, as shown in Figure 13. The figure shows the baseload and the total load with respect to FCFS and the proposed FLWCS schemes for four different parking capacities. The URPL (which is computed according to Equation (5)) is the threshold point representing the normal operational limit of the power grid. From figures, it can be seen that in all the four cases, both of the charging schemes follow the normal operational limits of the power grid. However, the efficient utilization of the AP and the total load profile with the proposed FLWCS is higher than the FCFS-based charging scheme. The temporal varying baseload, the operational constraints of the power grid, the different behaviors of EV owners, the battery capacities, and the required amount of charging have an effect on the QoE of EVs and thereby on the QoP of the parking lot. Considering the QoE until full battery capacity, the QoP in terms of satisfied QoE for the four different cases is shown in Figure 14. In view of the EV owner's requirements, the two schemes have different QoP in each case. For example, with the proposed FLWCS a greater number of EVs are able to get the charging opportunities and thereby improving the QoP performance than the FCFS-based scheme. In case-1, the EVs with satisfied QoE are about 76% and 68% and the EVs with unsatisfied QoE are about 24% and 32% with the proposed FLWCS and the FCFS-based charging scheme (Figure 14a).

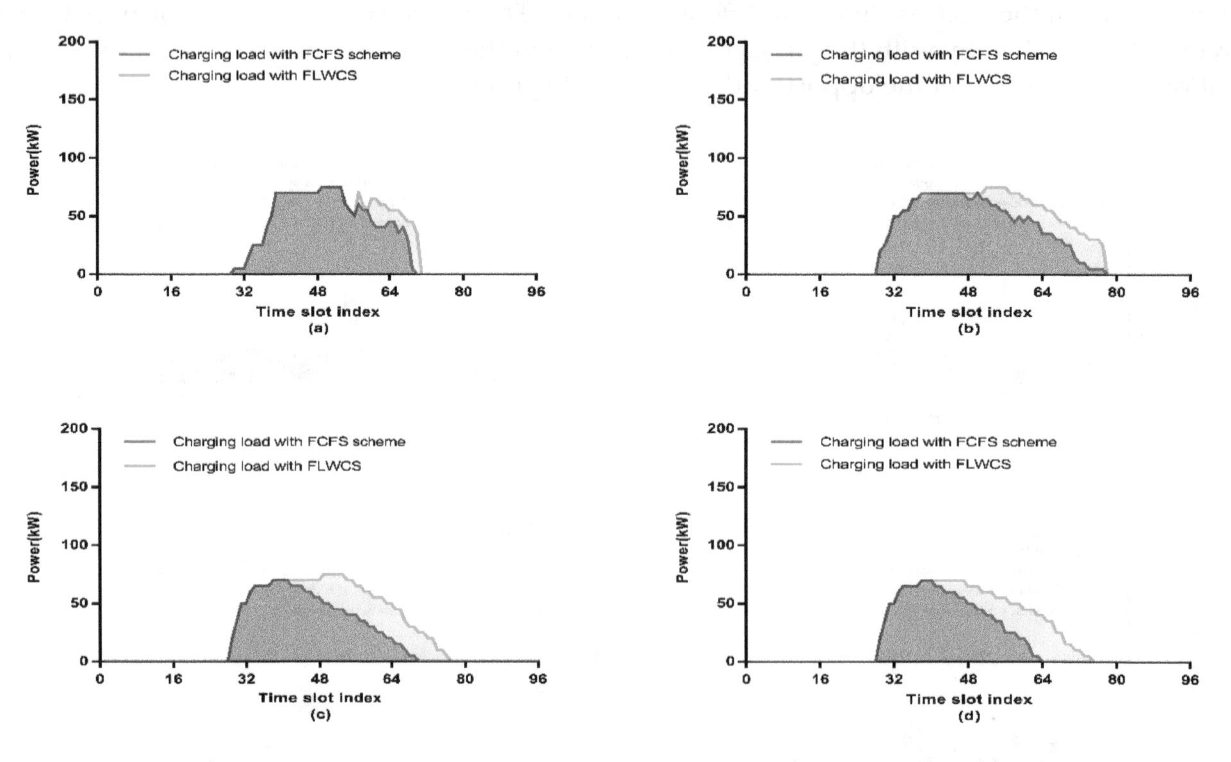

Figure 12. Charging load of EVs with FCFS and proposed FLWCS scheme in four different cases. (**a**) Number of parking spots is 50; (**b**) number of parking spots is 100; (**c**) number of parking spots is 150 and (**d**) number of parking spots is 200.

This implies that the proposed FLWCS has about 8% improved QoP than the FCFS-based scheme. By increasing the parking lot size from 50 to 100 parking spots in case-2, a degrading QoP performance was noted. The QoP is about 51% and 41% with respect to the proposed FLWCS and the FCFS-based charging schemes (Figure 14b). The performance was further analyzed by simulating scenarios of parking lots with 150 and 200 parking spots in case-3 and case-4 (Figure 14c,d). In case-3 and case-4, the QoP is about 43% and 38%, with the proposed FLWCS. In these cases, the QoP is about 31% and 24% with respect to the FCFS-based charging scheme. The results in these cases imply that the proposed FLWCS has about 12% and 14% higher QoP comparing to the FCFS-based charging scheme.

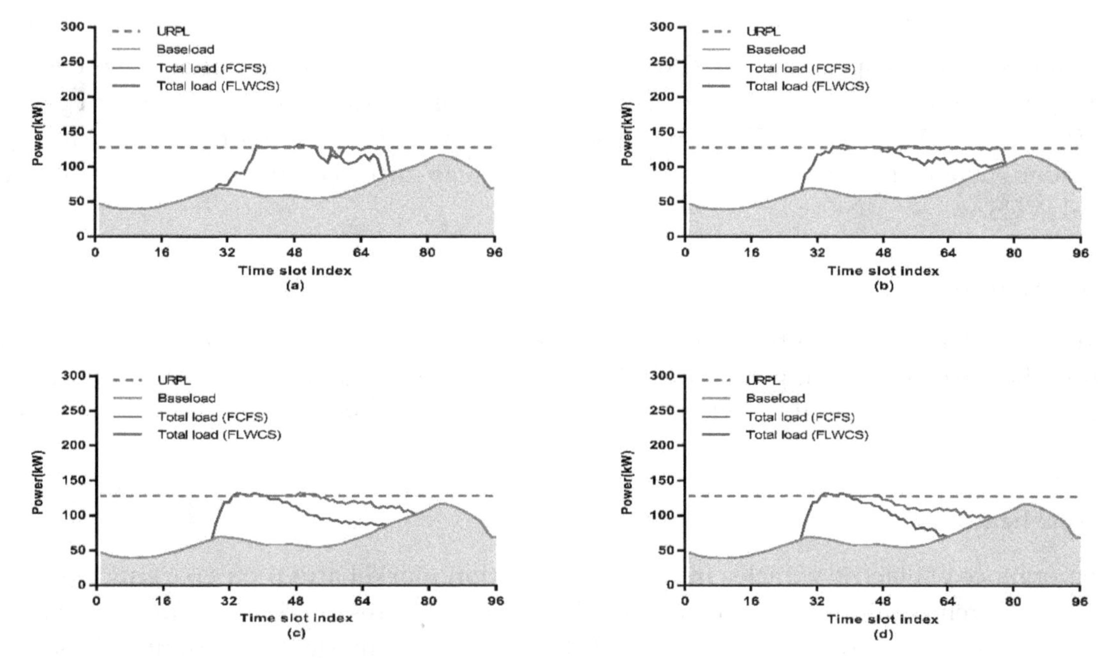

Figure 13. Total load with FCFS and proposed FLWCS scheme in four different cases. (**a**) Number of parking spots is 50; (**b**) number of parking spots is 100; (**c**) number of parking spots is 150 and (**d**) number of parking spots is 200.

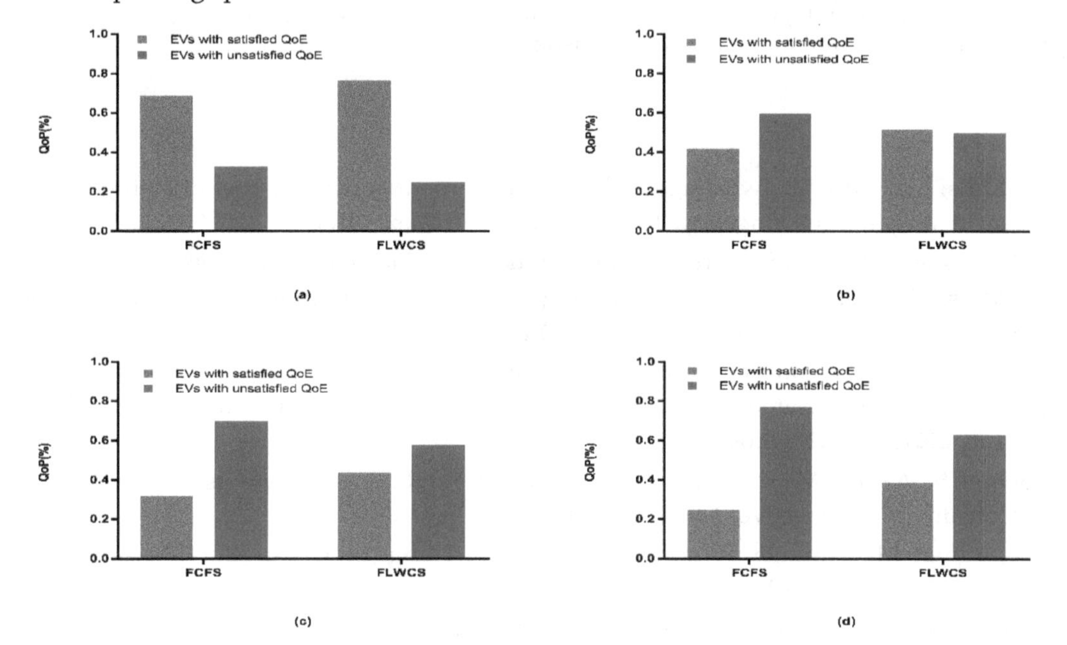

Figure 14. The quality of performance (QoP) with FCFS and proposed FLWCS in four different cases. (**a**) Number of parking spots is 50; (**b**) number of parking spots is 100; (**c**) number of parking spots is 150 and (**d**) number of parking spots is 200.

5. Conclusions

This study proposed a fuzzy logic weight-based charging scheme to distribute the charging power among the optimal number of EVs in such a way that maximizes the quality-of-performance under the operational constraints of the power grid. The developed fuzzy inference mechanism correlates different parameters such as state-of-charge, remaining parking duration and the available power into weighted values for each of the EVs. Once the weight values of all the EVs are known, their charging operations are controlled in each time slot such that the operational constraints of the power grid are respected. A java-based simulator was developed and tested for a parking lot with different parking capacities by considering four types of EVs with different penetration levels. The performance of the proposed FLWCS was analyzed against the conventional FCFS-based scheme in terms of QoP. The simulation result reveals that the proposed FLWCS has a significant performance over the conventional FCFS-based charging scheme. In more detail, increasing the parking spots to a certain number such as from 50 EVs to 200 EVs the QoP was improved by about 8% to 14%, with the proposed FLWCS.

Research limitations:

There is a tradeoff between the power grid requirements and EV owner's satisfaction and it is believed that the variable charging rate could be used to optimally analyze these requirements. In the future, the proposed scheme will be extended for more complex charging scenarios based on variable charging.

Research Implications:

The emergence of electric vehicles in the transportation market and their charging system offers a vast range of research possibilities in the field of electro-mobility research. Therefore, there is a need to study the socioeconomic implications of EV fleets by developing models for sustainable development such as social, environmental and market economics.

Author Contributions: Conceptualization, S.H.; Y.-C.K.; methodology, S.H., K.-B.L. and Y.-C.K. Software, S.H.; validation, M.A.A., Y.-C.K., K.-B.L.; writing—original draft preparation, S.H.; writing—review & editing, M.A.A., Y.-C.K. All authors have read and agreed to the published version of the manuscript.

References

1. Geng, B.; Mills, J.K.; Sun, D. Two-stage energy management control of fuel cell plug-in hybrid electric vehicles considering fuel cell longevity. *IEEE Trans. Veh. Technol.* **2011**, *61*, 498–508. [CrossRef]
2. Chan, C.C. The state of the art of electric, hybrid, and fuel cell vehicles. *Proc. IEEE* **2007**, *95*, 704–718. [CrossRef]
3. Geng, B.; Mills, J.K.; Sun, D. Two-stage charging strategy for plug-in electric vehicles at the residential transformer level. *IEEE Trans. Smart Grid* **2013**, *4*, 1442–1452. [CrossRef]
4. Azmat, M.; Kummer, S. Potential applications of unmanned ground and aerial vehicles to mitigate challenges of transport and logistics-related critical success factors in the humanitarian supply chain. *Asian J. Sustain. Soc. Responsib.* **2020**, *5*, 1–22. [CrossRef]
5. Wintersberger, S.; Azmat, M.; Kummer, S. Are We Ready to Ride Autonomous Vehicles? A Pilot Study on Austrian Consumers' Perspective. *Logistics* **2019**, *3*, 20. [CrossRef]
6. Vaidian, I.; Azmat, M.; Kummer, S. Impact of Internet of Things on Urban Mobility. In Proceedings of the Innovation Arabia, Dubai, UAE, 24–27 February 2019.
7. Lopes, J.A.P.; Soares, F.J.; Almeida, P.M.R. Integration of electric vehicles in the electric power system. *Proc. IEEE* **2010**, *99*, 168–183. [CrossRef]
8. Schroeder, A. Modeling storage and demand management in power distribution grids. *Appl. Energy* **2011**, *88*, 4700–4712. [CrossRef]
9. Muñoz-Villamizar, A.; Montoya-Torres, J.R.; Faulin, J. Impact of the use of electric vehicles in collaborative urban transport networks: A case study. *Transp. Res. Part D Transp. Environ.* **2017**, *50*, 40–54. [CrossRef]
10. Mu, Y.; Wu, J.; Jenkins, N.; Jia, H.; Wang, C. A spatial–temporal model for grid impact analysis of plug-in electric vehicles. *Appl. Energy* **2014**, *114*, 456–465. [CrossRef]
11. Moghaddam, Z.; Ahmad, I.; Habibi, D.; Phung, Q.V. Smart charging strategy for electric vehicle charging stations. *IEEE Trans. Transp. Electrif.* **2017**, *4*, 76–88. [CrossRef]
12. Shao, S.; Pipattanasomporn, M.; Rahman, S. Challenges of PHEV penetration to the residential distribution network. In Proceedings of the 2009 IEEE Power & Energy Society General Meeting, Calgary, AB, Canada, 26–30 July 2009; pp. 1–8.

13. Jiyao, A.; Guilin, W.; Wei, X. Improved results on Fuzzy H∞ filter design for TS Fuzzy systems. *Discret. Dyn. Nat. Soc.* **2010**, *2010*. [CrossRef]
14. An, J.; Li, T.; Wen, G.; Li, R. New stability conditions for uncertain TS fuzzy systems with interval time-varying delay. *Int. J. Control Autom. Syst.* **2012**, *10*, 490–497. [CrossRef]
15. Seddig, K.; Jochem, P.; Fichtner, W. Two-stage stochastic optimization for cost-minimal charging of electric vehicles at public charging stations with photovoltaics. *Appl. Energy* **2019**, *242*, 769–781. [CrossRef]
16. Perez-Diaz, A.; Gerding, E.; McGroarty, F. Coordination and payment mechanisms for electric vehicle aggregators. *Appl. Energy* **2018**, *212*, 185–195. [CrossRef]
17. Gnann, T.; Klingler, A.-L.; Kühnbach, M. The load shift potential of plug-in electric vehicles with different amounts of charging infrastructure. *J. Power Sources* **2018**, *390*, 20–29. [CrossRef]
18. Shukla, R.M.; Sengupta, S.; Patra, A.N. Smart plug-in electric vehicle charging to reduce electric load variation at a parking place. In Proceedings of the 2018 IEEE 8th Annual Computing and Communication Workshop and Conference (CCWC), Athens, Greece, 19–23 March 2018; pp. 632–638.
19. Luo, Y.; Zhu, T.; Wan, S.; Zhang, S.; Li, K. Optimal charging scheduling for large-scale EV (electric vehicle) deployment based on the interaction of the smart-grid and intelligent-transport systems. *Energy* **2016**, *97*, 359–368. [CrossRef]
20. Van-Linh, N.; Tuan, T.; Bacha, S.; Be, N. Charging strategies to minimize the peak load for an electric vehicle fleet. In Proceedings of the 40th Annual Conference of the IEEE Industrial Electronics Society, IECON, Dallas, TX, USA, 29 October–1 November 2014.
21. García-Villalobos, J.; Zamora, I.; Knezović, K.; Marinelli, M. Multi-objective optimization control of plug-in electric vehicles in low voltage distribution networks. *Appl. Energy* **2016**, *180*, 155–168. [CrossRef]
22. Fu, H.; Han, Y.; Wang, J.; Zhao, Q. A Novel Optimization of Plug-In Electric Vehicles Charging and Discharging Behaviors in Electrical Distribution Grid. *J. Electr. Comput. Eng.* **2018**, *2018*. [CrossRef]
23. Jiang, W.; Zhen, Y. A Real-Time EV Charging Scheduling for Parking Lots With PV System and Energy Store System. *IEEE Access* **2019**, *7*, 86184–86193. [CrossRef]
24. Park, J.; Sim, Y.; Lee, G.; Cho, D.-H. A fuzzy logic based electric vehicle scheduling in smart charging network. In Proceedings of the 2019 16th IEEE Annual Consumer Communications & Networking Conference (CCNC), Vegas, NV, USA, 11–14 January 2019; pp. 1–6.
25. Hariri, A.O.; Esfahani, M.M.; Mohammed, O. A cognitive price-based approach for real-time management of en-route electric vehicles. In Proceedings of the 2018 IEEE Transportation Electrification Conference and Expo (ITEC), Long Beach, CA, USA, 13–15 June 2018; pp. 922–927.
26. Karmaker, A.K.; Hossain, M.; Manoj Kumar, N.; Jagadeesan, V.; Jayakumar, A.; Ray, B. Analysis of Using Biogas Resources for Electric Vehicle Charging in Bangladesh: A Techno-Economic-Environmental Perspective. *Sustainability* **2020**, *12*, 2579. [CrossRef]
27. Chang, W.-Y. The state of charge estimating methods for battery: A review. *ISRN Appl. Math.* **2013**, *2013*. [CrossRef]
28. Dubey, A.; Santoso, S. Electric vehicle charging on residential distribution systems: Impacts and mitigations. *IEEE Access* **2015**, *3*, 1871–1893. [CrossRef]
29. Hussain, S.; Muhammad, F.; Kim, Y.-C. Communication Network Architecture based on Logical Nodes for Electric Vehicles. In Proceedings of the 2017 International Symposium on Information Technology Convergence, Shijiazhuang, China, 19–21 October 2017; pp. 321–326.
30. Nijhuis, M.; Gibescu, M.; Cobben, J. Valuation of measurement data for low voltage network expansion planning. *Electr. Power Syst. Res.* **2017**, *151*, 59–67. [CrossRef]
31. Rezaee, S.; Farjah, E.; Khorramdel, B. Probabilistic analysis of plug-in electric vehicles impact on electrical grid through homes and parking lots. *IEEE Trans. Sustain. Energy* **2013**, *4*, 1024–1033. [CrossRef]
32. Yagcitekin, B.; Uzunoglu, M. A double-layer smart charging strategy of electric vehicles taking routing and charge scheduling into account. *Appl. Energy* **2016**, *167*, 407–419. [CrossRef]
33. El-Bayeh, C.Z.; Mougharbel, I.; Saad, M.; Chandra, A.; Lefebvre, S.; Asber, D.; Lenoir, L. A novel approach for sizing electric vehicles Parking Lot located at any bus on a network. In Proceedings of the 2016 IEEE Power and Energy Society General Meeting (PESGM), Boston, MA, USA, 17–21 July 2016; pp. 1–5.
34. Jiyao, A.; Tang, J.; Yu, Y. Fuzzy multi-objective optimized with efficient energy and time-varying price for EV charging system. In Proceedings of the International Conference on Artificial Intelligence (ICAI), Las Vegas, NV, USA, 17–20 July 2017; pp. 47–53.

35. Li, Y.; Yang, Z.; Li, G.; Mu, Y.; Zhao, D.; Chen, C.; Shen, B. Optimal scheduling of isolated microgrid with an electric vehicle battery swapping station in multi-stakeholder scenarios: A bi-level programming approach via real-time pricing. *Appl. Energy* **2018**, *232*, 54–68. [CrossRef]
36. Zadeh, L.A. Fuzzy sets. *Inf. Control* **1965**, *8*, 338–353. [CrossRef]
37. An, J.; Hu, M.; Fu, L.; Zhan, J. A novel fuzzy approach for combining uncertain conflict evidences in the Dempster-Shafer theory. *IEEE Access* **2019**, *7*, 7481–7501. [CrossRef]
38. Hussain, S.; Ahmed, M.A.; Kim, Y.-C. Efficient Power Management Algorithm Based on Fuzzy Logic Inference for Electric Vehicles Parking Lot. *IEEE Access* **2019**, *7*, 65467–65485. [CrossRef]
39. Andrenacci, N.; Genovese, A.; Ragona, R. Determination of the level of service and customer crowding for electric charging stations through fuzzy models and simulation techniques. *Appl. Energy* **2017**, *208*, 97–107. [CrossRef]
40. Bai, Y.; Wang, D. Fundamentals of fuzzy logic control—Fuzzy sets, fuzzy rules and defuzzifications. In *Advanced Fuzzy Logic Technologies in Industrial Applications*; Springer: Berlin/Heidelberg, Germany, 2006; pp. 17–36.
41. Hussain, M. Fuzzy Relations. Master's Thesis, Blekinge Institute of Technology School of Engineering, Karlskrona, Sweden, 2010.
42. Vo, P.-N.; Detyniecki, M. Towards smooth monotonicity in fuzzy inference system based on gradual generalized modus ponens. In Proceedings of the 8th Conference of the European Society for Fuzzy Logic and Technology (EUSFLAT-13), Milano, Italy, 11–13 September 2013.
43. Lu, S.; Samaan, N.; Diao, R.; Elizondo, M.; Jin, C.; Mayhorn, E.; Zhang, Y.; Kirkham, H. Centralized and decentralized control for demand response. In Proceedings of the ISGT 2011, Anaheim, CA, USA, 17–19 January 2011; pp. 1–8.
44. Brodt-Giles, D. OpenEI An Open Energy Data and Information Exchange for International Audiences. Available online: https://openei.org/doe-opendata/dataset/open-energy-information-openei-org (accessed on 1 January 2020).
45. Mazidi, M.; Abbaspour, A.; Fotuhi-Firuzabad, M.; Rastegar, M. Optimal allocation of PHEV parking lots to minimize distribution system losses. In Proceedings of the 2015 IEEE Eindhoven PowerTech, Eindhoven, The Netherlands, 29 June–2 July 2015; pp. 1–6.
46. Fotuhi-Firuzabad, M.; Abbaspour, A.; Mazidi, M.; Rastegar, M. Optimal Allocation of PHEV Parking Lots to Minimize Distribution System Losses. *Int. J. Energy Power Eng.* **2015**, *9*, 843–848.
47. Tamura, S.; Kikuchi, T. V2G strategy for frequency regulation based on economic evaluation considering EV battery longevity. In Proceedings of the 2018 IEEE International Telecommunications Energy Conference (INTELEC), Turin, Italy, 7–11 October 2018; pp. 1–6.
48. Wang, Q.; Jiang, B.; Li, B.; Yan, Y. A critical review of thermal management models and solutions of lithium-ion batteries for the development of pure electric vehicles. *Renew. Sustain. Energy Rev.* **2016**, *64*, 106–128. [CrossRef]
49. Wang, Y.; Gao, Q.; Wang, G.; Lu, P.; Zhao, M.; Bao, W. A review on research status and key technologies of battery thermal management and its enhanced safety. *Int. J. Energy Res.* **2018**, *42*, 4008–4033. [CrossRef]
50. Kongjeen, Y.; Bhumkittipich, K. Impact of plug-in electric vehicles integrated into power distribution system based on voltage-dependent power flow analysis. *Energies* **2018**, *11*, 1571. [CrossRef]
51. Cingolani, P.; Alcalá-Fdez, J. jFuzzyLogic: A java library to design fuzzy logic controllers according to the standard for fuzzy control programming. *Int. J. Comput. Intell. Syst.* **2013**, *6*, 61–75. [CrossRef]
52. Nour, M.; Said, S.M.; Ali, A.; Farkas, C. Smart Charging of Electric Vehicles According to Electricity Price. In Proceedings of the Conference on Innovative Trends in Computer Engineering (ITCE), Tianjin, China, 17–19 September 2018; pp. 432–437.
53. Kim, Y.C.; Lee, D.E.; Lee, B.J.; Kim, Y.S.; Mukherjee, B. Dynamic channel reservation based on mobility in wireless ATM networks. *IEEE Commun. Mag.* **1999**, *37*, 47–51.

7

Integration of Electric Vehicles in Low-Voltage Distribution Networks Considering Voltage Management

Miguel Carrión [1,*], Rafael Zárate-Miñano [2] and Ruth Domínguez [1]

1 Department of Electrical Engineering, University of Castilla—La Mancha, 45071 Toledo, Spain; Ruth.Dominguez@uclm.es
2 Department of Electrical Engineering, University of Castilla—La Mancha, 13400 Almadén, Spain; Rafael.Zarate@uclm.es
* Correspondence: Miguel.Carrion@uclm.es.

Abstract: The expected growth of the number of electric vehicles can be challenging for planning and operating power systems. In this sense, distribution networks are considered the Achilles' heel of the process of adapting current power systems for a high presence of electric vehicles. This paper aims at deciding the maximum number of three-phase high-power charging points that can be installed in a low-voltage residential distribution grid. In order to increase the number of installed charging points, a mixed-integer formulation is proposed to model the provision of decentralized voltage support by electric vehicle chargers. This formulation is afterwards integrated into a modified AC optimal power flow formulation to characterize the steady-state operation of the distribution network during a given planning horizon. The performance of the proposed formulations have been tested in a case study based on the distribution network of La Graciosa island in Spain.

Keywords: charging points; electric vehicles; operation; planning; reactive power provision; voltage support

---

1. Introduction

A massive presence of electric vehicles may endanger the operation of distribution systems [1]. If a large number of electric vehicles is charged in a non-coordinated manner, undervoltage phenomena can happen, jeopardizing the stability of the distribution networks, [2]. In order to avoid this situation, different actions can be implemented. The easiest, but most aggressive manner to protect the distribution network operation, is to curtail the active power demand consumed by electricity end users when the normal operation of distribution networks is at danger. To prevent from this drastic measure, a wide number of smart charging procedures able to reduce the simultaneous charge of electric vehicles have been proposed during recent years [3,4]. Usually, these types of procedures need the presence of a central operator in charge of deciding at which time each electric vehicle may be charged or not. These procedures reduce the control of electric vehicle users over the initiation, duration and completion of the charging processes of their vehicles.

An alternative procedure to avoid undervoltage episodes consists in applying reactive power control mechanisms in the charging points of electric vehicles. Observe that the high resistance–reactance ratio in low-voltage distribution networks makes the reactive power support an effective tool for voltage management [5]. Traditionally, the voltage control is performed locally by injecting reactive power in those buses with voltage deviations. In this sense, the power rectifiers used in the charging points of electric vehicles are suitable to be upgraded to provide such a service [6,7]. Then, electric vehicle chargers may monitor locally the voltage at the charging point and provide the appropriate reactive power value based on a pre-established control law. Observe that this voltage control mechanism is totally decentralized and it is not required to change the charging preferences of electric vehicle users.

The active participation of electric vehicles in the operation of power systems has been extensively studied during recent years. The authors of [8] propose a number of electric vehicle charging algorithms considering explicitly the possible negative impacts on the transmission and distribution grids. A coordinated dispatch strategy for electric vehicles and renewable units at distribution level is provided in [9]. Reference [10] develops a probabilistic approach to assess the impact of electric vehicles on distribution grids considering a detailed modeling of the batteries. In reference [11], the provision of ancillary services by electric vehicles in a realistic case study is analyzed in detail. References [12,13] study the contribution of electric vehicles to the primary frequency response. Reference [14] presents a Markov decision problem that seeks to minimize the charging cost of a single electric vehicle that participates in the secondary frequency regulation. References [15,16] propose different charging procedures to minimize power losses.

The optimal planning of electric vehicle charging infrastructures has been a major concern of researchers [17]. The maximization of the penetration of electric vehicles has been studied in [18,19]. Considering that the power demanded by the charging points of electrical vehicles is quite high compared to typical consumption profiles of households, how to decide the placement of charging stations in the distribution grids is a relevant problem to solve by distribution network operators. Different approaches have been proposed to decide the optimal location of charging stations according to the needs of electric vehicle users and the particular design of distribution networks. For instance, reference [20] assumes the role of a distribution network planner aiming at expanding optimally the distribution network considering the presence of charging stations. The objective function of this problem seeks the minimization of investment and operation costs, and the maximization of the utilization of charging stations and the reliability of the distribution network. Reference [21] proposes a multi-year expansion procedure for distribution networks considering uncontrolled and smart charging. The authors of [22] propose a three-layered decision approach for deciding the location, capacity and operation policy of charging stations. Reference [23] takes the perspective of a private investor of charging stations and proposes a bi-level optimization model to maximize the expected profit obtained by the investor ensuring a certain degree of satisfaction of electric vehicle users. The optimal location of charging stations is formulated as a mixed-integer linear programming problem in [24]. In this work, the electricity demanded by electric vehicles is characterized using a novel capacitated-flow refueling location model. Finally, reference [25] proposes a capacity planning model of charging stations enforcing explicitly the reliable operation of the distribution network and the satisfaction of electric vehicle users in terms of accessibility to the charging services.

The provision of voltage support in distribution networks by electric vehicles has been also analyzed in the technical literature. Reference [26] studies different voltage support functions for electric vehicles in distribution grids with high electric vehicle penetration. The authors of [27] analyze the influence of reactive power support of electric vehicle chargers in low-voltage residential distribution grids. Reference [28] proposes a combined control scheme to improve the voltage profile in residential distribution networks considering battery storages and electric vehicles. A bidirectional charging control strategy of electric vehicles to simultaneously regulate the voltage and frequency has been developed in [29]. Reference [30] analyzes the reactive power support by electric vehicle charging stations. The authors of [31] study the capability and cost of providing reactive power service by electric vehicles. Different procedures are developed in [32] for managing the active and reactive power dispatch of a set of electric vehicles. Finally, reference [33] carries out a literature review of mathematical procedures for designing reactive power compensation in distribution networks.

This paper aims at deciding the maximum number of three-phase high-power charging points that can be installed in a low-voltage residential distribution grid. As stated in the European Roadmap on the Electrification of Road Transport [34], optimization tools are required to optimize the location of charging points and the development of the electricity network. It is assumed that the distribution network operator has to decide the maximum number of charging points that can be installed from a list of charging point requests ensuring a reliable operation of the network. We assume that charging points are private and they are installed in residential buildings. In order to increase the number of installed charging points, it is considered that charging points are able to provide reactive power support to maintain appropriate voltage levels. Unlike other mentioned works focused on the operation of distribution networks [26–32], the provision of reactive power support in this paper is considered from a planning perspective. Therefore, the number of charging points in the distribution network is not a known parameter, but a decision variable of the distribution network planner. Observe also that the analyzed problem is different from those focused on non-residential, large-scale charging stations [20–25]. As an example, in the case of domestic charging points, the location of charging points and the number of electric vehicles associated with each candidate charging point are know in advance. Finally, an optimal power flow problem including reactive power support constraints has been formulated in order to analyze the steady state operation of the distribution network considering the presence of new charging points.

2. Optimization Models

In this section, two optimization models are presented considering that electric vehicle chargers can provide voltage support by means of a reactive power-voltage magnitude control law. First, the mathematical formulation of this control law is provided in Section 2.1. The first optimization problem is described in Section 2.2 and intends to maximize the penetration level of electric vehicles in a given distribution network. The penetration level of electric vehicles can be defined as the maximum number of electric vehicle chargers that can be installed ensuring a reliable and secure operation of the distribution network. The second optimization model is provided in Section 2.3 and it is a modified version of the optimal power flow problem [35] that aims at simulating the operation of the distribution network considering a given number of electric vehicle chargers.

From a mathematical point of view, the formulation of the problem associated with the determination of the maximum number of charging points that can be installed in a distribution grid is challenging. The decision associated with the installation of a charging point is binary, to install or not to install, whereas the expressions modeling the power flow in the distribution network include the product of variables and trigonometric functions. As a result, the determination of the maximum number of charging points in a distribution network is a mixed-integer non-linear problem. Additionally, the modeling of voltage control by means of a reactive power-voltage droop curve is mathematically non-convex and requires the usage of additional binary variables.

The initial hypothesis and limitations of the proposed approach are the following:

1. A low-voltage residential distribution network at usage is considered.
2. The solution of the problem described in this work does not intend to overcome previous operation limitations of the analyzed distribution network.
3. The upgrade of distribution assets is not considered. Therefore, new investments in transformers, cables, capacitors and storages are not modeled in this work.
4. Since the proposed approach is focused on high-power rate chargers, only three-phase chargers are considered in this study.
5. Only domestic chargers are considered. The installation of large charging stations is not taken into account in this work.
6. The power factor value of electric vehicle chargers is a constant value.
7. The proposed approach does not consider the coordinated charge of electric vehicles and vehicle-to-grid capability.

Since an existing residential distribution network that operates properly under normal conditions is analyzed, and three-phase chargers do not cause additional unbalances between phases, a single-phase equivalent analysis has been adopted in this paper. Observe that single-phase equivalents are typically used in planning models. However, if desired, the proposed procedure can be straightforwardly modified to incorporate a three-phase modeling of the distribution network.

Considering that the focus of this paper is to consider reactive power support by electric vehicle chargers, investments in capacity banks or storages are not included in the proposed approach. However, observe that the inclusion of capacity banks, lines upgrading, etc., can be easily integrated.

Finally, the notation used throughout the rest of this section is included in the Appendix A for quick reference.

2.1. Reactive Power—Voltage Magnitude Control Law Formulation

Since distribution networks are traditionally associated with high resistance–reactance ratios, the effect of resistance is no longer negligible, and the assumptions taken for high-voltage systems are no longer valid. An effective procedure to mitigate voltage decrements is to reduce the active power consumption in the grid. This is a very effective tool, but implies the modification of the consumption patterns of electricity end-users [36]. For this reason, local reactive power injection mechanisms are typically used, since they do not need the installation of additional devices and exploit the capability of the inverters used in electric vehicle charging points.

Figure 1 represents graphically a typical reactive power-voltage magnitude control law for a given electric vehicle charger k located in bus n on day d and period t, [37]. The reactive power demanded by the charging point is denoted by q^{EV}_{kdt}, whereas the voltage magnitude is v_{ndt}. If the nodal voltage is low, $V^{\min}_n \leq v_{ndt} \leq V^{(1)}_n$, the reactive power demanded by the charger point is a negative value, $-Q^{\text{EV,max}}_k$, corresponding with the maximum reactive power that can be injected by the charger into the grid. As the voltage magnitude increases, $V^{(1)}_n \leq v_{ndt} \leq V^{(2)}_n$, the injection of reactive power decreases linearly. Finally, if the voltage magnitude is greater than $V^{(2)}_n$, $V^{(2)}_n \leq v_{ndt} \leq V^{\max}_n$, the reactive power injected by the charging point is equal to zero.

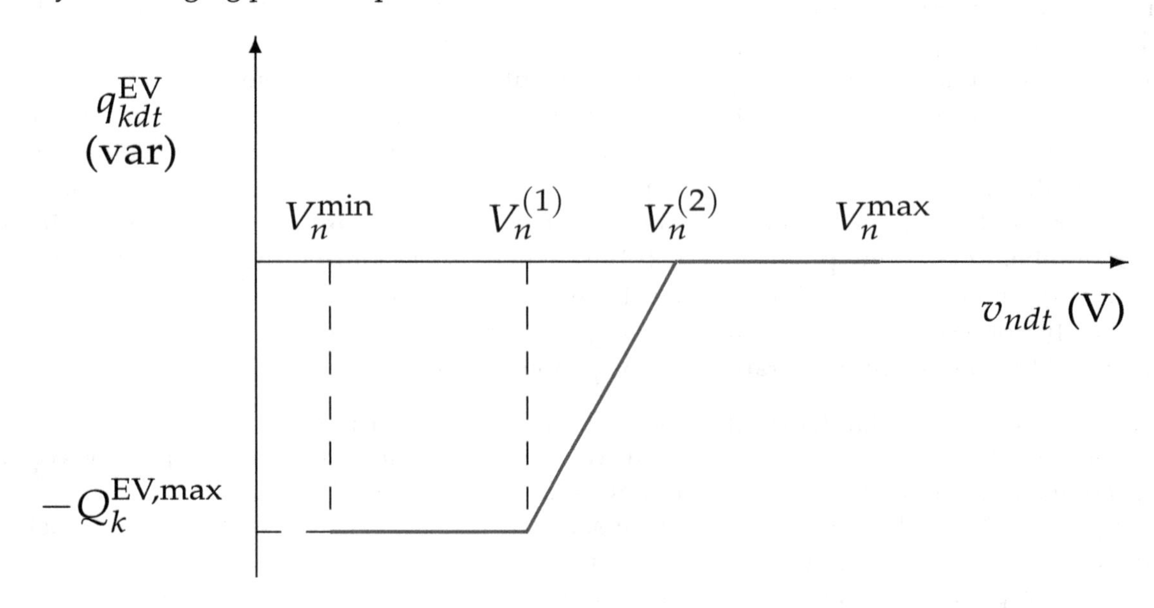

Figure 1. Reactive power-voltage magnitude droop curve.

Figure 1 can be mathematically expressed as follows:

$$q^{\text{EV}}_{kdt} = \left\{ \begin{array}{c} -Q^{\text{EV,max}}_k, \;\; V^{\min}_n \leq v_{ndt} \leq V^{(1)}_n \\ \alpha^{\text{EV}}_{kn} + \beta^{\text{EV}}_{kn} v_{ndt}, \; V^{(1)}_n \leq v_{ndt} \leq V^{(2)}_n \\ 0, \qquad V^{(2)}_n \leq v_{ndt} \leq V^{\max}_n \end{array} \right\}, \quad \forall k \in K_{ndt}, \forall n, \forall d, \forall t \tag{1}$$

where:

$$P_k^{\text{EV,max}} = S_k^{\text{EV,max}} \cos\phi_k, \quad \forall k \tag{2}$$

$$Q_k^{\text{EV,max}} = S_k^{\text{EV,max}} \sin\phi_k = P_k^{\text{EV,max}} \tan\phi_k, \quad \forall k \tag{3}$$

$$\alpha_{kn}^{\text{EV}} = Q_k^{\text{EV,max}} \left(\frac{-V_n^{(2)}}{V_n^{(2)} - V_n^{(1)}} \right), \quad \forall k, \forall n \tag{4}$$

$$\beta_{kn}^{\text{EV}} = \frac{Q_k^{\text{EV,max}}}{V_n^{(2)} - V_n^{(1)}}, \quad \forall k, \forall n \tag{5}$$

Symbols $S_k^{\text{EV,max}}$ and $\cos\phi_k$ denote the apparent power and the power factor of charger k, respectively. Equations (2) and (3) compute the maximum active and reactive powers associated with a charger for a given power factor, $\cos\phi_k$. It is assumed that the active power consumed by the charger is equal to the maximum value, but the reactive power is dependent on the nodal voltage magnitude, as defined in (1). Parameters α_{kn}^{EV} and β_{kn}^{EV} in Equations (4) and (5) are used to compute the reactive power injection q_{kdt}^{EV} associated with the voltage magnitude level v_{ndt} when the voltage takes values between $V_n^{(1)}$ and $V_n^{(2)}$, as represented in Figure 1. Observe that the reactive power in this interval is a linear expression of the voltage magnitude with intercept and slope equal to α_{kn}^{EV} and β_{kn}^{EV}, respectively.

Traditionally, electric vehicle chargers operate with power factor equal to 1. Therefore, if $\cos\phi_k = 1$, then $P_k^{\text{EV,max}} = S_k^{\text{EV,max}}$ and $Q_k^{\text{EV,max}} = 0$. However, hereinafter it is assumed a leading power factor, $\{\cos\phi_k < 1, \phi_k < 0\}$, and that electric vehicle chargers are able to inject reactive power as shown in Figure 1.

Observe that expression (1) cannot be incorporated directly into the formulation of an optimization problem. The reason of this is that the mathematical expression of q_{kdt}^{EV} is a piecewise function of the optimization variable representing nodal voltage magnitudes, v_{ndt}. The mathematical formulation of piecewise functions using mixed-integer linear expressions has been typically used in different power system problems, [38,39]. Therefore, constraints (6)–(14) are proposed to equivalently formulate (1). In this formulation, binary variables $y_{ndt}^{(1)}$ and $y_{ndt}^{(2)}$ are used to identify the block associated with the value of v_{ndt} in the piecewise function (1). In this manner, the voltage is in the first block if $\{y_{ndt}^{(1)}, y_{ndt}^{(2)}\} = \{1,0\}$; if $\{y_{ndt}^{(1)}, y_{ndt}^{(2)}\} = \{0,1\}$, the voltage is in the second block; finally, the voltage is in the third block if $\{y_{ndt}^{(1)}, y_{ndt}^{(2)}\} = \{0,0\}$. For the sake of clarity, Table 1 provides the rationale of variables $\{y_{ndt}^{(1)}, y_{ndt}^{(2)}\}$. In formulation (6)–(14), variable $v_{ndt}^{(j)}$ is equal to v_{ndt} if v_{ndt} is within block j, being equal to 0 otherwise. This equality is enforced through constraints (6)–(9). In these constraints, the values of binary variables $\{y_{ndt}^{(1)}, y_{ndt}^{(2)}\}$ are assigned depending on the value of v_{ndt}. For instance, if v_{ndt} is in the first block, it has to be satisfied that $V_n^{\min} \leq v_{ndt} \leq V_n^{(1)}$. To obtain this result, the only feasible solution is $y_{ndt}^{(1)} = 1$, $y_{ndt}^{(2)} = 0$, and $v_{ndt} = v_{ndt}^{(1)}$. A similar reasoning can be done for blocks 2 and 3. The value of q_{kdt}^{EV} is assigned through constraints (10). If $y_{ndt}^{(1)} = 1$ and $y_{ndt}^{(2)} = 0$, then $q_{kdt}^{\text{EV}} = -Q_{kn}^{\text{EV,max}}$, whereas $q_{kdt}^{\text{EV}} = 0$ if $y_{ndt}^{(1)} = 0$ and $y_{ndt}^{(2)} = 0$. Note that the combination $\{y_{ndt}^{(1)}, y_{ndt}^{(2)}\} = \{1,1\}$ is not allowed by constraints (11). Finally, if $y_{ndt}^{(1)} = 0$ and $y_{ndt}^{(2)} = 1$, v_{ndt} is in the second block of the droop curve, $v_{ndt} = v_{ndt}^{(2)}$, and q_{kdt}^{EV} should be expressed as $\alpha_{kn}^{\text{EV}} y_{ndt}^{(2)} + \beta_{kn}^{\text{EV}} v_{ndt}^{(2)} y_{ndt}^{(2)}$. However, this expression contains the non-linear product of variables $y_{ndt}^{(2)}$ and $v_{ndt}^{(2)}$. In order to preserve the linearity of the formulation, the auxiliary variable v_{ndt}^{aux} is used to equivalently represent the product of $v_{ndt}^{(2)}$ and $y_{ndt}^{(2)}$. The linearization of the product of a continuous and a binary variable has been previously done, for instance, in [40]. The assignment $v_{ndt}^{\text{aux}} = v_{ndt}^{(2)} y_{ndt}^{(2)}$ is linearly expressed through constraints (12) and (13). In these constraints, M is a big enough parameter ($M \geq V_n^{\max}$). If $y_{ndt}^{(1)} = 0$ and $y_{ndt}^{(2)} = 1$, then constraints (12) state that $v_{ndt}^{\text{aux}} = v_{ndt}^{(2)}$ and $q_{kdt}^{\text{EV}} = \alpha_{kn}^{\text{EV}} + \beta_{kn}^{\text{EV}} v_{ndt}^{(2)}$. On the other hand, if $y_{ndt}^{(2)} = 0$, then v_{ndt} is either in the first or third block of the piecewise function (1) and

$v^{aux}_{ndt} = 0$ by constraints (13). The performance of constraints (12) and (13) is described in Table 2. Finally, the binary nature of variables is stated in (14).

$$v_{ndt} = \sum_{j=1}^{3} v^{(j)}_{ndt}, \quad \forall n, \forall d, \forall t \tag{6}$$

$$V^{\min}_n y^{(1)}_{ndt} \leq v^{(1)}_{ndt} \leq V^{(1)}_n y^{(1)}_{ndt}, \quad \forall n, \forall d, \forall t \tag{7}$$

$$V^{(1)}_n y^{(2)}_{ndt} \leq v^{(2)}_{ndt} \leq V^{(2)}_n y^{(2)}_{ndt}, \quad \forall n, \forall d, \forall t \tag{8}$$

$$V^{(2)}_n (1 - y^{(1)}_{ndt} - y^{(2)}_{ndt}) \leq v^{(3)}_{ndt} \leq V^{\max}_n (1 - y^{(1)}_{ndt} - y^{(2)}_{ndt}), \quad \forall n, \forall d, \forall t \tag{9}$$

$$q^{EV}_{kdt} = -Q^{EV,\max}_k y^{(1)}_{ndt} + \alpha^{EV}_{kn} y^{(2)}_{ndt} + \beta^{EV}_{kn} v^{aux}_{ndt}, \quad \forall k \in K_n, \forall n, \forall d, \forall t \tag{10}$$

$$y^{(1)}_{ndt} + y^{(2)}_{ndt} \leq 1, \quad \forall n, \forall d, \forall t \tag{11}$$

$$-M(1 - y^{(2)}_{ndt}) \leq v^{aux}_{ndt} - v^{(2)}_{ndt} \leq M(1 - y^{(2)}_{ndt}), \quad \forall n, \forall d, \forall t \tag{12}$$

$$-M y^{(2)}_{ndt} \leq v^{aux}_{ndt} \leq M y^{(2)}_{ndt}, \quad \forall n, \forall d, \forall t \tag{13}$$

$$y^{(1)}_{ndt}, y^{(2)}_{ndt} \in \{0, 1\}, \quad \forall n, \forall d, \forall t \tag{14}$$

Table 1. Rationale of variables $\{y^{(1)}_{ndt}, y^{(2)}_{ndt}\}$.

# Block	$y^{(1)}_{ndt}$	$y^{(2)}_{ndt}$	v_{ndt}	q^{EV}_{kdt}
1	1	0	$V^{\min}_n \leq v_{ndt} \leq V^{(1)}_n$	$-Q^{EV,\max}_k$
2	0	1	$V^{(1)}_n \leq v_{ndt} \leq V^{(2)}_n$	$\alpha^{EV}_{kn} + \beta^{EV}_{kn} v_{ndt}$
3	0	0	$V^{(2)}_n \leq v_{ndt} \leq V^{\max}_n$	0

Table 2. Linearization of product $y^{(2)}_{ndt} v^{(2)}_{ndt}$.

$y^{(2)}_{ndt}$	**Constraint** (12) $-M(1-y^{(2)}_{ndt}) \leq v^{aux}_{ndt} - v^{(2)}_{ndt} \leq M(1-y^{(2)}_{ndt})$	**Constraint** (13) $-M(1-y^{(2)}_{ndt}) \leq v^{aux}_{ndt} - v^{(2)}_{ndt} \leq M(1-y^{(2)}_{ndt})$	$v^{aux}_{ndt} \left(= y^{(2)}_{ndt} v^{(2)}_{ndt}\right)$
0	$-M \leq v^{aux}_{ndt} - v^{(2)}_{ndt} \leq M$	$0 \leq v^{aux}_{ndt} \leq 0$	0
1	$0 \leq v^{aux}_{ndt} - v^{(2)}_{ndt} \leq 0$	$-M \leq v^{aux}_{ndt} \leq M$	$v^{(2)}_{ndt}$

2.2. Maximization of the Penetration Level of Electric Vehicles

The maximization of the penetration level of electric vehicles is formulated in this subsection. In this problem, a low-voltage distribution network operating under normal conditions is considered.

It is also assumed that the distribution network planner has a request list of high-power three-phase electric vehicles chargers that are desired to be installed in the distribution network.

Each charging point request is indexed by k and has information about the location of the charger in the distribution network and the nominal charge rate, $S^{EV,\max}_k$. The objective of the distribution network planner is to accept the maximum number of charger point requests ensuring the adequate operation of the network. For doing that, the worst case is analyzed, which means that the distribution network has to be capable of operating in normal conditions assuming a simultaneity factor equal to 1. In other words, it is assumed that the distribution network has to be capable of operating under normal operation limits when households consume the contracted peak power and all electric vehicles are simultaneously charging. Note that this strong requirement can be relaxed if desired to obtain a larger penetration level of electric vehicles. Finally, note that this problem is solved by a single time period, i.e., $D = \{1\}$, $T = \{1\}$. The formulation of this problem is the following:

$$\text{Maximize}_{\Theta_1} \sum_{k \in K} x_k \tag{15}$$

Subject to:

- Voltage magnitude limits

$$V_n^{\min} \leq v_{ndt} \leq V_n^{\max}, \quad \forall n \in N, \forall d \in D, \forall t \in T \tag{16}$$

- Distribution transformer capacity constraints

$$\sqrt{(p_{sdt}^{\mathrm{S}})^2 + (q_{sdt}^{\mathrm{S}})^2} \leq S^{\mathrm{S,max}}, \quad \forall s \in S, \forall d \in D, \forall t \in T \tag{17}$$

- Active and reactive power flow constraints

$$p_{nmdt}^{\mathrm{L}} = v_{ndt}^2 G_{nm} - v_{ndt} v_{mdt} (G_{nm} \cos(\theta_{ndt} - \theta_{mdt}) + B_{nm} \sin(\theta_{ndt} - \theta_{mdt})), \quad \forall \{n,m\} \in L, \forall d \in D, \forall t \in T \tag{18}$$

$$q_{nmdt}^{\mathrm{L}} = -v_{ndt}^2 B_{nm} - v_{ndt} v_{mdt} (G_{nm} \sin(\theta_{ndt} - \theta_{mdt}) - B_{nm} \cos(\theta_{ndt} - \theta_{mdt})), \quad \forall \{n,m\} \in L, \forall d \in D, \forall t \in T \tag{19}$$

$$\sqrt{(p_{nmdt}^{\mathrm{L}})^2 + (q_{nmdt}^{\mathrm{L}})^2} \leq S_{nm}^{\mathrm{L,max}}, \quad \forall \{n,m\} \in L, \forall d \in D, \forall t \in T \tag{20}$$

- Active and reactive power balance constraints

$$\sum_{s \in S_n} p_{sdt}^{\mathrm{S}} + \sum_{m \in N_n} p_{mndt}^{\mathrm{L}} = P_{ndt}^{\mathrm{D}} + \sum_{k \in K_{ndt}} P_{kdt}^{\mathrm{EV}} x_k, \quad \forall n \in N, \forall d \in D, \forall t \in T \tag{21}$$

$$\sum_{s \in S_n} q_{sdt}^{\mathrm{S}} + \sum_{m \in N_n} q_{mndt}^{\mathrm{L}} = Q_{ndt}^{\mathrm{D}} + \sum_{k \in K_{ndt}} q_{kdt}^{\mathrm{EV,S}}, \quad \forall n \in N, \forall d \in D, \forall t \in T \tag{22}$$

- Reactive power—voltage magnitude control constraints

$$-(1 - x_k) Q_k^{\mathrm{EV,max}} \leq q_{kdt}^{\mathrm{EV,S}} - q_{kdt}^{\mathrm{EV}} \leq (1 - x_k) Q_k^{\mathrm{EV,max}}, \quad \forall k \in K, \forall d \in D, \forall t \in T \tag{23}$$

$$-x_k Q_k^{\mathrm{EV,max}} \leq q_{kdt}^{\mathrm{EV,S}} \leq x_k Q_k^{\mathrm{EV,max}}, \quad \forall k \in K, \forall d \in D, \forall t \in T \tag{24}$$

$$\text{Constraints (6)–(14)} \tag{25}$$

where Θ_1 is the set of optimization variables in problem (15)–(25).

The objective function (15) represents the maximization of the number of accepted requests of electric vehicle chargers. The acceptance of a request is characterized by binary variable x_k that is equal to 1 if charger k is accepted, being equal to zero otherwise. Voltage magnitudes are bounded by constraints (16). Constraints (17) limit the power output of each distribution transformer s. Despite the fact that distribution networks are usually operated as radial systems, for the sake of generality, we have considered in this formulation that several distribution substations may feed the analyzed residential distribution system. The active and reactive power flows through distribution lines are expressed by constraints (18) and (19), respectively. As usual, distribution lines are represented by using the series impedance model. The capacity of the lines is bounded by constraints (20). Equations (21) and (22) establish the active and reactive power balances in each bus for each time period. These equations enforce the active and reactive power balances in each bus of the distribution network considering the nodal consumption and distribution line flows. Finally, constraints (23)–(25) formulate the reactive power-voltage magnitude control law for those selected chargers. Constraints (23) and (24) are used to state that the reactive power contribution of charger k, $q_{kdt}^{\mathrm{EV,S}}$, must be equal to q_{kdt}^{EV} if charger k is accepted ($x_k = 1$), being equal to zero if the charger is not selected ($x_k = 0$). Note that the value of q_{kdt}^{EV} depends on the nodal voltage magnitude v_{ndt}, as formulated in (6)–(14).

Problem (15)–(25), hereinafter defined as (P1), is a nonlinear mixed-integer programming model that can be solved using commercial software.

2.3. Operation of the Distribution Network Considering Voltage Management of Electric Vehicles

This subsection solves an optimal power flow (OPF) problem to simulate the steady-state operation of a distribution network considering that electric vehicles can provide voltage support. This problem can be solved sequentially to simulate the operation during a planning horizon. This problem is denoted as (P2) and is formulated as follows:

$$\begin{cases} \text{Minimize}_{\Theta_{2,dt}} \sum_{n \in N} (v^{+}_{ndt} + v^{-}_{ndt}) \end{cases} \tag{26}$$

- Voltage limits

$$V_n^{\min} - v^{-}_{ndt} \leq v_{ndt} \leq V_n^{\max} + v^{+}_{ndt}, \quad \forall n \in N \tag{27}$$

$$v^{-}_{ndt}, v^{+}_{ndt} \geq 0, \quad \forall n \in N \tag{28}$$

- Assignment of selected charging points

$$x_k = x_k^*, \quad \forall k \in K \tag{29}$$

$$\left. \text{Constraints (17)–(25),} \right\}, \forall d \in D, \forall t \in T \tag{30}$$

where $\Theta_{2,dt}$ comprises all optimization variables in problem (P2) for day d and period t.

The objective function of (P2) is formulated in (26) and consists in minimizing voltage deviations over upper and lower limits, $V_n^{\max}$ and $V_n^{\min}$. Since these voltage deviations are penalized in the objective function (26), they will be greater than zero only in the case in which the demand cannot be procured satisfying all the technical constraints of the distribution network. The voltage magnitude limits, including positive and negative deviations, are formulated in (27) and (28). Finally, the charging points considered in the formulation are specified through constraints (29), being x_k^* the optimal values of variables x_k obtained from solving problem (P1).

2.4. Tool Usage

The usage of the optimization models presented in this section is described below:

- Distribution network operators may determine the maximum number of charging points for electric vehicles by solving problem (P1). The most important input data needed to solve this problem are: (i) the technical data of the distribution network, (ii) the peak power contracted by each consumer, (iii) the technical characteristics of the candidate charging points, and (iv) the parameters describing the reactive power droop curve. The outputs of this problem are the set of candidate charging points that are accepted to be installed in the considered distribution network.
- The steady-state behavior of the distribution network considering reactive power injections and the set of installed charging points is characterized by solving problem (P2). This problem can be solved for a specified number of days and time periods. The input data of this problem are: (i) the technical data of the distribution network, (ii) the location and technical characteristics of each accepted charging point, (iii) the actual energy consumption per bus and period (considering electric vehicle demand), and (iv) the parameters describing the reactive power droop curve.

The main outputs of this problem are (i) voltages per bus and period and (ii) active and reactive power flows per line and period.

Figure 2 represents graphically the tool usage described above:

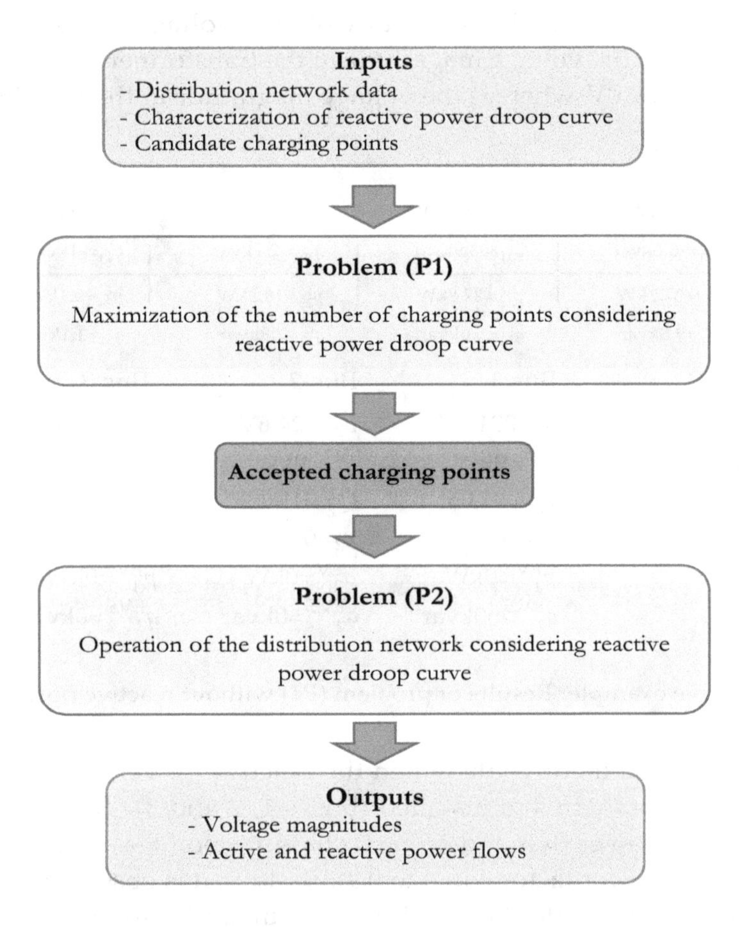

Figure 2. Tool usage.

3. Illustrative Example

In this section we solve an illustrative example to analyze the impact of considering the reactive power-voltage magnitude droop curve in the selection process of charging points. We assumed a radial distribution network where a distribution transformer feeds the demand of four buses. As stated in Section 2, a single-phase analysis was performed in this study. Each demand bus consumed 9 kW and 1 kva per phase, and had associated a charging point request of $S_k^{\text{EV,max}}$ =5 kVA per phase. The magnitudes of the impedance and resistance of each line branch were equal to 128.8 and 32.2 Ω, respectively. The nominal single phase voltage magnitude was 230 V. The minimum and maximum values of voltage magnitudes were 0.95 and 1.05 times the nominal value, i.e., 218.5 and 241.5 V, respectively. Parameters $V_n^{(1)}$ and $V_n^{(2)}$ describing the reactive power-voltage magnitude curve were equal to 224.25 and 230 V, respectively. The power factor of the charger points in the case considering the reactive power-voltage magnitude droop curve was equal to 0.9 (leading). Therefore, $P_k^{\text{EV,max}} = S_k^{\text{EV,max}} 0.9 = 4.5$ kW, whereas $Q_k^{\text{EV,max}} = P_k^{\text{EV,max}} \sqrt{\frac{1}{(0.9)^2} - 1} = 2.18$ kvar. The capacity of the lines, $S_{nm}^{\text{L,max}}$, was 100 kVA. Please note, that the values of the parameters selected for this case study are not meant to be representative of a realistic network and they are only used for illustrative purposes.

The main results obtained are depicted in Figures 3 and 4, respectively. These figures provide per phase values associated with voltage magnitudes, consumption of active and reactive power and number of installed charging points per bus, as well as active and reactive power flows per line.

For the sake of clarity, indices d and t have been omitted from the mathematical symbols included in both figures. Figure 3 provides the results obtained from solving problem (P1) without considering the reactive power droop curve. In other words, problem (15)–(22) was solved enforcing $q_{kdt}^{EV,S} = 0$. From Figure 3 we can conclude that the power transfer capability of the analyzed network was low and no charging points could be installed without violating voltage limits ($x_k = 0, \forall k \in \{1, \cdots, 4\}$). In this sense, we observed that the voltage magnitude in the transformer network bus was greater than the nominal value, 240.7 > 230.0V, whereas the voltage magnitude at the final bus, 4, was equal to the lower voltage limit, 218.5V.

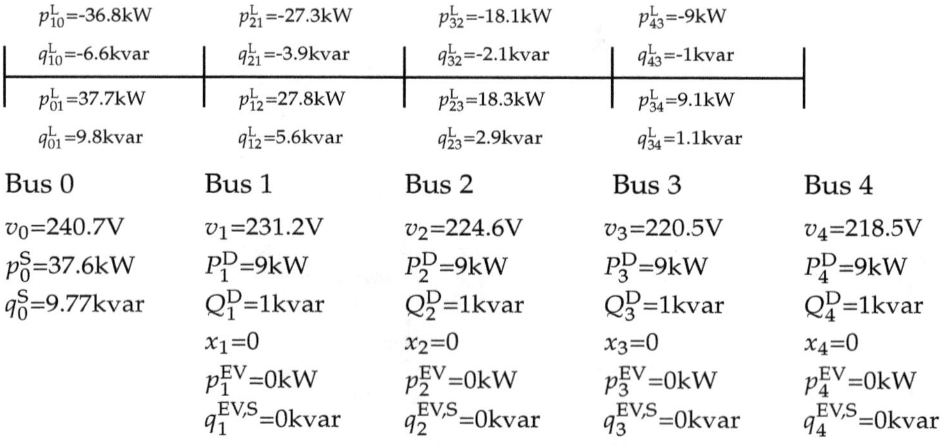

p_{10}^L=-36.8kW	p_{21}^L=-27.3kW	p_{32}^L=-18.1kW	p_{43}^L=-9kW	
q_{10}^L=-6.6kvar	q_{21}^L=-3.9kvar	q_{32}^L=-2.1kvar	q_{43}^L=-1kvar	
p_{01}^L=37.7kW	p_{12}^L=27.8kW	p_{23}^L=18.3kW	p_{34}^L=9.1kW	
q_{01}^L=9.8kvar	q_{12}^L=5.6kvar	q_{23}^L=2.9kvar	q_{34}^L=1.1kvar	
Bus 0	Bus 1	Bus 2	Bus 3	Bus 4
v_0=240.7V	v_1=231.2V	v_2=224.6V	v_3=220.5V	v_4=218.5V
p_0^S=37.6kW	P_1^D=9kW	P_2^D=9kW	P_3^D=9kW	P_4^D=9kW
q_0^S=9.77kvar	Q_1^D=1kvar	Q_2^D=1kvar	Q_3^D=1kvar	Q_4^D=1kvar
	x_1=0	x_2=0	x_3=0	x_4=0
	p_1^{EV}=0kW	p_2^{EV}=0kW	p_3^{EV}=0kW	p_4^{EV}=0kW
	$q_1^{EV,S}$=0kvar	$q_2^{EV,S}$=0kvar	$q_3^{EV,S}$=0kvar	$q_4^{EV,S}$=0kvar

Figure 3. Illustrative example: Results of problem (P1) without reactive power droop curve.

Figure 4 provides the obtained results when the reactive power droop curve was considered. In this case, three charging points were installed (buses 1, 2 and 3). Since the power factor of the charger was 0.9, the active power demand of each charging point was reduced from 5kW (power factor equal to 1) to 4.5 kW (power factor equal to 0.9). Besides, it is observed that reactive power was injected from the charging points into the grid if voltage magnitudes are lower than $V^{(2)} = 230.0$ V. This was the case of buses 2 and 3, with voltage magnitudes equal to 224.2 and 220.5 V, respectively.

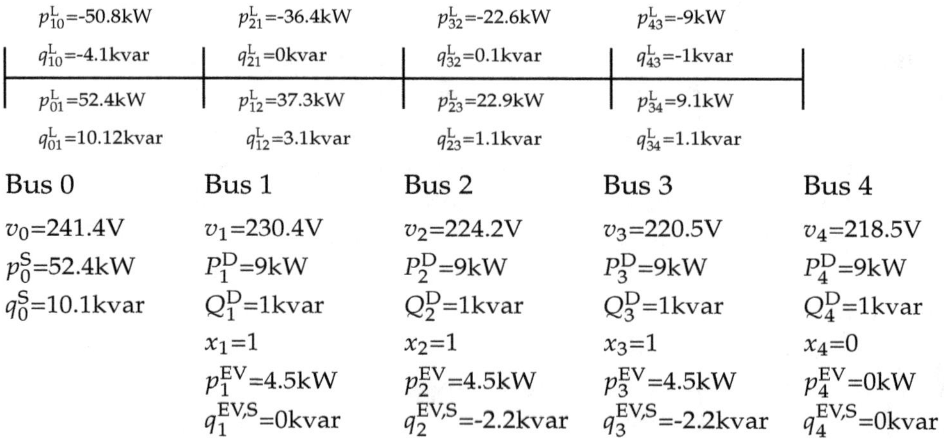

p_{10}^L=-50.8kW	p_{21}^L=-36.4kW	p_{32}^L=-22.6kW	p_{43}^L=-9kW	
q_{10}^L=-4.1kvar	q_{21}^L=0kvar	q_{32}^L=0.1kvar	q_{43}^L=-1kvar	
p_{01}^L=52.4kW	p_{12}^L=37.3kW	p_{23}^L=22.9kW	p_{34}^L=9.1kW	
q_{01}^L=10.12kvar	q_{12}^L=3.1kvar	q_{23}^L=1.1kvar	q_{34}^L=1.1kvar	
Bus 0	Bus 1	Bus 2	Bus 3	Bus 4
v_0=241.4V	v_1=230.4V	v_2=224.2V	v_3=220.5V	v_4=218.5V
p_0^S=52.4kW	P_1^D=9kW	P_2^D=9kW	P_3^D=9kW	P_4^D=9kW
q_0^S=10.1kvar	Q_1^D=1kvar	Q_2^D=1kvar	Q_3^D=1kvar	Q_4^D=1kvar
	x_1=1	x_2=1	x_3=1	x_4=0
	p_1^{EV}=4.5kW	p_2^{EV}=4.5kW	p_3^{EV}=4.5kW	p_4^{EV}=0kW
	$q_1^{EV,S}$=0kvar	$q_2^{EV,S}$=-2.2kvar	$q_3^{EV,S}$=-2.2kvar	$q_4^{EV,S}$=0kvar

Figure 4. Illustrative example: Results of problem (P1) with reactive power droop curve.

Table 3 provides the values of the variables $y_{ndt}^{(1)}$ and $y_{ndt}^{(2)}$ for each bus of the network in which a charging point was installed. Observe that only bus 1 had a voltage magnitude greater than $V_n^{(2)}$, which results in $y_{1dt}^{(1)} = y_{1dt}^{(2)} = 0$ and $q_{kdt}^{EV,S} = 0$ by constraints (6)–(14). On the other hand, voltages in buses 2 and 3 were lower than $V_n^{(1)}$, which resulted in $y_{1dt}^{(1)} = 1, y_{1dt}^{(2)} = 0$ and $q_{kdt}^{\text{EV,S}} = -Q_k^{\text{EV,max}} = -2.18$ kvar. Observe that a negative value of $q_{kdt}^{\text{EV,S}}$ indicates a reactive power injection on the bus where charger k was located (constraint (22)).

Table 3. Illustrative example: Variables $\{y_{ndt}^{(1)}, y_{ndt}^{(2)}\}$.

# Bus	$y_{ndt}^{(1)}$	$y_{ndt}^{(2)}$	v_{ndt} (V)	$q_{kdt}^{\text{EV,S}}$ (kvar)
1	0	0	230.4 $\left(v_{1dt} > 230.00 = V_1^{(2)}\right)$	0
2	1	0	224.2 $\left(v_{2dt} < 224.25 = V_2^{(1)}\right)$	−2.18
3	1	0	220.5 $\left(v_{3dt} < 224.25 = V_3^{(1)}\right)$	−2.18

4. Case Study

This section analyzes a case study based on the distribution system of La Graciosa, Canary Islands, Spain. La Graciosa is a small island of 27 km^2 with a population around 660 people [41]. This case study is focused on a district that is fed by a single 400-kVA power transformer, [42]. This district comprises 26 buildings that are connected by 26 lines.

4.1. Input Data

Figure 5 shows the location of the district and the buildings. The distribution network is represented in Figure 6.

The demands associated with the considered buildings were obtained from the first 26 residential demands provided in [43], which are downloadable in [44]. In this case study we assumed that there were no charging points in the district. In order to analyze a case in which a high number of chargers are demanded, it was assumed that 44 charging points were requested, that is the number of electric vehicles associated with the first 26 households in [43]. The power factor of residential demands was equal to 0.95 (lagging).

The main input data describing the considered distribution network are provided in Table 4. All loads were assumed to have a three-phase grid connection and a nominal neutral-to-phase voltage magnitude equal to 230 V. The minimum and maximum values that voltage magnitudes could take were 0.95 and 1.05 times the nominal value, i.e., 218.5 and 241.5 V, respectively. Parameters $V_n^{(1)}$ and $V_n^{(2)}$ describing the reactive power-voltage magnitude curve were equal to 224.25 and 230 V, respectively. The power factor of the charging points in the case in which the reactive power-voltage magnitude droop curve was considered was 0.95 (leading).

The performances of problems (P1) and (P2) were analyzed by considering two typical three-phase chargers with rate powers equal to 11 and 22 kVA.

All mathematical cases were solved using GAMS and Dicopt 12.6.111 in a linux-based server of four 3.0 GHz processors and 250 GB of RAM. Due to the complexity of mixed-integer non-linear programs, it is convenient to solve problems (P1) and (P2) starting from a so-called warm start solution. For doing that, problems (P1) and (P2) were solved neglecting the reactive power injections (6)–(14) and binary variables $y_{ndt}^{(1)}$ and $y_{ndt}^{(2)}$. Afterwards, initial values of variables $y_{ndt}^{(1)}$ and $y_{ndt}^{(2)}$ could be assigned as a function of the resulting voltage magnitudes in the previous problem. Therefore, for the warm start solution, if $v_{ndt} \leq V_n^{(1)}$, then $\{y_{ndt}^{(1)} = 1, y_{ndt}^{(2)} = 0\}$; if $V_n^{(1)} \leq v_{ndt} \leq V_n^{(2)}$, then $\{y_{ndt}^{(1)} = 0, y_{ndt}^{(2)} = 1\}$; and $v_{ndt} \geq V_n^{(2)}$, then $\{y_{ndt}^{(1)} = 0, y_{ndt}^{(2)} = 0\}$.

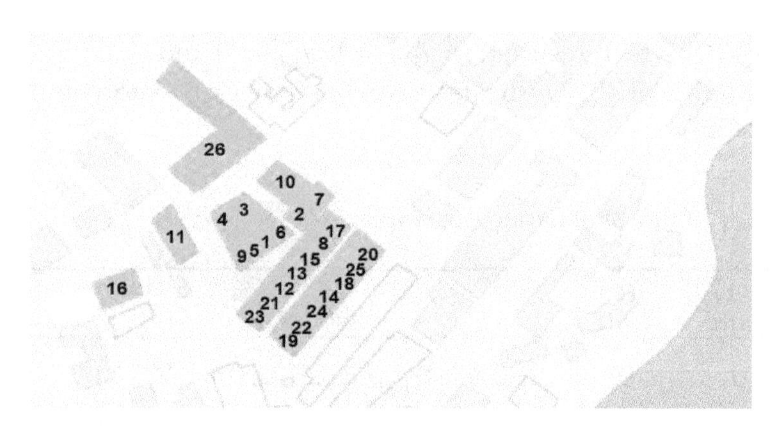

Figure 5. Case study: Location of the distribution network.

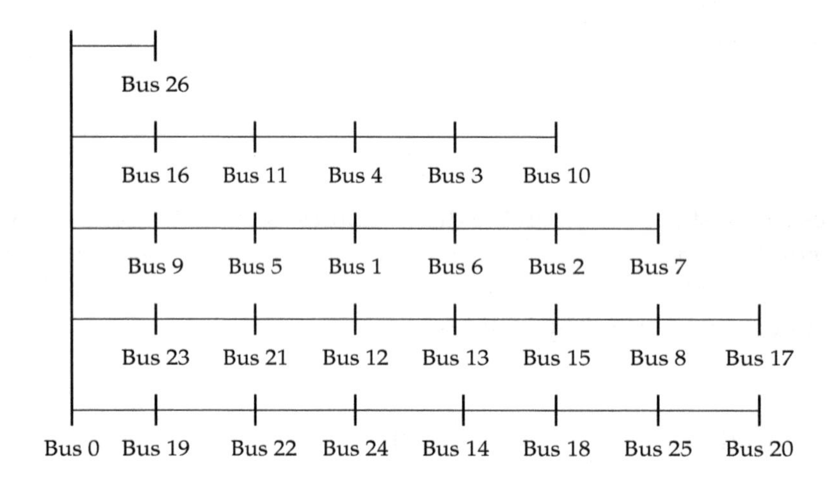

Figure 6. Case study: Distribution network [42].

Table 4. Case study: input data.

Branch	Length (m)	Bus	Peak Demand (kW)	Annual Demand (MWh)	Contracted Power (kW)	Requested Chargers (#)
0-26	65	1	4.99	6.88	5.75	1
0-16	18	2	6.67	10.02	6.90	2
16-11	34	3	6.29	9.62	6.90	1
11-4	19	4	6.10	7.38	6.90	1
4-3	16	5	6.18	11.41	6.90	1
3-10	15	6	6.34	9.24	6.90	1
0-9	50	7	7.00	9.35	8.05	2
9-5	8	8	9.01	13.03	9.20	2
5-1	8	9	5.39	6.36	5.75	1
1-6	8	10	6.10	8.47	6.90	2
6-2	10	11	6.49	7.97	6.90	2
2-7	7	12	6.11	11.98	6.90	1
0-23	70	13	5.03	9.99	5.75	1
23-21	10	14	6.15	7.71	6.90	2

Table 4. *Cont.*

Branch	Length (m)	Bus	Peak Demand (kW)	Annual Demand (MWh)	Contracted Power (kW)	Requested Chargers (#)
21-12	10	15	6.79	9.51	6.90	2
12-13	10	16	6.20	7.80	6.90	1
13-15	10	17	4.81	8.36	5.75	1
15-8	10	18	6.72	10.03	6.90	3
8-17	10	19	4.49	5.94	5.75	1
0-19	95	20	6.17	7.86	6.90	1
19-22	10	21	6.89	7.78	6.90	2
22-24	10	22	5.91	8.39	6.90	1
24-14	10	23	6.07	9.74	6.90	3
14-18	10	24	7.20	10.45	8.05	2
18-25	10	25	7.46	11.50	8.05	2
25-20	10	26	6.18	7.40	6.90	1

4.2. Results: Maximization of Number of Charging Points

This section analyzes the performance of problem (P1) on the distribution network described above. In order to analyze the advantages of considering the reactive power-voltage magnitude droop curve, all cases have been also solved without considering the reactive power droop curve. Besides, with the aim of analyzing the influence of the distribution line parameters, two different sets of data were used to characterize the distribution lines, namely Z_1 and Z_2. In Z_1, it was assumed that the impedance of each line was equal to 0.551 + 0.089i Ω/m [27], whereas it was equal to 0.716 + 0.089i Ω/m for Z_2. Observe that the relationship resistance/reactance of Z_1 was equal to 6.1, whereas it was equal to 8.6 for Z_2. Observe that high resistance/reactance ratios were typical of low-voltage distribution networks. Considering the values of the impedances in cases Z_1 and Z_2, the power transfer capability of the network would be higher for Z_1 than for Z_2. In this manner, it was expected that the voltage droop along the lines in case Z_2 would be higher than that in case Z_1.

Table 5 provides the results obtained from solving problem (P1) in terms of the number of charging points and the voltage magnitude in each bus of the system for each case. In total, eight different cases have been solved in this section. The average number of constraints, continuous and binary variables was equal to 734, 191 and 93, respectively. Each case was solved in a time smaller than 0.2 s. From the results provided in Table 5 we can observe that the number of charging points resulting from the model considering the reactive power droop curve was always greater than those obtained in the case without reactive power droop curve. If reactive power injections were considered, the number of installed charging points were up to 8.3% and 10.7% higher for 11 and 22 kVA chargers, respectively. As expected, the number of installed charging points strongly depended on the nominal power of the charger (11/22 kVA) and on the parameters of the lines (Z_1, Z_2). Focusing on the model with reactive power droop curve and Z_1, we observed that the number of charging points decreased from 26, for a 11 kVA charger, to 21, for a 22 kVA charger. For Z_2, the number of installed charging points was lower and it was equal to 21 for a 11 kVA charger, and 15 for a 22 kVA charger. It was also observed that the voltage magnitudes at some ending buses (7, 17 and 20) were close to the minimum allowed voltage magnitude, 218.5 V. These voltage values indicated that the installation of additional charging points in these lines was not feasible.

Table 5. Case study: Selection of charging points and voltage magnitudes.

	Without Reactive Power Droop Curve								With Reactive Power Droop Curve							
	11kVA-Z_1		11kva-Z_2		22kVA-Z_1		22kVA-Z_2		11kVA-Z_1		11kVA-Z_2		22kVA-Z_1		22kVA-Z_2	
Bus	x_k	v (V)	x_k	v (V)	x_k	v (V)	x_k	v (V)	x_k	v (V)	x_k	v (V)	x_k	v (V)	x_k	v (V)
0	0	241.5	0	241.5	0	241.5	0	241.5	0	241.5	0	241.5	0	241.5	0	241.5
26	1	241.3	1	241.2	1	241.2	1	241.1	1	241.3	1	241.2	1	241.2	1	241.1
16	1	236.5	1	234.9	1	234.0	1	233.0	1	236.7	1	235.2	1	233.4	1	233.4
11	2	234.3	2	232.0	2	230.6	2	229.4	2	234.5	2	232.4	2	229.7	2	230.0
4	1	231.5	1	228.4	1	226.9	1	225.7	1	231.9	1	229.0	1	225.2	1	226.5
3	1	229.1	1	225.2	1	223.8	1	223.3	1	229.6	1	226.0	1	221.4	1	224.2
10	2	227.5	2	223.1	1	222.2	0	222.8	2	228.0	2	224.1	2	218.7	0	223.7
9	1	240.1	1	239.9	1	239.8	1	239.7	1	240.0	1	240.0	1	239.8	1	239.7
5	1	233.1	1	232.3	1	231.7	1	232.2	1	232.1	1	232.5	1	231.9	1	231.7
1	1	227.9	1	227.0	1	226.4	0	228.3	1	225.9	1	227.3	1	226.9	0	227.3
6	1	224.4	0	224.0	0	224.1	0	225.3	1	221.5	0	224.2	0	224.5	0	224.3
2	0	223.1	0	222.3	0	222.8	0	223.6	1	219.3	0	222.6	0	223.2	0	222.6
7	0	222.1	0	221.0	0	221.8	0	222.4	0	218.5	0	221.3	0	222.2	0	221.3
23	3	240.0	3	239.8	3	239.4	3	239.5	3	240.0	3	239.8	3	239.5	3	239.2
21	2	233.0	2	231.7	2	230.5	1	232.7	2	232.8	2	232.0	2	230.9	2	230.3
12	1	228.2	1	226.7	1	225.8	0	228.8	1	227.8	1	227.0	1	226.2	0	226.4
13	1	224.9	0	223.6	0	223.3	0	225.7	1	224.3	0	223.9	0	223.8	0	223.3
15	0	223.0	0	221.1	0	221.4	0	223.3	0	222.0	0	221.4	0	221.9	0	220.8
8	0	221.7	0	219.4	0	220.1	0	221.6	0	220.7	0	219.7	0	220.6	0	219.1
17	0	221.2	0	218.8	0	219.6	0	221.0	0	220.2	0	219.1	0	220.1	0	218.5
19	1	240.6	1	240.6	1	240.5	1	240.5	1	240.6	1	240.5	1	240.4	1	240.4
22	1	233.4	1	233.6	1	233.4	1	233.6	1	232.7	1	232.5	1	231.6	1	232.4
24	2	227.7	1	228.5	1	228.6	0	229.6	2	226.3	2	226.3	2	225.1	0	227.3
14	1	224.4	0	225.4	0	226.2	0	226.6	2	222.2	0	223.2	0	222.7	0	224.3
18	0	222.5	0	223.1	0	224.4	0	224.3	0	220.4	0	220.9	0	220.8	0	221.9
25	0	221.2	0	221.4	0	223.1	0	222.6	0	219.1	0	219.3	0	219.5	0	220.2
20	0	220.6	0	220.7	0	222.5	0	221.8	0	218.5	0	218.5	0	218.9	0	219.4
$\sum_k x_k$	24		20		19		14		26		21		21		15	

4.3. 30-Days Simulation

In this subsection we solve problem (P2) to analyze the steady-state operation of the distribution network during a planning horizon of 30 days divided into 10-min periods. For doing that, we simulate the operation of the distribution network considering 11 and 22 kVA chargers using the distribution line parameters denoted by Z_1. The aggregated hourly demand is depicted in Figure 7.

The energy charged in each 10-min period by each electric vehicle was generated from the charging profile L1 included in [43]. Based on these data, the daily energy charged and the starting hour of the charging for each electric vehicle were characterized as random variables. Therefore, a Kernel density distribution has been fitted for each random variable and for each electric vehicle, as done in [45]. The resulting daily probability distributions for each of the 41 electric vehicles are represented in Figure 8. Figure 8a indicates that the average daily energy charged per vehicle was less than 10 kWh, whereas Figure 8b shows that the charge of the vehicles usually started in the middle part of the day.

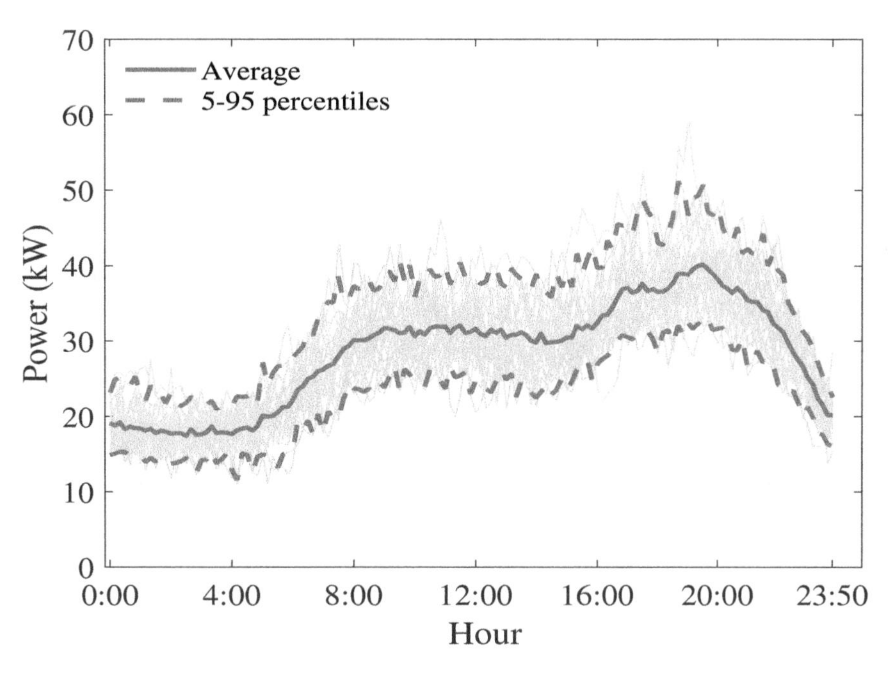

Figure 7. Case study: aggregated residential demand.

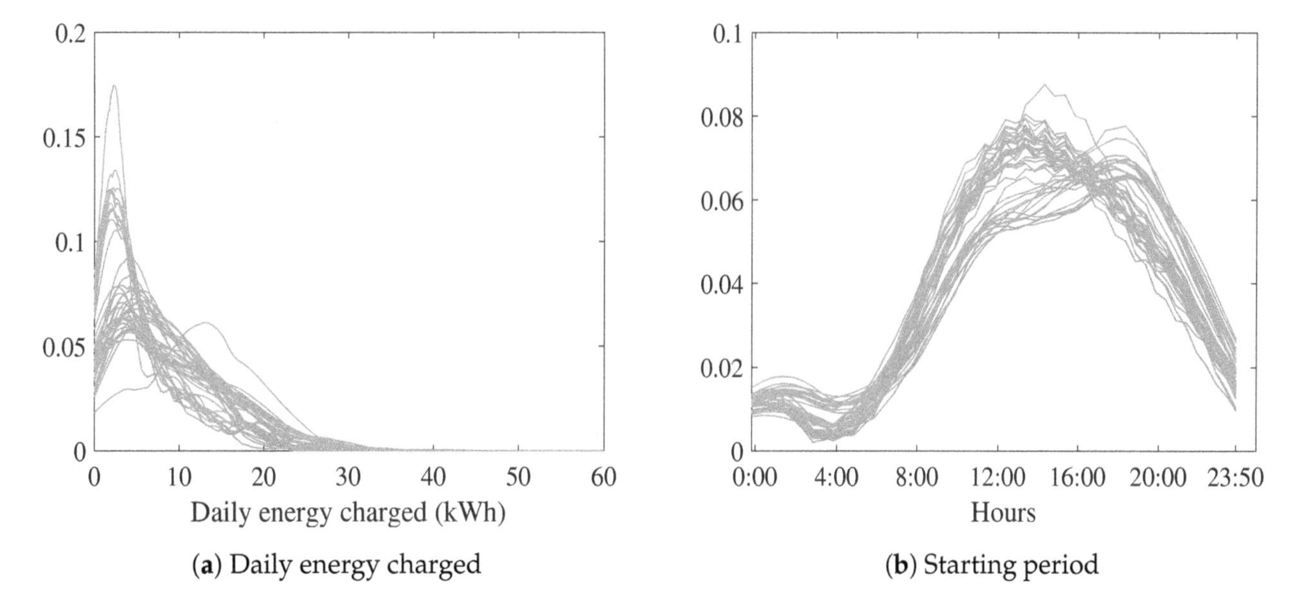

Figure 8. Case study: probability distributions of the daily energy charged and the starting hour of the charging per vehicle.

Using the probability distributions represented in Figure 8, different charging profiles have been randomly generated for each day considering 11 and 22 kVA chargers and 10-min time periods. In order to analyze the influence of increasing the charging demand of electric vehicles, two additional cases were generated considering that the energy demanded by the electric vehicles was 50% higher. The cases generated from the energy values represented in Figure 8a are denoted by 11 kVA-E_1 and 22 kVA-E_1, whereas cases with higher energy demanded are denoted by 11 kVA-E_2 and 22 kVA-E_2, respectively. As an example, the average energy charged per vehicle and day in case 11 kVA-E_1 was 8.5 kWh resulting, for a typical electric vehicle consumption rate of 0.2 kWh/km, in a daily distance driven equal to 42.5 km. The resulting charging profiles are provided in Figure 9.

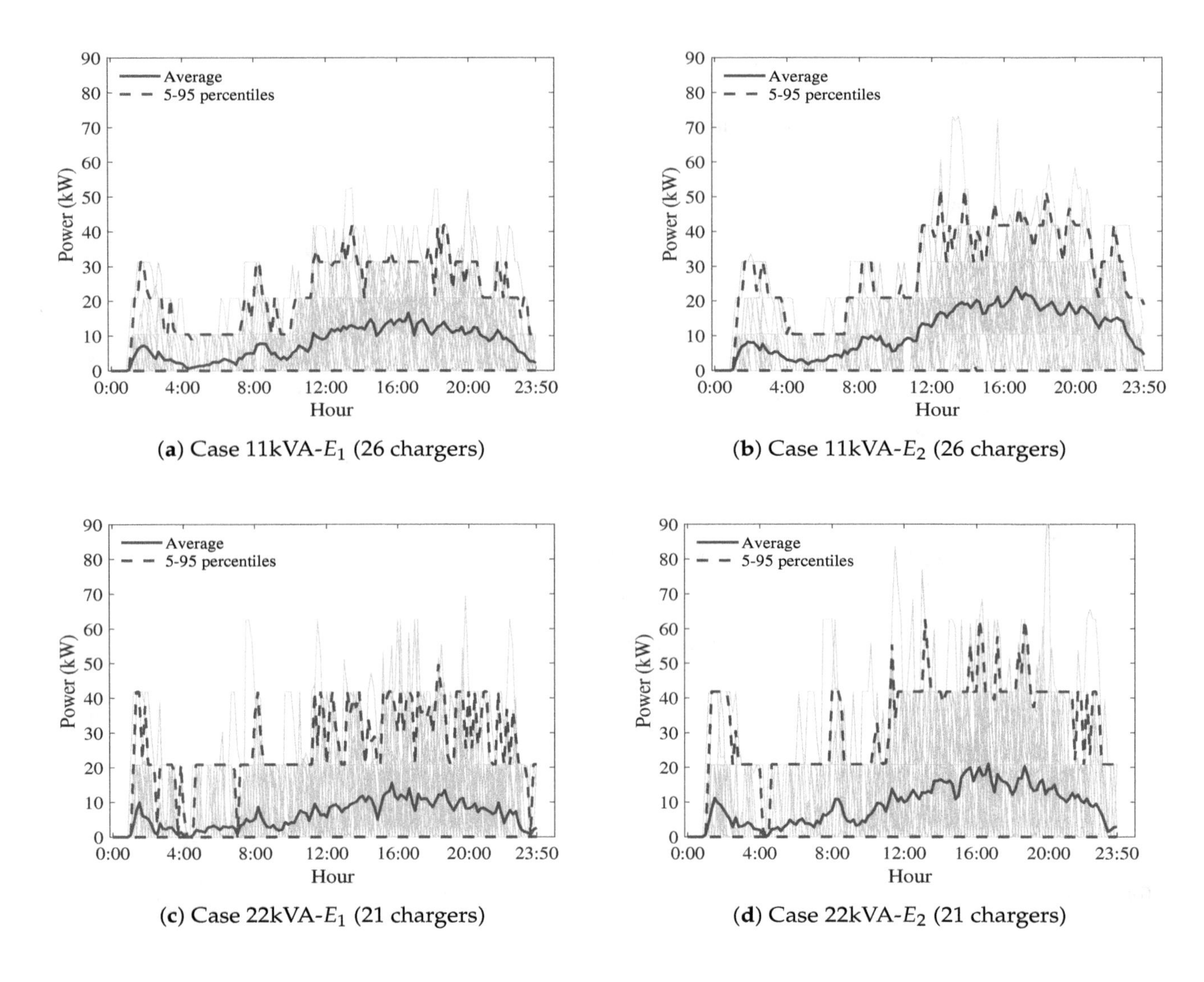

(**a**) Case 11kVA-E_1 (26 chargers) (**b**) Case 11kVA-E_2 (26 chargers)

(**c**) Case 22kVA-E_1 (21 chargers) (**d**) Case 22kVA-E_2 (21 chargers)

Figure 9. Case study: aggregated charging profiles.

Because of the variation of the demand during the day, three different voltage levels were considered in the distribution transformer bus in order to provide voltage support: 231 V from 0:00 to 6:00 h and from 23:00 to 23:50 h; 232 V from 17:00 to 21:00 h; and 231.5 V for the rest of hours.

Considering cases with and without reactive power provision, a total number of $30 \times 144 \times 2 \times (2+2) = 34,560$ instances of problem (P2) have been solved in this section. The average number of constraints, continuous and binary variables was equal to 734, 245 and 34, respectively. Each single instance was solved in a time smaller than 0.02 s.

Table 6 provides the minimum ($\underline{v}$) and mean ($\hat{v}$) voltage magnitudes and the sum of the reactive power injected ($\sum q$) per bus and case during the 30-day simulation. It was verified that voltage magnitudes were always greater than the lower limit, 218.5 V, and none voltage violations have been observed in the analyzed cases. As expected, voltage magnitudes were lower in those buses that were located furthest from the distribution transformer. In general, voltages decreased as the rate power of the charger and the energy charged increased. However, it was noticed that effect of increasing the power of the charger was more relevant than the growth of the energy demanded. For instance, from case 11kVA-E_1 to case 22kVA-E_1, the average minimum voltage decreased 0.3 V, whereas this difference was 0.2 V if cases 11 kVA-E_1 and 11 kVA-E_2 were compared.

Table 6 shows that the reactive power injected was higher in those chargers located far from the distribution transformer. In this sense, the reactive power injected in the closest buses to the transformer (buses 9, 19, 23 and 26) is always 0. Additionally, the injection of reactive power grew as the rate power of the charger and the energy charged increased. For instance, a total amount of 382.8 kvarh was injected into the grid in case 11 kVA-E_1. This value corresponded to an average value of 14.7 kvarh per charger. This value increased 25.2% and 56.1% in cases 22 kVA-E_1 and 11kVA-E_2, respectively.

Table 6. Case study: voltage magnitudes per bus (V) and reactive power injections (kvarh).

	11kVA-E_1			11kVA-E_2			22kVA-E_1			22kVA-E_2		
Bus	$\underline{v}$	$\hat{v}$	$\sum q$	$\underline{v}$	$\hat{v}$	$\sum q$	$\underline{v}$	$\hat{v}$	$\sum q$	$\underline{v}$	$\hat{v}$	$\sum q$
0	231.0	231.4	0.0	231.0	231.4	0.0	231.0	231.4	0.0	231.0	231.4	0.0
26	230.9	231.4	0.0	230.9	231.4	0.0	230.7	231.4	0.0	230.7	231.4	0.0
16	229.8	231.1	0.0	229.8	231.1	0.0	229.0	231.1	0.5	228.8	231.1	0.7
11	229.0	231.0	1.1	229.0	230.9	3.3	228.0	231.0	10.3	227.5	230.9	16.9
4	227.9	230.8	5.1	227.9	230.7	9.2	226.6	230.8	18.8	226.0	230.7	30.3
3	227.2	230.6	9.6	227.2	230.5	15.7	225.4	230.6	27.3	225.1	230.5	43.2
10	226.6	230.5	30.1	226.4	230.4	46.6	225.0	230.5	71.5	224.0	230.4	111.6
9	230.6	231.3	0.0	230.6	231.3	0.0	230.6	231.3	0.0	230.6	231.3	0.0
5	228.3	230.6	7.1	228.3	230.5	11.1	227.9	230.7	21.3	227.7	230.6	33.6
1	225.9	230.1	17.2	225.5	230.0	26.5	224.6	230.2	36.1	224.6	230.2	54.9
6	223.2	229.7	49.4	222.8	229.6	75.3	223.4	229.9	0.0	223.4	229.9	0.0
2	222.0	229.5	55.2	221.5	229.4	78.5	222.6	229.7	0.0	222.6	229.7	0.0
7	221.7	229.4	0.0	221.1	229.2	0.0	221.9	229.6	0.0	221.9	229.5	0.0
23	230.7	231.3	0.0	230.7	231.3	0.0	230.7	231.3	0.0	230.6	231.3	0.0
21	228.3	230.5	8.9	228.1	230.5	14.5	226.6	230.5	27.3	226.6	230.5	45.9
12	225.5	229.9	43.8	225.5	229.8	66.5	224.4	229.9	83.4	224.0	229.8	126.5
13	223.7	229.4	28.2	223.7	229.3	42.4	223.9	229.5	0.0	223.0	229.4	0.0
15	222.8	229.1	0.0	222.8	229.0	0.0	223.5	229.2	0.0	222.2	229.1	0.0
8	222.5	228.9	0.0	222.5	228.8	0.0	223.3	228.9	0.0	221.8	228.9	0.0
17	222.4	228.8	0.0	222.4	228.7	0.0	223.1	228.8	0.0	221.6	228.8	0.0
19	230.8	231.4	0.0	230.8	231.4	0.0	230.8	231.4	0.0	230.7	231.4	0.0
22	228.2	230.7	4.2	227.2	230.7	8.1	226.9	230.8	15.4	226.9	230.7	24.9
24	225.4	230.2	39.9	224.4	230.1	67.0	224.5	230.3	75.5	224.3	230.2	114.3
14	223.0	229.8	82.9	222.5	229.6	131.7	223.8	229.9	0.0	223.4	229.9	0.0
18	222.4	229.5	0.0	222.2	229.4	0.0	223.1	229.7	0.0	222.7	229.6	0.0
25	221.9	229.3	0.0	221.9	229.2	0.0	222.5	229.5	0.0	222.0	229.4	0.0
20	221.9	229.2	0.0	221.9	229.1	0.0	222.1	229.4	0.0	221.9	229.3	0.0

As an example, Figure 10 represents the results obtained for bus 10 during the first day in case 11kVA-E_1 in terms of voltage magnitudes, active power consumed by the household and the electric vehicles, and the injected reactive power. As indicated in Table 5, two electric vehicles chargers were installed in bus 10. As observed in Figure 10c, one vehicle started the charging at 8:10 am, and the other at 11:40 am. The total energy consumed by the house, without considering electric vehicles, was 8.9 kWh, whereas the sum of the energy charged by two electric vehicles was only 3.5 kWh. However, the highest active power peaks corresponded to the periods in which the vehicles were being charged. The high active power demanded by the electric vehicle chargers had a high impact on voltage magnitudes. Figure 10a shows that voltage drops occurred when electric vehicles were charging. Since voltages were under the nominal value, 230 V, reactive power was injected into the grid, as observed in Figure 10d. Finally, Figure 10b shows that the voltage magnitudes in the case in which the reactive power droop was considered are higher than those in the case in which the reactive power droop was not accounted for.

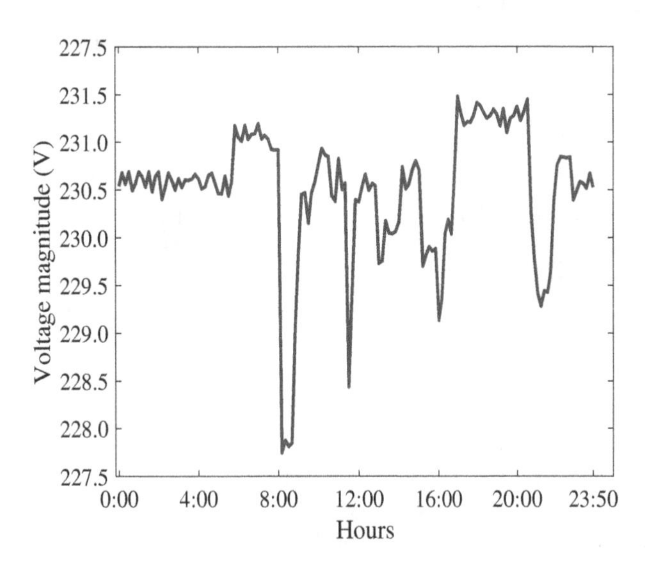

(**a**) Voltage magnitudes

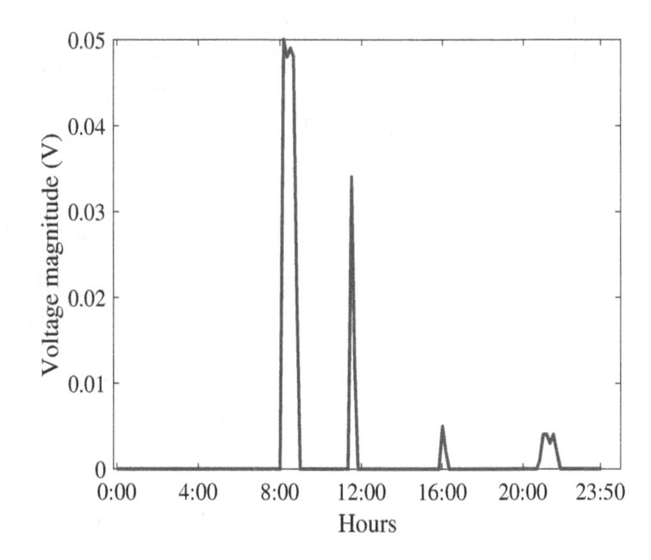

(**b**) Voltage magnitude difference with respect to the case without reactive power droop

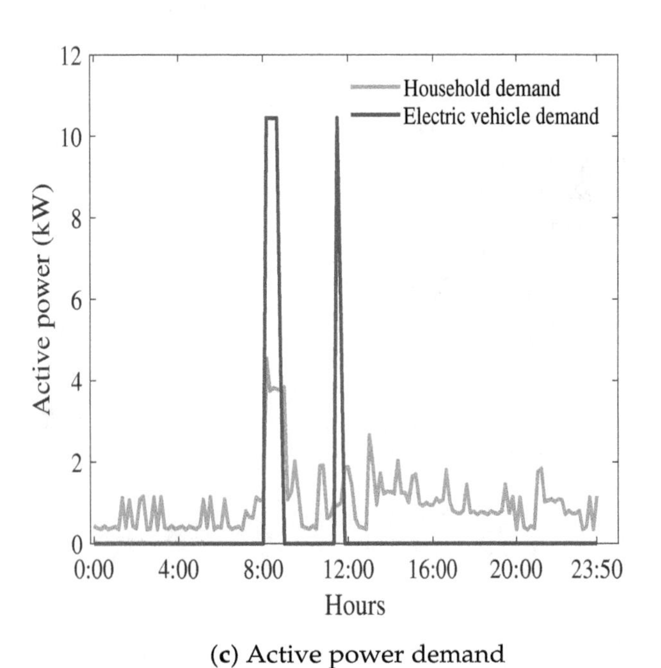

(**c**) Active power demand

(**d**) Reactive power injection of electric vehicles

Figure 10. Case study: Results bus 10 (11 kVA-E_1).

Finally, Figure 11 represents the average values and difference between voltage magnitudes in cases 11 kVA-E_1 and 22 kVA-E_1 using models with and without reactive power injection. For the sake of conciseness, a set of selected buses corresponding with the last buses of each branch were considered. The positive values of this difference indicated that voltages in cases with reactive power injection were higher than those in cases without reactive power injection. It was observed that the average voltage difference was larger in those periods with higher demand (between 12 and 22 h).

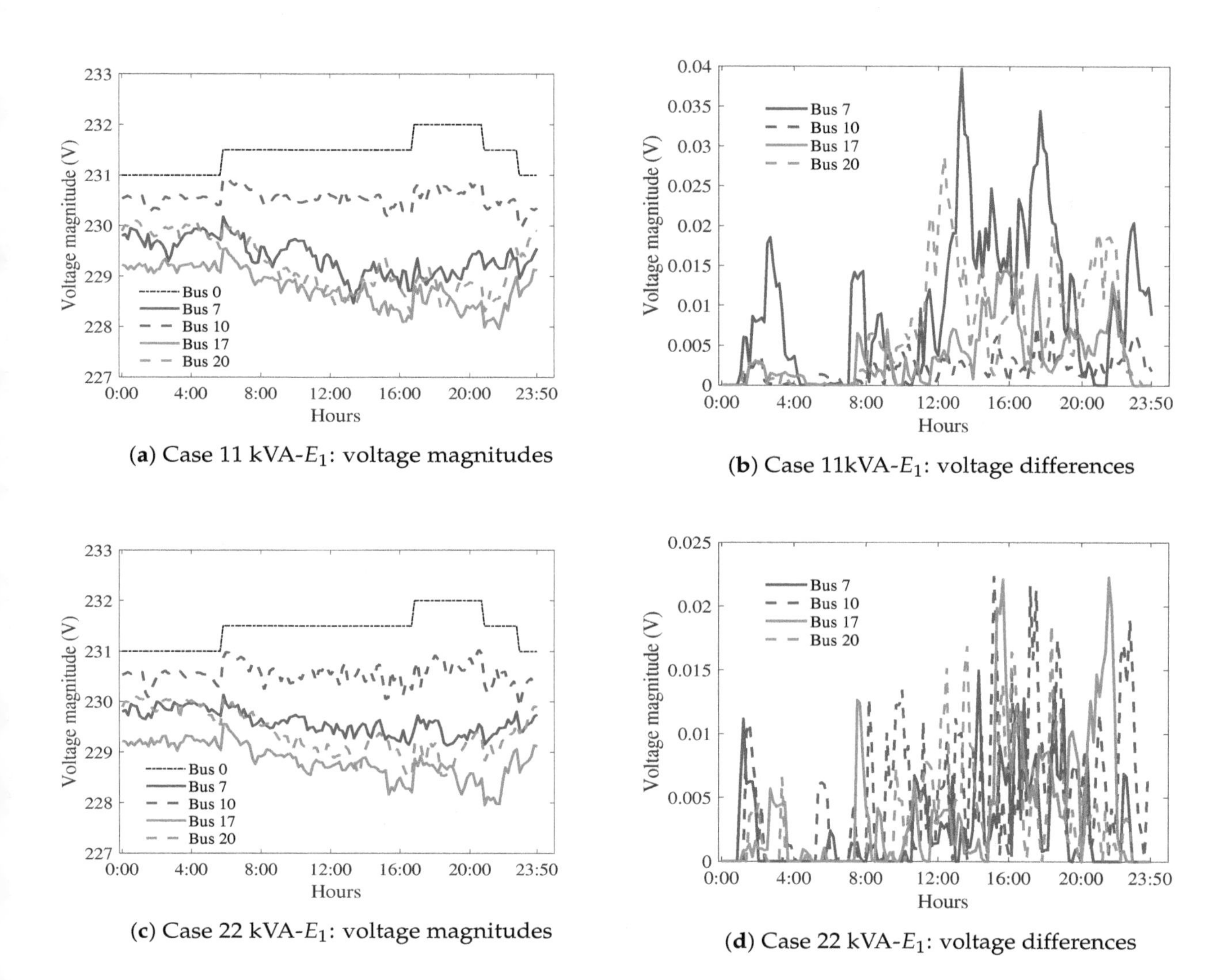

(**a**) Case 11 kVA-E_1: voltage magnitudes

(**b**) Case 11kVA-E_1: voltage differences

(**c**) Case 22 kVA-E_1: voltage magnitudes

(**d**) Case 22 kVA-E_1: voltage differences

Figure 11. Case study: comparison of voltage magnitudes .

5. Conclusions

This paper has presented a non-linear mixed-integer formulation for deciding the maximum number of charging points of electric vehicles that can be installed in a low-voltage distribution network. In order to increase the number of charging points, it has been considered that charging points are able to provide voltage support by injecting reactive power if voltage magnitudes are lower than a specified value. Therefore, the reactive power-voltage magnitude control is activated locally and it does not require the presence of a central operator. Additionally, an optimal power flow formulation has been provided to simulate the steady-state operation of a distribution network considering that electric vehicles can provide voltage support.

The proposed model is tested on a case study based on La Graciosa distribution network. Based on the numerical results presented in the case study, it has been observed that the number of installed charging points is higher if reactive power injections are considered. These increases are up to 8.3% and 10.7% for 11 and 22 kVA chargers, respectively. It has been also observed that the parameters of the distribution lines have a high influence on the number of installed chargers. For instance, if the resistance of the lines decreases 50%, the number of installed chargers increases 19.2% and 21.0% for 11 and 22 kVA chargers, respectively. It has been verified through 30-day simulations that the voltage magnitudes resulting from the model including reactive power injections are greater than the specified lower voltage limit. It has been also observed that the effect of increasing the power of the charger is more relevant than the growth of the energy demanded in terms of voltage decrease.

Research is currently underway to optimize the power factor assigned to each charger in order to maximize the number of installed charging points.

Author Contributions: M.C. proposed the core idea, performed the simulations and exported the results. M.C., R.Z.-M. and R.D. contributed to the design of the models and the writing of this manuscript. All authors have read and agreed to the published version of the manuscript.

Acknowledgments: The authors of this work would like to acknowledge and dedicate this paper to Spanish health professionals that have put their lives at risk and worked hard against COVID-19 while this paper was written.

Appendix A. Notation

The notation used throughout the paper is included below for quick reference. Observe that the definitions of symbols included in this appendix pretend to be general enough to be used in the two optimization models formulated in this paper. Unless otherwise indicated, all symbols refer to per-phase values.

Sets and Indices

D Set of days, indexed by d
K Set of electric vehicle chargers, indexed by k
K_{ndt} Set of electric vehicles chargers at usage in bus n, day d and period t.
L Set of distribution lines
N Set of buses, indexed by n and m
N_n Set buses connected to bus n, indexed by n and m
S Set of distribution transformers, indexed by s
S_n Set of distribution transformers in bus n, indexed by s
T Set of time periods, indexed by t

Parameters

B_{nm} Susceptance of the line linking buses n and m
G_{nm} Conductance of the line linking buses n and m
M Large enough parameter
P^{D}_{ndt} Active power demand in bus n, day d and period t
$P^{\text{EV,max}}_{kdt}$ Active power demanded by charger k on day d and period t
$P^{\text{EV,max}}_{k}$ Active power rate of charger k
Q^{D}_{ndt} Reactive power demand in bus n, day d and period t
$Q^{\text{EV,max}}_{k}$ Maximum reactive power that can be consumed by charger k
Q^{EV}_{kdt} Reactive power consumed by charger k on day d and period t
$S^{\text{EV,max}}_{k}$ Rate charge power of charger k
$S^{\text{L,max}}_{nm}$ Capacity of the line linking buses n and m
$S^{\text{S,max}}_{s}$ Capacity of distribution transformer s
$V^{(1)}_{n}$ First breakpoint of the reactive power-voltage magnitude droop curve of bus n
$V^{(2)}_{n}$ Second breakpoint of the reactive power-voltage magnitude droop curve of bus n
$V^{\max}_{n}$ Upper limit of the voltage magnitude of bus n
$V^{\min}_{n}$ Lower limit of the voltage magnitude of bus n

$\alpha_{kn}^{\mathrm{EV}}$ Auxiliary parameter used to formulate the reactive power-voltage magnitude droop curve of charger k in bus n
β_{kn}^{EV} Auxiliary parameter used to formulate the reactive power-voltage magnitude droop curve of charger k in bus n
ϕ_k Power factor angle of charger k

Variables

p_{nmdt}^{L} Active power flow through line linking buses n and m on day d and period t
p_{sdt}^{S} Active power supplied by distribution transformer s on day d and period t
q_{kdt}^{EV} Reactive power that must be consumed by charger k on day d and period t, if the charging point is accepted
$q_{kdt}^{\mathrm{EV,S}}$ Reactive power consumed by charger k on day d and period t
q_{nmdt}^{L} Reactive power flow through line linking buses n and m on day d and period t
q_{sdt}^{S} Reactive power supplied by distribution transformer s on day d and period t
v_{ndt} Voltage magnitude of bus n on day d and period t
v_{ndt}^{+} Positive voltage magnitude deviation in bus n on day d and period t
v_{ndt}^{-} Negative voltage magnitude deviation in bus n on day d and period t
v_{ndt}^{aux} Auxiliary continuous variable used to linearize the bilinear product of variables v_{ndt} and $\{y_{ndt}^{(1)}, y_{ndt}^{(2)}\}$
$v_{ndt}^{(j)}$ Auxiliary continuous variable associated with block-j of the voltage magnitude of bus n on day d and period t
x_k Binary variable that is equal to 1 if the request for charging point k is accepted
$y_{ndt}^{(1)}$ Auxiliary binary variable used to formulate the reactive power-voltage magnitude droop curve of bus n on day d and period t
$y_{ndt}^{(2)}$ Auxiliary binary variable used to formulate the reactive power-voltage magnitude droop curve of bus n on day d and period t
θ_{ndt} Voltage angle of bus n on day d and period t

References

1. Melhorn, A.C.; McKenna, K.; Keane, A.; Flynn, D.; Dimitrovski, A. Autonomous plug and play electric vehicle charging scenarios including reactive power provision: A probabilistic load flow analysis. *IET Gener. Transm. Distrib.* **2017**, *11*, 768–775. [CrossRef]
2. Graham, R.L.; Francis, J.; Bogacz, R.J. *Challenges and Opportunities of Grid Modernization and Electric Transportation*; U.S. Department of Energy: Washington DC, USA, 2017.
3. Knezović, K.; Marinelli, M.; Zecchino, A.; Andersen, P.B.; Traeholt, C. Supporting involvement of electric vehicles in distribution grids: Lowering the barriers for a proactive integration. *Energy* **2017**, *134*, 458–468. [CrossRef]
4. Arias, N.B.; Hashemi, S. Distribution System Services Provided by Electric Vehicles: Recent Status, Challenges, and Future Prospects. *IEEE Trans. Intell. Transp. Syst.* **2019**, *20*, 4277–4296. [CrossRef]
5. Gómez Expósito, A.; Conejo, A.J.; Cañizares, C. (Eds.) *Electric Energy Systems: Analysis and Operation*; The Electric Power Engineering Series; CRC Press: Boca Raton, FL, USA, 2009.
6. Demirok, E.; Casado-González, P.; Frederiksen, K.H.; Sera, D.; Rodríguez, P.; Teodorescu, R. Local reactive power control methods for overvoltage prevention of distributed solar inverters in low-voltage grids. *IEEE J. Photovolt.* **2011**, *1*, 174–182. [CrossRef]
7. Sortomme, E.; Negash, A.I.; Venkata S.S.; Kirschen, D.S. Voltage dependent load models of charging electric vehicles. In Proceedings of the IEEE PES General Meeting, Vancouver, BC, Canada, 21–25 July 2013; pp. 1–5.
8. Munoz, E.R.; Razeghi, G.; Zhang, L.; Jabbari, F. Electric vehicle charging algorithms for coordination of the grid and distribution transformer levels. *Energy* **2016**, *113*, 930–942. [CrossRef]

9. Wang, L.; Sharkh, S.; Chipperfield, A. Optimal coordination of vehicle-to-grid batteries and renewable generators in a distribution system. *Energy* **2016**, *113*, 1250–1264. [CrossRef]
10. Valsera-Naranjo, E.; Sumper, A.; Villafafila-Robles, R.; Martínez-Vicente, D. Probabilistic Method to Assess the Impact of Charging of Electric Vehicles on Distribution Grids. *Energies* **2012**, 1503–1531. [CrossRef]
11. Ciechanowicz, D.; Knoll, A.; Osswald, P.; Pelzer, D. Towards a business case for vehicle-to-Grid: Maximizing profits in ancillary service markets. In *Plug in Electric Vehicles in Smart Grids. Power Systems*; Rajakaruna, S., Shahnia, F., Ghosh, A., Eds.; Springer: Singapore, 2015.
12. Mu, Y.; Wu, J.; Ekanayake, J.; Jenkins, N.; Jia, H. Primary frequency response from electric vehicles in the Great Britain power system. *IEEE Trans. Smart Grid* **2013**, *4*, 1142–1150. [CrossRef]
13. Carrión, M.; Domínguez, R.; Cañas-Carretón M.; Zárate-Miñano R. Scheduling isolated power systems considering electric vehicles and primary frequency response. *Energy* **2019**, *168*, 1192–1207. [CrossRef]
14. Donadee, J.; Ilic, M.D. Stochastic optimization of grid to vehicle frequency regulation capacity bids. *IEEE Trans. Smart Grid* **2014**, *5*, 1061–1069. [CrossRef]
15. Clement-Nyns, K.; Driesen, J. The impact of charging plug-in hybrid electric vehicles on a residential distribution grid. *IEEE Trans. Power Syst.* **2010**, *25*, 371–380. [CrossRef]
16. Sortomme, E.; Hindi, M.M.; MacPherson, S.D.J.; Venkata, S.S. Coordinated charging of plug-in hybrid electric vehicles to minimize distribution system losses. *IEEE Trans. Smart Grid* **2011**, *2*, 198–205. [CrossRef]
17. Abdalrahman, A.; Zhuang, W. A Survey on PEV Charging Infrastructure: Impact Assessment and Planning. *Energies* **2017**, *10*, 1650. [CrossRef]
18. Humayd, A.S.B.; Bhattacharya, K. A Novel Framework for Evaluating Maximum PEV Penetration into Distribution Systems. *IEEE Trans. Smart Grid* **2018**, *9*, 2741–2751. [CrossRef]
19. Zárate-Miñano, R.; Flores Burgos, A.; Carrión, M. Analysis of different modeling approaches for integration studies of plug-in electric vehicles. *Int. J. Electr. Power Energy Syst.* **2020**, *114*, 105398. [CrossRef]
20. Xiang, Y.; Yang, W.; Liu, J.; Li, F. Multi-objective distribution network expansion incorporating electric vehicle charging stations. *Energies* **2016**, *9*, 909. [CrossRef]
21. Zeng, B.; Feng, J.; Zhang, J.; Liu, Z. An optimal integrated planning method for supporting growing penetration of electric vehicles in distribution systems. *Energy* **2017**, *126*, 273–284. [CrossRef]
22. Kong, C.; Jovanovic, R.; Bayram, I.S.; Devetsikiotis, M. A hierarchical optimization model for a network of electric vehicle charging stations. *Energies* **2017**, *10*, 675. [CrossRef]
23. Liu, Y.; Xiang, Y.; Tan, Y.; Wang, B.; Liu, J.; Yang, Z. Optimal Allocation Model for EV Charging Stations Coordinating Investor and User Benefits. *IEEE Access* **2018**, *6*, 36039–36049. [CrossRef]
24. Zhang, H.; Moura, S.J.; Hu, Z.; Song, Y. PEV Fast-Charging Station Siting and Sizing on Coupled Transportation and Power Networks. *IEEE Trans. Smart Grid* **2018**, *9*, 2595–2605. [CrossRef]
25. Abdalrahman, A.; Zhuang, W. QoS-Aware Capacity Planning of Networked PEV Charging Infrastructure. *IEEE Open J. Veh. Technol.* **2020**, *1*, 116–129. [CrossRef]
26. Huang, S.; Pillai, J.R.; Bak-Jensen, B.; Thøgersen, P. Voltage support from electric vehicles in distribution grid. In Proceedings of the 15th European Conference on Power Electronics and Applications (EPE), Lille, France, 2–6 September 2013; pp. 1–8.
27. Leemput, N.; Geth, F.; Van Roy, J.; Büscher, J.; Driesen, J. Reactive power support in residential LV distribution grids through electric vehicle charging. *Sustain. Energy Grids Netw.* **2015**, *3*, 24–35. [CrossRef]
28. Behravesh, V.; Keypour, R.; Akbari Foroud, A. Control strategy for improving voltage quality in residential power distribution network consisting of roof-top photovoltaic-wind hybrid systems, battery storage and electric vehicles. *Sol. Energy* **2019**, *182*, 80–95. [CrossRef]
29. Mohammadi, F.; Nazri, G.-A.; Saif, M. A Bidirectional Power Charging Control Strategy for Plug-in Hybrid Electric Vehicles. *Sustainability* **2019**, *11*, 4317. [CrossRef]
30. Kesler, M.; Kisacikoglu, M.C.; Tolbert, L.M. Vehicle-to-grid reactive power operation using plug-in electric vehicle bidirectional offboard charger. *IEEE Trans. Ind. Electron.* **2014**, *61*, 6778–6784. [CrossRef]
31. Mojdehi, M.N.; Fardad, M.; Ghosh, P. Technical and economical evaluation of reactive power service from aggregated EVs. *Electr. Power Syst. Res.* **2016**, *133*, 132–141. [CrossRef]
32. Wang, J.; Bharati, G.R.; Paudyal, S.; Ceylan, O.; Bhattarai, B.P.; Myers, K.S. Coordinated electric vehicle charging with reactive power support to distribution grids. *IEEE Trans. Ind. Informat.* **2018**, *15*, 54–63. [CrossRef]

33. Stanelyte, D.; Radziukynas, V. Review of Voltage and Reactive Power Control Algorithms in Electrical Distribution Networks. *Energies* **2020**, *13*, 58. [CrossRef]
34. ERTRAC; EPoSS; ETIP SNET. European Roadmap Electrification of Road Transport. 2017. Available online: https://egvi.eu/wp_content/uploads/2018/01/ertrac_electrificationroadmap2017.pdf (accessed on 5 July 2020).
35. Carpentier, J. Contribution to the economic dispatch problem. *Bull. Soc. Fr. Electr.* **1962**, *8*, 431–447.
36. Yang, G.; Marra, F.; Juamperez, M.; Kjaer, S.B.; Hashemi, S.; Østergaard, J.; Ipsen, H.H.; Frederiksen, K.H. Voltage rise mitigation for solar PV integration at LV grids. *J. Mod. Power Syst. Clean Energy* **2015**, *3*, 411–421 [CrossRef]
37. Pukhrem, S.; Basu, M.; Conlon, M.F.; Sunderland, K. Enhanced network voltage management techniques under the proliferation of rooftop solar PV installation in low-voltage distribution network. *IEEE J. Emerg. Sel. Top. Power Electron.* **2017**, *5*, 681–694. [CrossRef]
38. Arroyo, J.M.; Conejo, A.J. Optimal response of a thermal unit to an electricity spot market. *IEEE Trans. Power Syst.* **2000**, *15*, 1098–1104. [CrossRef]
39. Conejo, A.J.; Arroyo, J.M.; Contreras, J.; Villamor, F.A. Self-scheduling of a hydro producer in a pool-based electricity market. *IEEE Trans. Power Syst.* **2002**, *17*, 1265–1271. [CrossRef]
40. Carrión, M.; Dvorkin, Y.; Pandžić, H. Primary Frequency Response in Capacity Expansion with Energy Storage. *IEEE Trans. Power Syst.* **2018**, *33*, 1824–1835. [CrossRef]
41. Asensio, M.; Meneses de Quevedo, P.; Muñoz-Delgado, G.; Contreras, J. Joint Distribution Network and Renewable Energy Expansion Planning considering Demand Response and Energy Storage—Part II: Numerical Results. *IEEE Trans. Smart Grid* **2018**, *9*, 667–675. [CrossRef]
42. Sánchez, J.; Pavón, M.C.; Romero, L.; Guerrero, M.C.; Álvarez, S. Potential for exploiting the synergies between buildings through DSM approaches. Case study: La Graciosa Island. *Energy Convers. Manag.* **2019**, *194*, 199–216. [CrossRef]
43. Muratori, M. Impact of uncoordinated plug-in electric vehicles charging on residential power demand. *Nat. Energy* **2018**, *3*, 193–201. [CrossRef]
44. Muratori, M. Impact of Uncoordinated Plug-in Electric Vehicle Charging on Residential Power Demand—Supplementary Data. Available online: https://data.nrel.gov/submissions/69 (accessed on 25 May 2020).
45. Carrión, M. Determination of the selling price offered by electricity suppliers to electric vehicle users. *IEEE Trans. Smart Grids* **2019**, *10*, 6655–6666. [CrossRef]

8

Optimal Battery Storage Participation in European Energy and Reserves Markets

Kristina Pandžić [1,†], Ivan Pavić [2,†], Ivan Andročec [3,†] and Hrvoje Pandžić [2,*,†]

1 Croatian TSO (Hrvatski Operator Prijenosnog Sustava d.o.o.—HOPS), Zagreb 10000, Croatia; kristina.pandzic@hops.hr
2 Department of Energy and Power Systems, Faculty of Electrical Engineering and Computing, University of Zagreb, Zagreb 10000, Croatia; ivan.pavic@fer.hr
3 Hrvatska Elektroprivreda d.d., Zagreb 10000, Croatia; ivan.androcec@hep.hr
* Correspondence: hrvoje.pandzic@fer.hr
† These authors contributed equally to this work.

Abstract: Battery energy storage is becoming an important asset in modern power systems. Considering the market prices and battery storage characteristics, reserve provision is a tempting play fields for such assets. This paper aims at filling the gap by developing a mathematically rigorous model and applying it to the existing and future electricity market design in Europe. The paper presents a bilevel model for optimal battery storage participation in day-ahead energy market as a price taker, and reserve capacity and activation market as a price maker. It uses an accurate battery charging model to reliably represent the behavior of real-life lithium-ion battery storage. The proposed bilevel model is converted into a mixed-integer linear program by using the Karush–Kuhn–Tucker optimality conditions. The case study uses real-life data on reserve capacity and activation costs and quantities in German markets. The reserves activation quantities and activation prices are modeled by a set of credible scenarios in the lower-level problem. Finally, a sensitivity analysis is conducted to comprehend to what extent do battery storage bidding prices affect its overall profit.

Keywords: battery storage; day-ahead market; reserve market; optimal scheduling

---

1. Introduction

The European power sector is characterized by an ongoing liberalization and integration of national markets into one common marketplace. After the successful introduction of national electricity exchanges, followed by their coupling, the focus switched to the provision of ancillary services. The frequency reserves, as fairly location-independent services, were first in line to be governed by the market laws. Most of the European systems already have well-organized reserve markets, but their harmonization, which is the foundation for the integrated European reserve markets, is yet to be initiated. Reserve markets will use the same cross-border interconnection capacities as the energy market, and therefore these two markets must be co-optimized. The most recent European Union energy package incorporates detailed rules on how the reserve markets are to be organized, co-optimized and coupled, forming a cornerstone for all future reserve market research [1].

The reserve markets, depending on the type of reserve and different countries' regulations, are organized as either single-stage capacity-only markets or two-stage capacity and activation markets. The former type includes only capacity auction where the reserve providers' bids consist of capacity volume (in MW) and price (€ /MW). Using a merit order list (MOL), the the transmission system operator (TSO) accepts the cheapest bids until the required capacity is reached. A reserve provider must take into account the potential activated energy cost within its capacity bid as it is usually not separately remunerated (it could also be remunerated based on a regulated price). Such capacity is activated based on uniform price or some other rule. Usually, the frequency containment reserve (FCR) and sometimes automatic frequency restoration reserve (aFRR) are modeled this way. The latter type, along with the capacity procurement, includes the activated energy auction as well. A reserve provider's bid consist of energy volume (in MWh) and price (in € /MW). Using the MOL, energy offers are activated when needed. Such pricing activates the cheapest units first and therefore yields lower overall cost. Usually, manual restoration and replacement reserves and often automatic restoration reserves are modeled this way.

Both stages can be modeled in either a pay-as-bid or marginal pricing manner. In the current German secondary reserve market, both the capacity and the activated energy are priced as pay-as-bid [2,3]. However, the PICASSO project published a report with a conclusion that the pricing of aFRR activated energy in a future European-wide aFRR activation platform will be guided by the marginal pricing rule [4], which is adopted in this paper as well.

The capacity of the installed battery storage worldwide was around 10 GWh in 2017 [5]. In Germany alone, as one of the leaders in battery installations, in 2018 the capacity of home storage systems was around 930 MWh and large storage systems around 550 MWh [5]. The capacity of industrial storage systems is hard to estimate due to a lack of information. It is estimated that by the end of 2030 the battery capacity would rise to 181–421 GWh worldwide [5]. Most of the large storage systems operate in FCR markets. The FCR markets, in developed countries such as Germany and UK, are coming close to saturation, but new revenue streams are unlocking such as grid deferral and aFRR markets [6].

Coupling of national reserve markets and their co-optimization with energy markets creates new possibilities for battery storage as they could sell their services cross-border and position themselves in multiple markets. The battery storage as a fully flexible resource must be able to simultaneously bid in both the energy and reserve markets and must maintain its state-of-energy (SOE) within the allowed, i.e., feasible, range. Energy markets include a large number of different units, both capacity- and technology-wise, and its size is considerably larger then one battery storage. For example, the French power system had the minimum demand of 30.4 GW in 2018 during the summer and the peak demand of 96.6 GW during the winter [7]. Battery storage impact on such large market is negligible and therefore it can be seen as a price taker. However, reserve markets are smaller in size. For example, the German aFRR market has total demand of above 2 GW, while German FCR market is somewhat higher than 0.5 GW [8]. The battery storage trading on those markets should be modeled as price maker as its behavior could affect the prices.

In this paper, a novel battery storage scheduling algorithm for joint participation on energy and reserve market is designed and validated on a realistic test case. The battery storage acts as a price taker in the day-ahead energy market and as a price maker in the reserves market. Such algorithms are deemed to be the backbone for future battery scheduling in the large coupled and co-optimized energy and reserve markets in Europe. The focus of the paper is on aFRR markets as they are becoming a new source of revenue for the battery storage systems. However, the developed algorithm can easily be adjusted for other types of reserves.

2. Literature Review and Contributions

Depending on its capacity with respect to the total system load, energy storage can be considered too small to affect market prices, i.e., price taker, or to have a sufficient capacity to alter the market outcomes, thus becoming a price maker. Some early studies model the energy storage as a price taker, which means the prices in the models are known upfront [9,10].

Arbitrage alone might not be sufficient to justify the investment cost of energy storage. The authors in [11] prove that large-scale energy storage will dampen the price difference between on- and off-peak hours when performing arbitrage. It hereby reduces the profit it can make in the energy market, suggesting that energy storage should be used for ancillary services as well.

In [12], the authors model a profit-seeking price-taker energy storage that participates in energy and reserve day-ahead market and energy hour-ahead market. Stochastic unit commitment is used to derive scenarios for the cost of power and reserve in the hour-ahead market, as well as the actual reserve activation quantities. The uncertain parameters arise from the wind power plant output uncertainty. An optimal energy storage bidding model considering day-ahead energy, spinning reserve and regulation markets is presented in [13]. The price-taker energy storage considers uncertainties of predicted market prices and energy deployment in spinning reserve and regulation markets. The optimal bidding schedule is secured against realization of uncertainties using robust optimization framework.

Optimal bidding strategies are studied for battery energy storage systems in the reserve market with battery aging constraints in [14,15]. On the other hand, [16] combines power from unpredictable wind and photovoltaic sources with energy storage in the day-ahead electricity market using a stochastic two-stage programming environment, where the first stage is the day-ahead market, while the second stage simulates the balancing market using multiple scenario sets with historical data. An interested reader may find a comprehensive overview of operating models of energy storage is available in [17].

The Alberta Electric System Operator (AESO) compared sequential clearing of the energy and reserve market with their co-optimization and concluded that co-optimization was more cost-efficient then sequential clearing [18]. Authors in [19] propose a model that co-optimizes energy and reserve market for a combined cycle plant using a mixed-integer linear program (MILP). Paper [20] proposes a nonlinear model for co-optimization of energy and reserves in competitive electricity markets including nonlinear constraints such as power flow losses, unreliability and generation repair time. The authors in [21] clarify two approaches used in the literature to formulate the reserve requirements. The first one is by pre-defining the necessary reserve requirements using ad-hoc rules, such as the 3 + 5% rule [22], and setting the reserve requirements as parameters in the optimization problem. The second approach incorporates the power balance and transmission constraints both at the day-ahead and the balancing stage. These approaches are studied and evaluated in the MISO (Midwestern Independent System Operator) system in [23]. Another model that proposes an optimal dispatch of the energy and reserve capacity, but considering uncertain demand, is presented in [24]. The effects of co-optimized and individual clearing of the energy and reserve markets are investigated.

Despite a large body of literature focused on either theoretical or US-market based participation of energy storage, there are very few papers that replicate the operation of European markets and integrate them in a rigorous and scientific framework. One of the pivotal papers in modeling battery storage providing primary frequency response in the European setting is [25]. The presented optimization problem and the case study is focused and based on data for the German market. German energy and reserves market was also targeted in [26], where the pay-as-bid feature as well as longer time steps for providing reserve (4–12 h) was adopted. German aFRR market was the main topic in papers [27,28]. The former paper tackles the aFRR activation duration and price forecasting while the latter one deals with the bidding process in the German energy and aFRR reserve markets. The model in the paper [28] creates bids for storage to participate in the aFRR market based on price and activation forecasts meaning that it does not observe energy storage as a price forming factor but as a price taker.

With respect to the examined literature, this paper aims at filling the gap by combining a mathematically rigorous mathematical model with application to the existing and future electricity markets currently designed in Europe. Contribution of the paper is threefold. First, we develop a bilevel model for optimal battery storage participation in day-ahead energy market as a price taker, and reserve capacity and activation market as a price maker. Conceptually, this paper is an alternative to the approach of price maker algorithms for the German aFRR presented in [27,28]. As opposed to the majority of the literature that uses a generic energy storage model, we use an accurate battery charging model to reliably represent the behavior of actual battery storage. The proposed bilevel model is converted into a mixed-integer linear program by using the Karush–Kuhn–Tucker (KKT) optimality conditions. Second, we use real-life data on reserve capacity and activation costs and quantities to bring relevant conclusions. The reserves activation quantities and, consequently, the activation price is modeled by a set of credible scenarios. Thirdly, we provide a sensitivity analysis to comprehend to what extent do battery storage bidding prices affect its overall profit.

In the following chapter we first define the indices, parameters and variables used in the model and then present the model itself. The KKT optimality conditions and linearization technique are also presented. In Section 4 we present a case study based on the German market. This section also includes a sensitivity analysis for different sets of battery storage bidding prices. Finally, the relevant conclusions are drawn in the final section.

3. Mathematical Formulation

3.1. Nomenclature

Sets:

I	Set of generation units, indexed by i.
J	Set of battery charging curve linear parts, indexed by j.
S	Set of reserve activation scenarios, indexed by s.
T	Set of time periods, indexed by t.

Parameters:

$C_i^{\mathrm{a}\downarrow}$	Generator i down reserve activation price (€ /MWh).
$C_i^{\mathrm{a}\uparrow}$	Generator i up reserve activation price (€ /MWh).
$C^{\mathrm{b,a}\downarrow}$	Battery storage down reserve activation price (€ /MWh).
$C^{\mathrm{b,a}\uparrow}$	Battery storage up reserve activation price (€ /MWh).
$C^{\mathrm{b,cap}\downarrow}$	Battery storage down reserve capacity price (€ /MW).
$C^{\mathrm{b,cap}\uparrow}$	Battery storage up reserve capacity price (€ /MW).
$C_i^{\mathrm{cap}\downarrow}$	Generator i down reserve capacity price (€ /MW).
$C_i^{\mathrm{cap}\uparrow}$	Generator i up reserve capacity price (€ /MW).
$G_{t,i}^{\downarrow}$	Generator i maximum down reserve capacity (MW).
$G_{t,i}^{\uparrow}$	Generator i maximum up reserve capacity (MW).
F_j	Maximum amount of energy that can be charged at specific state-of-energy breakpoint R_j as a portion of SOE.
P	Battery storage maximum charging and discharging power (MW).
R_j	Capacity of each state-of-energy segment j as a portion of the maximum state-of-energy SOE.
$R_t^{\mathrm{cap}\downarrow}$	Required down reserve capacity (MW).
$R_t^{\mathrm{cap}\uparrow}$	Required up reserve capacity (MW).
$R_{t,s}^{\mathrm{a}\downarrow}$	Activated down reserve energy (MWh).
$R_{t,s}^{\mathrm{a}\uparrow}$	Activated up reserve energy (MWh).
η^{ch}	Battery storage charging efficiency.
η^{dis}	Battery storage discharging efficiency.
λ_t^{da}	Day-ahead market price (€ /MW).

Variables:

$g_{t,i,s}^{\mathrm{a}\uparrow}$	Generator i activated down energy (MWh).
$g_{t,i,s}^{\mathrm{a}\downarrow}$	Generator i activated up energy (MWh).
$g_{t,i}^{\mathrm{cap}\downarrow}$	Generator i down capacity reserved quantity (MW).
$g_{t,i}^{\mathrm{cap}\uparrow}$	Generator i up capacity reserved quantity (MW).
$\overline{q}_t^{\downarrow}$	Battery storage down reserve capacity bid (MW).
$\overline{q}_t^{\uparrow}$	Battery storage up reserve capacity bid (MW).
$q_{t,s}^{\mathrm{a}\downarrow}$	Battery storage activated down reserve quantity in scenario s (MWh).

$q_{t,s}^{\mathrm{a}\uparrow}$	Battery storage activated up reserve quantity in scenario s (MWh).
$q_t^{\mathrm{cap}\downarrow}$	Battery storage down reserved capacity (MW).
$q_t^{\mathrm{cap}\uparrow}$	Battery storage up reserved capacity (MW).
q_t^{ch}	Battery storage charging quantity (MW).
q_t^{dis}	Battery storage discharging quantity (MW).
$soe_{t,s}$	Battery storage state-of-energy (MWh).
$\lambda_{t,s}^{\mathrm{a}\downarrow}$	Down reserve activation clearing price in scenario s (€ /MWh).
$\lambda_{t,s}^{\mathrm{a}\uparrow}$	Up reserve activation clearing price in scenario s (€ /MWh).
$\lambda_t^{\mathrm{cap}\downarrow}$	Down reserve capacity clearing price (€ /MW).
$\lambda_t^{\mathrm{cap}\uparrow}$	Up reserve capacity clearing price (€ /MW).

3.2. Initial Problem Formulation

The proposed battery storage optimal bidding problem is formulated as follows:

$$\underset{\Xi^{\mathrm{UL}}}{\text{Maximize}} \quad \sum_{t\in\mathcal{T}} \left[\lambda_t^{\mathrm{da}}(q_t^{\mathrm{dis}} - q_t^{\mathrm{ch}}) + \left(\lambda_t^{\mathrm{cap}\uparrow}\cdot q_t^{\mathrm{cap}\uparrow} + \lambda_t^{\mathrm{cap}\downarrow}\cdot q_t^{\mathrm{cap}\downarrow}\right) + \left(\lambda_{t,s}^{\mathrm{a}\uparrow}\cdot q_{t,s}^{\mathrm{a}\uparrow} + \lambda_{t,s}^{\mathrm{a}\downarrow}\cdot q_{t,s}^{\mathrm{a}\downarrow}\right)\right] \tag{1}$$

subject to:

$$0 \le q_t^{\mathrm{ch}} \le \frac{\Delta soe_t}{\Delta t\cdot \eta^{\mathrm{ch}}}, \quad \forall t \tag{2}$$

$$0 \le q_t^{\mathrm{dis}} \le P\cdot\eta^{\mathrm{dis}}, \quad \forall t \tag{3}$$

$$q_t^{\mathrm{ch}} - q_t^{\mathrm{dis}} + \overline{q}_t^{\downarrow} \le \frac{\Delta soe_{t,s}}{\Delta t\cdot\eta^{\mathrm{ch}}}, \quad \forall t,s \tag{4}$$

$$-q_t^{\mathrm{ch}} + q_t^{\mathrm{dis}} + \overline{q}_t^{\uparrow} \le P\cdot\eta^{\mathrm{dis}}, \quad \forall t \tag{5}$$

$$soe_{t,s} = soe_{t-1,s} + \Delta t\cdot q_t^{\mathrm{ch}}\cdot\eta^{\mathrm{ch}} + q_{t,s}^{\mathrm{a}\downarrow}\cdot\eta^{\mathrm{ch}} - \Delta t\cdot q_t^{\mathrm{dis}}/\eta^{\mathrm{dis}} - q_{t,s}^{\mathrm{a}\uparrow}/\eta^{\mathrm{dis}}, \quad \forall t,s \tag{6}$$

$$0 \le soe_{t,s} - \Delta t\cdot\overline{q}_t^{\uparrow}, \quad \forall t,s \tag{7}$$

$$soe_{t,s} + \Delta t\cdot\overline{q}_t^{\downarrow} \le SOE, \quad \forall t,s \tag{8}$$

$$soe_{t,s} = \sum_{j=1}^{J-1} soe_{t,j,s}, \quad \forall t \tag{9}$$

$$0 \le soe_{t,j,s} \le (R_{j+1} - R_j)\cdot SOE, \quad \forall t,j,s \tag{10}$$

$$\Delta soe_{t,s} = F_1\cdot SOE + \sum_{j=1}^{J-1}\frac{F_{j+1}-F_j}{R_{j+1}-R_j}\cdot soe_{t-1,j,s}, \quad \forall t,s \tag{11}$$

where $\Xi_{UL} = \{q_t^{\mathrm{ch}}, q_t^{\mathrm{dis}}, q_t^{\mathrm{cap}\uparrow}, q_t^{\mathrm{cap}\downarrow}, q_{t,s}^{\mathrm{a}\downarrow}, q_{t,s}^{\mathrm{a}\uparrow}, soe_{t,s}, soe_{t,j,s}, \Delta soe_{t,s}\}$.

Battery storage in objective function (1) draws benefits from three streams. The first part is the day-ahead market, where it performs energy arbitrage as a price taker. The battery storage can be either discharged, q_t^{dis}, or charged, q_t^{ch}, at the day-ahead market price λ_t^{da}. The second part is the capacity reservation market. Since this market is much smaller than the day-ahead market, battery storage is modeled as a price maker, i.e., the up and down capacity reservation prices $\lambda_t^{\mathrm{cap}\uparrow}$ and $\lambda_t^{\mathrm{cap}\downarrow}$ are dual variables whose values are decided in the lower-level problem considering the battery's bids. The final part of the objective function (1) displays the benefits of both the up and down reserve activation $q_{t,s}^{a\uparrow}$ and $q_{t,s}^{a\downarrow}$ at prices $\lambda_{t,s}^{a\uparrow}$ and $\lambda_{t,s}^{a\downarrow}$, respectively.

Constraints (2) and (3) limit the day-ahead charging and discharging power. The fact that the battery charging ability reduces with high state-of-energy values is considered by limiting the battery charging power in (2) by the maximum amount of energy the battery can charge in a single time-step, $\Delta soe_{t,s}$, divided by the length of the time-step to convert energy to power. On the other hand, the discharging battery ability in (3) is constant regardless of the state-of-energy. Constraints (4) and (5) impose charging and discharging limits to down and up reserve bids so the charging and discharging battery capacity is not exceeded. Down reserve in (4) can be provided by increasing the charging power from the day-ahead stage (in this case q_t^{ch} is positive and q_t^{dis} is zero) or by reducing or fully stopping the discharging power from the day-ahead stage and possibly starting to charge instead (in this case q_t^{ch} is zero and q_t^{dis} is positive). Similarly, up reserve in (5) can be provided by reducing the day-ahead charging power and/or increasing the day-ahead discharging power. Equation (6) calculates the state-of-energy per each reserve activation scenario. Since q_t^{ch} and q_t^{dis} are power quantities, they are multiplied by an appropriate time step duration Δt. Since the day-ahead market is on an hourly basis, q_t^{ch} and q_t^{dis} are multiplied by 1. The reserve activation quantities $q_{t,s}^{a\downarrow}$ and $q_{t,s}^{a\uparrow}$ are energy quantities, the same as the state-of-energy $soe_{t,s}$. Constraints (7) and (8) provide the lower and upper bounds on the battery state-of-energy considering the reserve activations per scenario and the bid reserve quantities. This ensures that regardless of the reserve activation scenarios the state-of-energy will remain within the given bounds. Constraints (9)–(11) calculate the amount of energy the battery can charge in a time-step, $\Delta soe_{t,s}$. To describe the nonlinear battery charging curve, a piecewise approximation given in Figure 1 is used. This curve shows the amount of energy a lithium-ion battery can withdraw from the grid depending on its current state-of-energy. The given picewise linear approximation divides the state-of-energy in multiple segments, $soe_{t,j,s}$, constituting the actual battery state-of-energy $soe_{t,s}$. These segments are used in (11) to calculate the amount of energy the battery can charge in time period t. Further details on this procedure are available in [29].

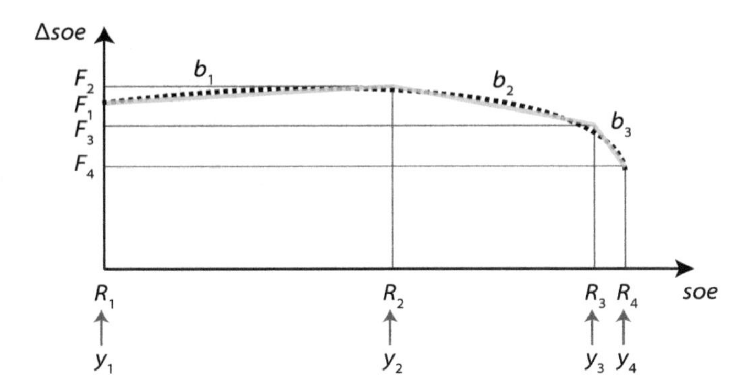

Figure 1. Piecewise linear approximation of an *soe*–Δ*soe* function.

The battery scheduling problem (1) is subject to the following lower-level problem (corresponding dual variables related to each constraint are listed after a colon):

$$\underset{\Xi^{LL}}{\text{Minimize}} \quad \sum_{t\in\mathcal{T}} \left[\sum_{i\in\mathcal{I}} C_i^{cap\uparrow} \cdot g_{t,i}^{cap\uparrow} + C^{b,cap\uparrow} \cdot q_t^{cap\uparrow} + \sum_{i\in\mathcal{I}} C_i^{cap\downarrow} \cdot g_{t,i}^{cap\downarrow} + C^{b,cap\downarrow} \cdot q_t^{cap\downarrow} \right] +$$
$$\sum_{t\in\mathcal{T}} \sum_{s\in\mathcal{S}} \pi_s \cdot \left[\sum_{i\in\mathcal{I}} C_i^{a\uparrow} \cdot g_{t,i,s}^{a\uparrow} + C^{b,a\uparrow} \cdot q_{t,s}^{a\uparrow} + \sum_{i\in\mathcal{I}} C_i^{a\downarrow} \cdot g_{t,i,s}^{a\downarrow} + C^{b,a\downarrow} \cdot q_{t,s}^{a\downarrow} \right] \tag{12}$$

subject to:

$$\sum_{i\in\mathcal{I}} g_{t,i}^{cap\uparrow} + q_t^{cap\uparrow} \geq R_t^{cap\uparrow}, \quad \forall t \ : \lambda_t^{cap\uparrow} \tag{13}$$

$$\sum_{i \in \mathcal{I}} g_{t,i}^{\text{cap}\downarrow} + q_t^{\text{cap}\downarrow} \geq R_t^{\text{cap}\downarrow}, \quad \forall t : \lambda_t^{\text{cap}\downarrow} \tag{14}$$

$$-\sum_{i \in \mathcal{I}} g_{t,i,s}^{\text{a}\uparrow} - q_{t,s}^{\text{a}\uparrow} + R_{t,s}^{\text{a}\uparrow} = 0, \quad \forall t,s : \lambda_{t,s}^{\text{a}\uparrow} \tag{15}$$

$$-\sum_{i \in \mathcal{I}} g_{t,i,s}^{\text{a}\downarrow} - q_{t,s}^{\text{a}\downarrow} + R_{t,s}^{\text{a}\downarrow} = 0, \quad \forall t,s : \lambda_{t,s}^{\text{a}\downarrow} \tag{16}$$

$$g_{t,i}^{\text{cap}\uparrow} \leq G_{t,i}^{\uparrow}, \quad \forall t,i : \psi_{t,i}^{\uparrow} \tag{17}$$

$$g_{t,i}^{\text{cap}\downarrow} \leq G_{t,i}^{\downarrow}, \quad \forall t,i : \psi_{t,i}^{\downarrow} \tag{18}$$

$$g_{t,i,s}^{\text{a}\uparrow} \leq g_{t,i}^{\text{cap}\uparrow} \cdot \Delta t, \quad \forall t,i,s : \kappa_{t,i,s}^{\uparrow} \tag{19}$$

$$g_{t,i,s}^{\text{a}\downarrow} \leq g_{t,i}^{\text{cap}\downarrow} \cdot \Delta t, \quad \forall t,i,s : \kappa_{t,i,s}^{\downarrow} \tag{20}$$

$$q_t^{\text{cap}\uparrow} \leq \overline{q}_t^{\uparrow}, \quad \forall t : \zeta_t^{\uparrow} \tag{21}$$

$$q_t^{\text{cap}\downarrow} \leq \overline{q}_t^{\downarrow}, \quad \forall t : \zeta_t^{\downarrow} \tag{22}$$

$$q_{t,s}^{\text{a}\uparrow} \leq q_t^{\text{cap}\uparrow} \cdot \Delta t, \quad \forall t,s : \nu_{t,s}^{\uparrow} \tag{23}$$

$$q_{t,s}^{\text{a}\downarrow} \leq q_t^{\text{cap}\downarrow} \cdot \Delta t, \quad \forall t,s : \nu_{t,s}^{\downarrow} \tag{24}$$

$$g_{t,i,s}^{\text{a}\uparrow}, g_{t,i,s}^{\text{a}\downarrow} \geq 0, \quad \forall t,i,s : \alpha_{t,i,s}^{\uparrow}, \alpha_{t,i,s}^{\downarrow} \tag{25}$$

$$q_{t,s}^{\text{a}\uparrow}, q_{t,s}^{\text{a}\downarrow} \geq 0, \quad \forall t,s : \beta_{t,s}^{\uparrow}, \beta_{t,s}^{\downarrow} \tag{26}$$

$$g_{t,i}^{\text{cap}\uparrow}, g_{t,i}^{\text{cap}\downarrow} \geq 0, \quad \forall t,i : \gamma_{t,i}^{\uparrow}, \gamma_{t,i}^{\downarrow} \tag{27}$$

$$q_t^{\text{cap}\uparrow}, q_t^{\text{cap}\downarrow} \geq 0, \quad \forall t : \delta_t^{\uparrow}, \delta_t^{\downarrow} \tag{28}$$

where $\Xi_{LL} = \{g_{t,i}^{\text{cap}\uparrow}, g_{t,i}^{\text{cap}\downarrow}, g_{t,i,s}^{\text{a}\uparrow}, g_{t,i,s}^{\text{a}\downarrow}, q_t^{\text{cap}\uparrow}, q_t^{\text{cap}\downarrow}, q_{t,s}^{\text{a}\uparrow}, q_{t,s}^{\text{a}\downarrow}\}$.

The lower-level problem objective function (12) is the maximization of the social welfare, which includes minimizing the cost of both generators' and the battery's up and down capacity reservation as well as its activation per scenario. Constraints (13) and (14) impose the up and down required reserve capacity volumes, while Equations (15) and (16) decide on the contribution of each asset (generators and the battery storage) to up and down reserve activation per scenario. Up and down generators' cleared reserve capacities are restricted by their offered capacities in (17) and (18), while the generators' activated quantities are limited by their reserved capacities in (19) and (20). The same is achieved for the battery with constraints (21)–(24). Finally, nonnegativity of the lower-level variables is imposed in (25)–(28). The dual variables listed after a colon in constraints (13)–(28) indicate if those constraints are binding or not. Dual variables of constraints (13)–(16) take values of marginal cost for up capacity reservation, down capacity reservation, up capacity activation and down capacity activation, respectively, and are used in the upper-level problem to determine the profitability of the battery storage operation. The remaining dual variables defined for constraints (17)–(28) indicate how much this constraint worsen the objective function. If the value of a dual variable is zero, this constraint does not affect the objective function value, i.e., it is not binding.

Problem (1)–(2) is a bilevel problem and cannot be solved directly. Thus, the lower-level problem needs to be replaced by its equivalent constraints. We use Karush–Kuhn–Tucker optimality conditions to convert the initial bilevel problem into a mixed-integer linear program (MILP). An interested reader may find details on this mathematical technique in [30].

3.3. KKT Conditions of the Lower-Level Problem

The dual objective function:

$$\text{Maximize } -\sum_{t\in\mathcal{T}} \overline{q}_t^{\uparrow}\cdot\zeta_t^{\uparrow} - \sum_{t\in\mathcal{T}} \overline{q}_t^{\downarrow}\cdot\zeta_t^{\downarrow} + \sum_{t\in\mathcal{T}} R_t^{\text{cap}\uparrow}\cdot\lambda_t^{\text{cap}\uparrow} + \sum_{t\in\mathcal{T}} R_t^{\text{cap}\downarrow}\cdot\lambda_t^{\text{cap}\downarrow}$$
$$+\sum_{t\in\mathcal{T}}\sum_{s\in\mathcal{S}} R_{t,s}^{\text{a}\uparrow}\cdot\lambda_{t,s}^{\text{a}\uparrow} + \sum_{t\in\mathcal{T}}\sum_{s\in\mathcal{S}} R_{t,s}^{\text{a}\downarrow}\cdot\lambda_{t,s}^{\text{a}\downarrow} - \sum_{t\in\mathcal{T}}\sum_{i\in\mathcal{I}} G_{t,i}^{\uparrow}\cdot\psi_{t,i}^{\uparrow} - \sum_{t\in\mathcal{T}}\sum_{i\in\mathcal{I}} G_{t,i}^{\downarrow}\cdot\psi_{t,i}^{\downarrow} \quad (29)$$

Dual constraints and stationarity conditions:

$$-\sum_{s\in\mathcal{S}} \kappa_{t,i,s}^{\uparrow} + C_i^{\text{cap}\uparrow} - \lambda_t^{\text{cap}\uparrow} - \gamma_{t,i}^{\uparrow} + \psi_{t,i}^{\uparrow} = 0, \quad \forall t,i \quad (30)$$

$$-\sum_{s\in\mathcal{S}} \kappa_{t,i,s}^{\downarrow} + C_i^{\text{cap}\downarrow} - \lambda_t^{\text{cap}\downarrow} - \gamma_{t,i}^{\downarrow} + \psi_{t,i}^{\downarrow} = 0, \quad \forall t,i \quad (31)$$

$$\pi_s\cdot C_i^{\text{a}\uparrow} - \lambda_{t,s}^{\text{a}\uparrow} - \alpha_{t,i,s}^{\uparrow} + \kappa_{t,i,s}^{\uparrow} = 0, \quad \forall t,i,s \quad (32)$$

$$\pi_s\cdot C_i^{\text{a}\downarrow} - \lambda_{t,s}^{\text{a}\downarrow} - \alpha_{t,i,s}^{\downarrow} + \kappa_{t,i,s}^{\downarrow} = 0, \quad \forall t,i,s \quad (33)$$

$$-\sum_{s\in\mathcal{S}} \nu_{t,s}^{\uparrow} - \delta_t^{\uparrow} + \zeta_t^{\uparrow} - \lambda_t^{\text{cap}\uparrow} + C^{\text{b,cap}\uparrow} = 0, \quad \forall t \quad (34)$$

$$-\sum_{s\in\mathcal{S}} \nu_{t,s}^{\downarrow} - \delta_t^{\downarrow} + \zeta_t^{\downarrow} - \lambda_t^{\text{cap}\downarrow} + C^{\text{b,cap}\downarrow} = 0, \quad \forall t \quad (35)$$

$$-\beta_{t,s}^{\uparrow} - \lambda_{t,s}^{\text{a}\uparrow} + \nu_{t,s}^{\uparrow} + \pi_s\cdot C^{\text{b,a}\uparrow} = 0, \quad \forall t,s \quad (36)$$

$$-\beta_{t,s}^{\downarrow} - \lambda_{t,s}^{\text{a}\downarrow} + \nu_{t,s}^{\downarrow} + \pi_s\cdot C^{\text{b,a}\downarrow} = 0, \quad \forall t,s \quad (37)$$

Complementarity slackness:

$$(-\sum_{i\in\mathcal{I}} g_{t,i}^{\text{cap}\uparrow} - q_t^{\text{cap}\uparrow} + R_t^{\text{cap}\uparrow}) \perp \lambda_t^{\text{cap}\uparrow}, \quad \forall t \quad (38)$$

$$(-\sum_{i\in\mathcal{I}} g_{t,i}^{\text{cap}\downarrow} - q_t^{\text{cap}\downarrow} + R_t^{\text{cap}\downarrow}) \perp \lambda_t^{\text{cap}\downarrow}, \quad \forall t \quad (39)$$

$$(g_{t,i}^{\text{cap}\uparrow} - G_{t,i}^{\uparrow}) \perp \psi_{t,i}^{\uparrow}, \quad \forall t,i \quad (40)$$

$$(g_{t,i}^{\text{cap}\downarrow} - G_{t,i}^{\downarrow}) \perp \psi_{t,i}^{\downarrow}, \quad \forall t,i \quad (41)$$

$$(-g_{t,i}^{\text{cap}\uparrow}\cdot\Delta t + g_{t,i,s}^{\text{a}\uparrow}) \perp \kappa_{t,i,s}^{\uparrow}, \quad \forall t,i,s \quad (42)$$

$$(-g_{t,i}^{\text{cap}\downarrow} \cdot \Delta t + g_{t,i,s}^{\text{a}\downarrow}) \perp \kappa_{t,i,s}^{\downarrow}, \quad \forall t, i, s \tag{43}$$

$$(q_t^{\text{cap}\uparrow} - \overline{q}_t^{\uparrow}) \perp \zeta_t^{\uparrow}, \quad \forall t \tag{44}$$

$$(q_t^{\text{cap}\downarrow} - \overline{q}_t^{\downarrow}) \perp \zeta_t^{\downarrow}, \quad \forall t \tag{45}$$

$$(-q_t^{\text{cap}\uparrow} \cdot \Delta t + q_{t,s}^{\text{a}\uparrow}) \perp \nu_{t,s}^{\uparrow}, \quad \forall t, s \tag{46}$$

$$(-q_t^{\text{cap}\downarrow} \cdot \Delta t + q_{t,s}^{\text{a}\downarrow}) \perp \nu_{t,s}^{\downarrow}, \quad \forall t, s \tag{47}$$

$$-g_{t,i,s}^{\text{a}\uparrow} \perp \alpha_{t,i,s}^{\uparrow}, \quad \forall t, i, s \tag{48}$$

$$-g_{t,i,s}^{\text{a}\downarrow} \perp \alpha_{t,i,s}^{\downarrow}, \quad \forall t, i, s \tag{49}$$

$$-q_{t,s}^{\text{a}\uparrow} \perp \beta_{t,s}^{\uparrow}, \quad \forall t, s \tag{50}$$

$$-q_{t,s}^{\text{a}\downarrow} \perp \beta_{t,s}^{\downarrow}, \quad \forall t, s \tag{51}$$

$$-g_{t,i}^{\text{cap}\uparrow} \perp \gamma_{t,i}^{\uparrow}, \quad \forall t, i \tag{52}$$

$$-g_{t,i}^{\text{cap}\downarrow} \perp \gamma_{t,i}^{\downarrow}, \quad \forall t, i \tag{53}$$

$$-q_t^{\text{cap}\uparrow} \perp \delta_t^{\uparrow}, \quad \forall t \tag{54}$$

$$-q_t^{\text{cap}\downarrow} \perp \delta_t^{\downarrow}, \quad \forall t \tag{55}$$

where all dual variables are nonnegative, but $\lambda_{t,s}^{\text{a}\uparrow}$ and $\lambda_{t,s}^{\text{a}\downarrow}$, which are unrestricted.

The equivalent mixed-integer nonlinear program is (1), (30)–(55). The nonlinearity comes from multiplications of the upper-level variables (cleared battery-related quantities) and lower-level dual variables representing up and down reserve capacity reservation and activation. These are linearized using some of the KKT conditions and the strong duality equation as follows. First, the term $\lambda_t^{\text{cap}\uparrow} \cdot q_t^{\text{cap}\uparrow}$ is rewritten using KKT condition (34):

$$\lambda_t^{\text{cap}\uparrow} \cdot q_t^{\text{cap}\uparrow} = -\sum_{s \in \mathcal{S}} \nu_{t,s}^{\uparrow} \cdot q_t^{\text{cap}\uparrow} - \delta_t^{\uparrow} \cdot q_t^{\text{cap}\uparrow} + \zeta_t^{\uparrow} \cdot q_t^{\text{cap}\uparrow} + C^{\text{b,cap}\uparrow} \cdot q_t^{\text{cap}\uparrow} \tag{56}$$

Considering (46) and (54), Equation (56) is equal to:

$$\lambda_t^{\text{cap}\uparrow} \cdot q_t^{\text{cap}\uparrow} = -\sum_{s \in \mathcal{S}} \nu_{t,s}^{\uparrow} \cdot q_{t,s}^{\text{a}\uparrow} + \zeta_t^{\uparrow} \cdot q_t^{\text{cap}\uparrow} + C^{\text{b,cap}\uparrow} \cdot q_t^{\text{cap}\uparrow} \tag{57}$$

In a similar way, using (35), (47) and (55), we obtain the following equivalence:

$$\lambda_t^{\text{cap}\downarrow} \cdot q_t^{\text{cap}\downarrow} = -\sum_{s \in \mathcal{S}} \nu_{t,s}^{\downarrow} \cdot q_{t,s}^{\text{a}\downarrow} + \zeta_t^{\downarrow} \cdot q_t^{\text{cap}\downarrow} + C^{\text{b,cap}\downarrow} \cdot q_t^{\text{cap}\downarrow} \tag{58}$$

The term related to up reserve activation can be rewritten using (36):

$$\lambda^{a\uparrow}_{t,s} \cdot q^{a\uparrow}_{t,s} = -\beta^{\uparrow}_{t,s} \cdot q^{a\uparrow}_{t,s} + \nu^{\uparrow}_{t,s} \cdot q^{a\uparrow}_{t,s} + \pi_s \cdot C^{b,a\uparrow} \cdot q^{a\uparrow}_{t,s} \tag{59}$$

where $\beta^{\uparrow}_{t,s} \cdot q^{a\uparrow}_{t,s} = 0$ follows directly from (50). In a similar fashion, using (37) and (51) we obtain:

$$\lambda^{a\downarrow}_{t,s} \cdot q^{a\downarrow}_{t,s} = \nu^{\downarrow}_{t,s} \cdot q^{a\downarrow}_{t,s} + \pi_s \cdot C^{b,a\downarrow} \cdot q^{a\downarrow}_{t,s} \tag{60}$$

Finally, combining the obtained equalities (57)–(60) with the strong duality equality (The strong duality theorem states that, under certain conditions which are satisfied for linear optimization problems such as the one at hand, optimal solutions to the primal and the associated dual problem yield the same objective value [30].) (12) = (29), we obtain the following linear objective function of the upper-level problem:

$$\begin{gathered}
\underset{\Xi^{UL}}{\text{Maximize}} \quad \sum_{t\in\mathcal{T}} \Big[\lambda^{da}_t (q^{dis}_t - q^{ch}_t) + \\
\left(C^{b,cap\uparrow} \cdot q^{cap\uparrow}_t + C^{b,cap\downarrow} \cdot q^{cap\downarrow}_t\right) + \sum_{s\in\mathcal{S}} \left(\pi_s \cdot C^{b,a\uparrow} \cdot q^{a\uparrow}_{t,s} + \pi_s \cdot C^{b,a\downarrow} \cdot q^{a\downarrow}_{t,s}\right) - \\
\left(\sum_{i\in\mathcal{I}} C^{cap\uparrow}_i \cdot g^{cap\uparrow}_{t,i} + C^{b,cap\uparrow} \cdot q^{cap\uparrow}_t + \sum_{i\in\mathcal{I}} C^{cap\downarrow}_i \cdot g^{cap\downarrow}_{t,i} + C^{b,cap\downarrow} \cdot q^{cap\downarrow}_t\right) - \\
\sum_{s\in\mathcal{S}} \pi_s \cdot \left(\sum_{i\in\mathcal{I}} C^{a\uparrow}_i \cdot g^{a\uparrow}_{t,i,s} + C^{b,a\uparrow} \cdot q^{a\uparrow}_{t,s} + \sum_{i\in\mathcal{I}} C^{a\downarrow}_i \cdot g^{a\downarrow}_{t,i,s} + C^{b,a\downarrow} \cdot q^{a\downarrow}_{t,s}\right) + \\
\Big(R^{cap\uparrow}_t \cdot \lambda^{cap\uparrow}_t + R^{cap\downarrow}_t \cdot \lambda^{cap\downarrow}_t + \\
\sum_{s\in\mathcal{S}} R^{a\uparrow}_{t,s} \cdot \lambda^{a\uparrow}_{t,s} + \sum_{s\in\mathcal{S}} R^{a\downarrow}_{t,s} \cdot \lambda^{a\downarrow}_{t,s} - \sum_{i\in\mathcal{I}} G^{\uparrow}_{t,i} \cdot \psi^{\uparrow}_{t,i} - \sum_{i\in\mathcal{I}} G^{\downarrow}_{t,i} \cdot \psi^{\downarrow}_{t,i}\Big)\Big]
\end{gathered} \tag{61}$$

The final MILP formulation is (61) subject to constraints (2)–(11), (13)–(28), (30)–(55), where the orthogonal constraints (38)–(55) are easily linearized using the *big M* method.

4. Case Study

4.1. Input Data

The proposed model is tested on real data streaming from 1 May 2020. The day-ahead market prices, shown in Table 1 were taken from the German electricity exchange—EPEX, while the capacity and energy bids were gathered from an online German platform for balancing reserves auctions—Regelleistung.net. The former dataset is a series of 24 day-ahead prices, while the latter dataset for automatic frequency restoration reserve (for up and down reserve separately) consists of six 4-h periods, each of them including the following: total aFRR up/down volume and series of volume–price pairs (capacity price–capacity volume–energy price). The first stage in the auction is arranging the capacity price–capacity volume pairs in an ascending order by price, where all bids up to the total required volume (shown in Table 2) are accepted. Energy prices are used in the second stage in real-time when the TSO activates the accepted reserve providers. It arranges energy price–capacity volume pairs in an ascending order by price and all the bids up to total required energy are activated. For each 4-h period there are up to several hundred bids and many of them are identical both in terms of capacity and energy prices. To ease the computational efforts, we clustered similar ones and obtained between 30 and 90 total bids per timestep. Figure 2 shows the up reserve bids of the generators in the system. All the generators bid up a capacity reservation at zero €/MW (flat blue line), while the up reservation activation bids range from 36 to 2550 €/MWh (orange curve shows the activation bids sorted in ascending order). Figure 3 shows the down reserve bids of the generators in the system. As opposed to the up reserve, the down reserve capacity price is zero only for app. 500 €/MW, while the price of reservation of the remaining down reserve volume increases up to 8.6 €/MW (monotonically increasing blue curve). The corresponding activation prices are indicated with the orange curve. To minimize the operating cost (12), the system operator will activate the cheapest down reserve, i.e., the lowest values of the orange curve.

The data used to test and validate our model is taken from the German auction (www.regelleistung.net) and power system websites (www.smard.de) to accurately define one arbitrary chosen day. For the bids, real data for this specific day accounted for, on average, 283 and 333 bids over all bidding periods for the up and down reserve, respectively. In total in one day, there were 3697 bids for both up and down reserve during all bidding periods. A large number of those bids had the same values for both capacity and energy price or had the same number for one of those features and very similar for the other. To relieve the computational burden, but preserve the same level of accuracy, we aggregated those similar bids (in both features) and obtained on average 64 and 72 bids over all biding periods for the up and down reserve, respectively. This is in total 818 bids in one day for both up and down reserve during all bidding periods. This is still a very high number of bids even though the number of modeled bids was decreased by 88%. However, the accuracy of the case study remained untouched. When it comes to scenarios of activated aFRR, we used 10 scenarios as it is a sufficient number to validate the stochastic nature of the activation. Further increase in the number of scenarios would reduce computational efficiency for very low gains in the captured uncertainty.

Table 1. Day-ahead market prices (λ_t^{da}) on 1 May 2020.

Hour	Price (€/MWh)	Hour	Price (€/MWh)	Hour	Price (€/MWh)	Hour	Price (€/MWh)
1	5.5	7	2.54	13	0.35	19	18.99
2	5.35	8	1.50	14	−2.04	20	23.50
3	3.82	9	−1.57	15	−2.06	21	28.43
4	2.63	10	−2.43	16	−0.04	22	26.88
5	1.56	11	−2.89	17	1.95	23	20.91
6	2.46	12	−2.47	18	7.88	24	16.00

Table 2. Required up ($R_t^{cap\uparrow}$) and down ($R_t^{cap\downarrow}$) reserve per 4-hour periods on 1 May 2020.

	Hours 1–4	Hours 5–8	Hours 9–12	Hours 13–16	Hours 17–20	Hours 21–24
Up reserve (MW)	2359	2334	2355	2344	2357	2360
Down reserve (MW)	2247	2295	2338	2354	2316	2303

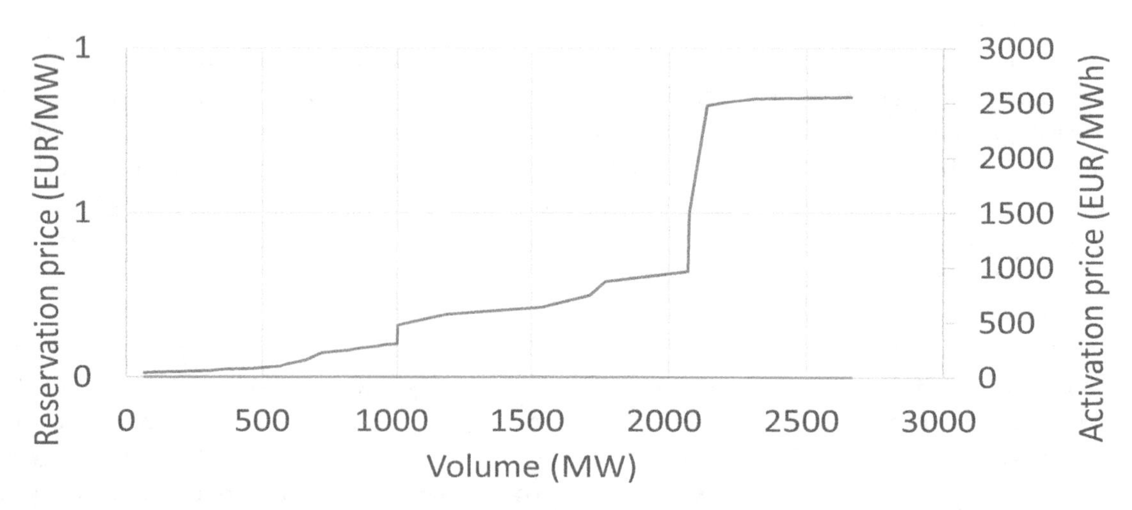

Figure 2. Up capacity reservation ($\lambda_t^{cap\uparrow}$) and activation ($\lambda_{t,s}^{a\uparrow}$) bids.

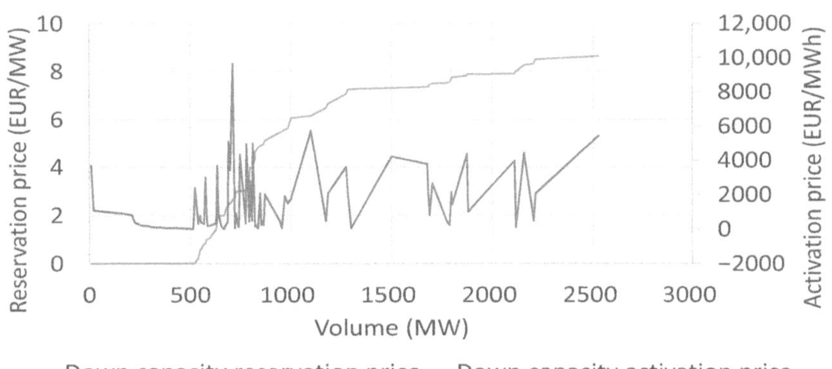

Figure 3. Down capacity reservation ($\lambda_t^{cap\downarrow}$) and activation ($\lambda_{t,s}^{a\downarrow}$) bids.

Energy prices are used in the second stage in real-time when the TSO activates the accepted reserve providers. It arranges energy price–capacity volume pairs in an ascending order by price and all the bids up to total required energy are activated.

A strategic battery storage (energy capacity 50 MWh; power capacity 50 MW; charging efficiency 1; discharging efficiency 0.82) is then added to the mentioned merit order lists. The system operator in the second stage of the reserve allocation process takes the energy bids, arranges them by price (ascending for up reserve, and descending for down reserve) and activates them one by one until satisfying the balancing energy request at a specific moment. The request for the total activated energy is modeled as an uncertain parameter through scenarios. In the case study, we used the quarter-hour activated aFRR balancing energies taken from the German electricity data transparency platform www.smard.de. The quarter-hours were summarized to an hourly resolution to match the hourly resolution of our model. Note that the same data was also used in papers [27,28]. The data for ten days streaming form May 1 to 10 May 2020 were taken as ten scenarios in our case study. The up and down reserve activation data are shown in Figures 4 and 5. To elaborate, each historical day (with all its hourly values) is shown as one scenario with a probability of 10%. The figures indicate a quite low activated volume, rarely surpassing 400 MWh, as compared to the reserved quantities from Table 2. Those scenarios affect our model results twofold: through the amount of activated reserve and through the price cleared for the activated reserve. In the case of batteries, the amount of activated reserve is relevant for securing a feasible state-of-energy evolution through time. It means that the state-of-energy boundaries will be satisfied regardless of which scenarios are actually realized. The price of activated reserves affects the profitability of reserve provision. The price maker models can be created in a way that their forecasted price is dependent on the activation scenarios as well, but they can not take into account the effect of the battery on the aFRR activated energy price formation.

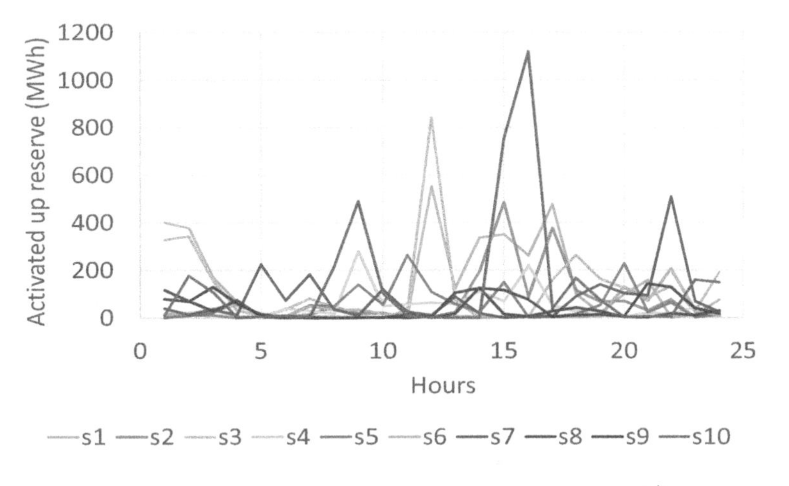

Figure 4. Up reserve activation per scenario ($R_{t,s}^{a\uparrow}$).

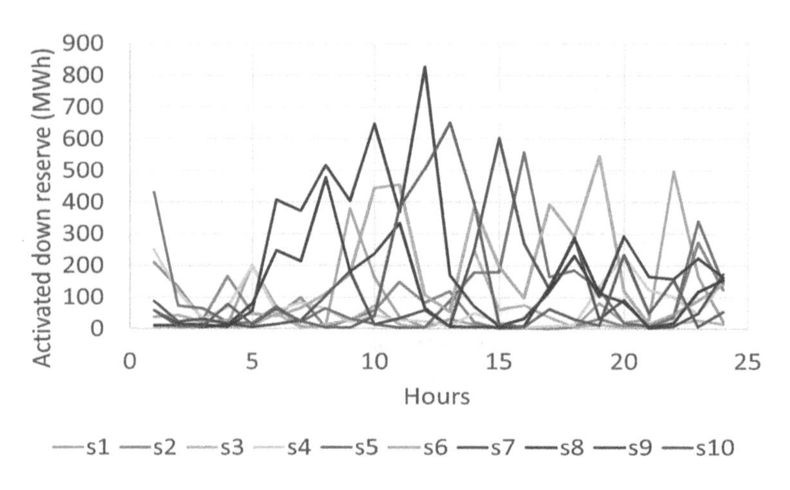

Figure 5. Down reserve activation per scenario ($R^{a\downarrow}_{t,s}$).

In the results of the case study, presented in the following subsection, we first analyze the results of the battery storage placing all four of its bids, i.e., up reserve capacity, down reserve capacity, up reserve activation and down reserve activation, at zero price. Note that, due to the marginal pricing, the battery storage will receive the marginal price that can only be better or equal to the one it bid price. After this analysis, we provide a sensitivity analysis with different values that the battery storage bids for the up reserve capacity, down reserve capacity, up reserve activation and down reserve activation.

4.2. Results

The maximum profit battery storage can achieve using the given input data is €22,171.61. While the revenue from providing down reserve capacity is quite high, €6724.47, the revenue from providing up reserve capacity is much lower, €21.03. On the other hand, the activation revenues are similar, €8506.66 for up reserve and €7291.02 for down reserve. The revenue in the day-ahead market is negative €371.57, as the battery storage primarily uses it to charge the energy later used for reserve activation. Figure 6 shows the battery storage day-ahead schedule along with the cleared up and down reserve capacities. Positive values represent the battery charging process, while negative ones the battery discharging process. In the day-ahead market, the battery storage generally charges during the night hours. It occasionally discharges (during hours 6, 8, 9, 12–15 and 19), but never over 18 MW. Provision of up reserve capacity (when activated, the battery discharges), never breaks 18 MW neither. It is significantly lower in volume than the down reserve capacity provision, which reaches 28 MW in hour 15. In some hours, e.g., 15, the system operator reserved both up and down reserve capacity from the battery. The activated amounts will differ based on the reserve activation scenario. For a more detailed explanation of the energy storage reserve activation please consult section 2.2 in [31].

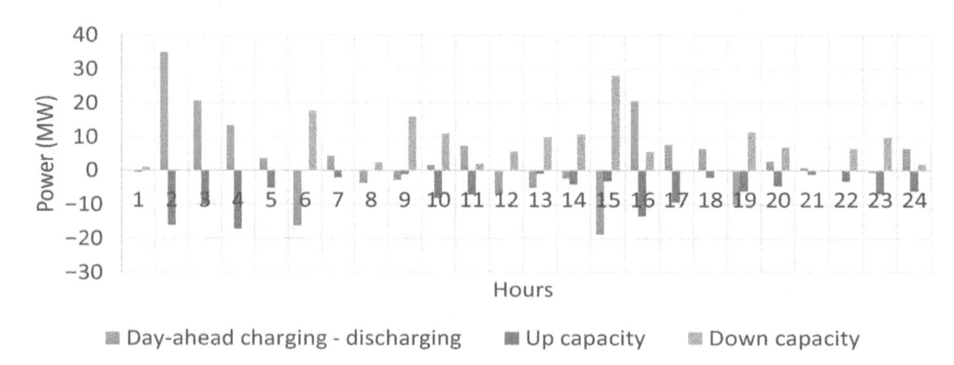

Figure 6. Battery charging day-ahead schedule ($q^{ch}_t - q^{dis}_t$) and up/down cleared reserve capacity q^{dis}_t and q^{ch}_t.

Figure 7 shows the propagation of the battery storage state-of-energy throughout the day for each scenario. Although the day-ahead schedule is the same, the activation direction (up or down) and the amount of activated reserve differs. For instance, in the 15-th hour, the battery storage reserves both up (3 MW) and down (28 MW) capacity. In scenario 3 we have 28 MW activated in the down direction and 1 MW in the up direction, while scenario 7 does not activate any up reserve, but activates 28 MW of down reserve. Since the modeled reserve is aFRR (15-minutes duration), a scenario can have activated both up and down reserve in the same hour (detailed visualization is available in Figures 8 and 9). In all scenarios, the battery storage is quite depleted in hour 15 and charges at 20.6 MW in the day-ahead market in hour 16. In the same hour, five out of ten scenarios provide 5.5 MW of down reserve (compare to Figure 6), enabling the battery storage to further charge in those scenarios (this is seen in Figure 7 as the ensemble of five scenarios with higher values of state-of-energy in hour 16). On the other hand, in the remaining five scenarios the battery activates 13.3 MW of up reserve, which reduces the charging effect from the day-ahead market, and consequently the battery receives less overall charge in those scenarios (this is seen in Figure 7 as the ensemble of five scenarios with lower values of state-of-energy in hour 16). Figure 7 is also useful to illustrate that the ending state-of-energy is highly dependent on the reserve activation scenario and ranges from 6 MWh for scenario 3 to 41.3 MWh for scenario 8.

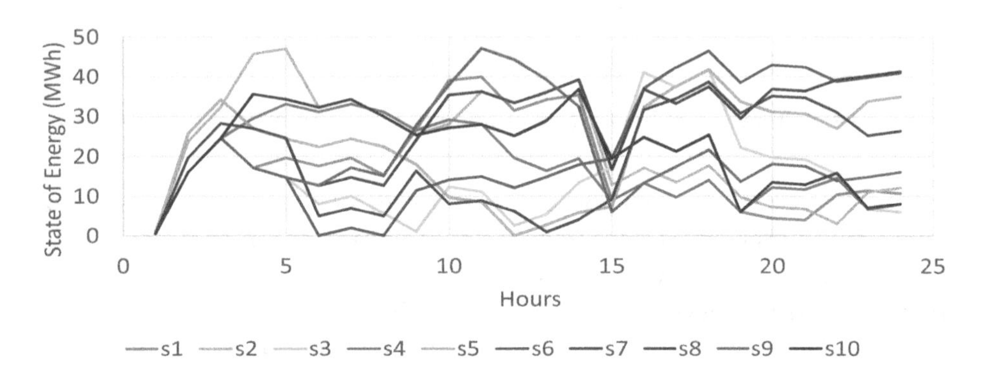

Figure 7. Propagation of the battery storage state-of-energy ($soe_{t,s}$) per scenario.

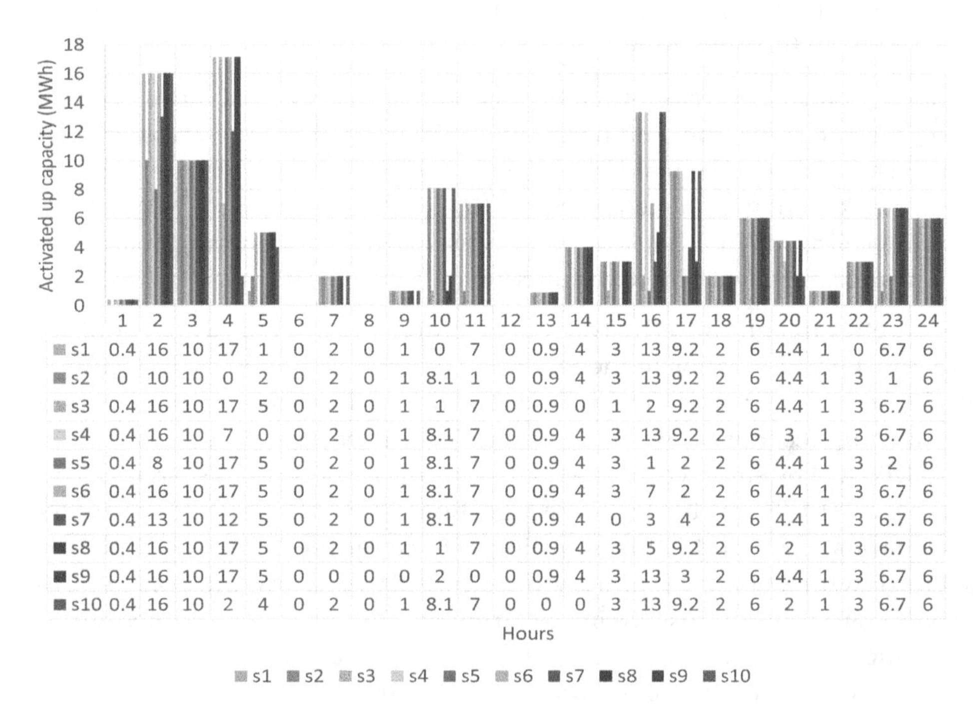

	1	2	3	4	5	6	7	8	9	10	11	12	13	14	15	16	17	18	19	20	21	22	23	24
s1	0.4	16	10	17	1	0	2	0	1	0	7	0	0.9	4	3	13	9.2	2	6	4.4	1	0	6.7	6
s2	0	10	10	0	2	0	2	0	1	8.1	1	0	0.9	4	3	13	9.2	2	6	4.4	1	3	1	6
s3	0.4	16	10	17	5	0	2	0	1	1	7	0	0.9	0	1	2	9.2	2	6	4.4	1	3	6.7	6
s4	0.4	16	10	7	0	0	2	0	1	8.1	7	0	0.9	4	3	13	9.2	2	6	3	1	3	6.7	6
s5	0.4	8	10	17	5	0	2	0	1	8.1	7	0	0.9	4	3	1	2	2	6	4.4	1	3	2	6
s6	0.4	16	10	17	5	0	2	0	1	8.1	7	0	0.9	4	3	7	2	2	6	4.4	1	3	6.7	6
s7	0.4	13	10	12	5	0	2	0	1	8.1	7	0	0.9	4	0	3	4	2	6	4.4	1	3	6.7	6
s8	0.4	16	10	17	5	0	2	0	1	1	7	0	0.9	4	3	5	9.2	2	6	2	1	3	6.7	6
s9	0.4	16	10	17	5	0	0	0	0	2	0	0	0.9	4	3	13	3	2	6	4.4	1	3	6.7	6
s10	0.4	16	10	2	4	0	2	0	1	8.1	7	0	0	0	3	13	9.2	2	6	2	1	3	6.7	6

Figure 8. Activation of the battery storage up reserve ($q^{a\uparrow}_{t,s}$) per scenario.

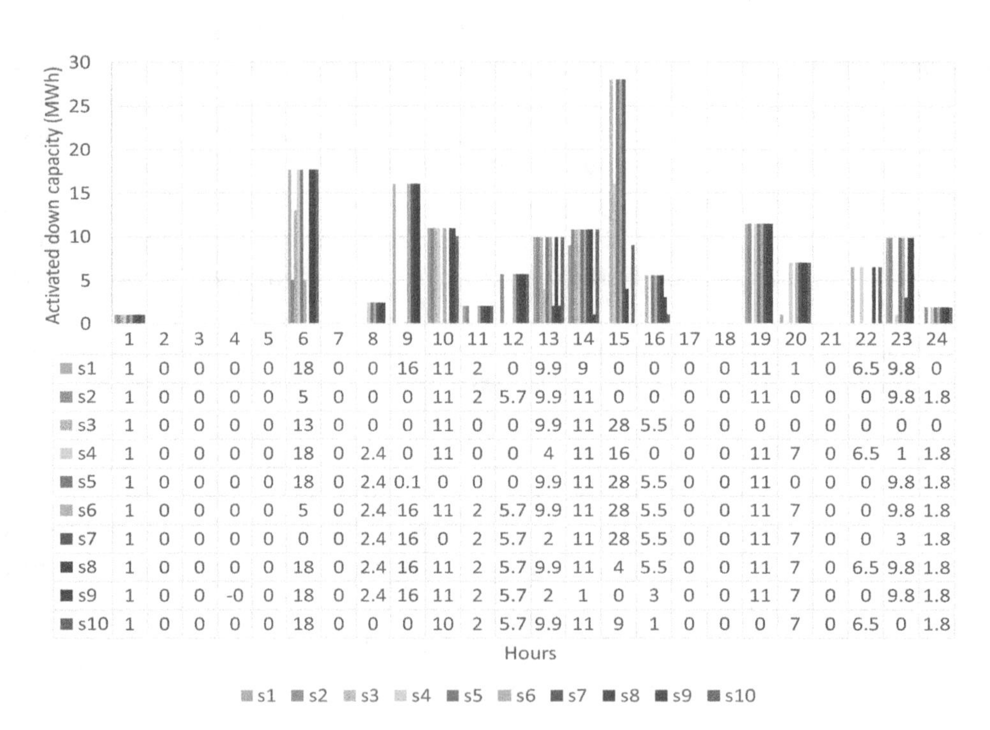

	1	2	3	4	5	6	7	8	9	10	11	12	13	14	15	16	17	18	19	20	21	22	23	24
s1	1	0	0	0	0	18	0	0	16	11	2	0	9.9	9	0	0	0	0	11	1	0	6.5	9.8	0
s2	1	0	0	0	0	5	0	0	0	11	2	5.7	9.9	11	0	0	0	0	11	0	0	0	9.8	1.8
s3	1	0	0	0	0	13	0	0	0	11	0	0	9.9	11	28	5.5	0	0	0	0	0	0	0	0
s4	1	0	0	0	0	18	0	2.4	0	11	0	0	4	11	16	0	0	0	11	7	0	6.5	1	1.8
s5	1	0	0	0	0	18	0	2.4	0.1	0	0	0	9.9	11	28	5.5	0	0	11	0	0	0	9.8	1.8
s6	1	0	0	0	0	5	0	2.4	16	11	2	5.7	9.9	11	28	5.5	0	0	11	7	0	0	9.8	1.8
s7	1	0	0	0	0	0	0	2.4	16	0	2	5.7	2	11	28	5.5	0	0	11	7	0	0	3	1.8
s8	1	0	0	0	0	18	0	2.4	16	11	2	5.7	9.9	11	4	5.5	0	0	11	7	0	6.5	9.8	1.8
s9	1	0	0	-0	0	18	0	2.4	16	11	2	5.7	2	1	0	3	0	0	11	7	0	0	9.8	1.8
s10	1	0	0	0	0	18	0	0	0	10	2	5.7	9.9	11	9	1	0	0	0	7	0	6.5	0	1.8

Figure 9. Activation of the battery storage down reserve ($q^{a\downarrow}_{t,s}$) per scenario.

Activation of the energy storage up and down reserves per scenario are visualized and listed in Figures 8 and 9. The numbers in the tables beneath these figures should be read column-by-column. In the first hour, the up reserve is fully activated (0.4 MW) in 9 out of 10 scenarios (Figure 8) and only in scenario s2 the battery up reserve remains inactive. The most noticeable property of the battery storage up reserve provision is having the activated capacity equal to the reserved capacity in the majority of scenarios. The lowest number of scenarios with fully activated up reserve occurs in hour 16, when only five scenarios experience full activation. Similar properties are observed for down reserve activation shown in Figure 9, where the lowest number of scenarios with fully activated reserve takes place in hour 15 with four full activations.

Generally, such uniform behavior of the battery storage reserve activation increases its utilization, i.e., the revenue of reserve activation, and harmonizes the state-of-energy across all scenarios. Since the last term in objective function (1) considers the weighed activation revenue, if the actual up reserve activation price in a certain hour of a scenario with 10% probability is €50/MWh, the value of the dual variable $\lambda^{a\uparrow}_{t,s}$ would be €5/MWh. This is a direct consequence of scenario probability π_s multiplying the activation costs in lower-level objective function (12).

To provide a better insight into the role of the battery storage in the overall reserve activation process, Tables 3 and 4 provide ratios of the reserve activation provided by the battery storage and the overall activated reserve for up and down direction. In the first hour, scenarios significantly vary in terms of the activated up reserve (Table 3). For scenario 1 the battery storage provides only 0.1% out of the activated 400 MWh. The same volume of battery's up activation in scenario 5 consists of 20% of the overall up reserve (0.4/2 MWh). In the second hour, the battery provides up to 16 MWh of the up reserve. In scenarios 2, 5, 6 and 7 this is sufficient to cover the entire required up reserve volume. When it comes to down reserve, the battery does not provide any portion in hours 2–5 (Table 4). In hour 9, it does not provide any reserve in scenarios that require low volumes, but it becomes active once the volumes increase (scenarios 1 and 6–9). This is because the down reserve activation prices of certain generators are negative (see the orange curve in Figure 3) and those are prioritized in the activation phase over the battery storage whose activation price is zero.

Table 3. Volume of up reserve activation provided by the battery per scenario as a portion of the overall activated reserve (rounded to an integer unless close to zero), in MWh.

Hour	s1	s2	s3	s4	s5	s6	s7	s8	s9	s10
1	0.4/400	0/0	0.4/327	0.4/8	0.4/2	0.4/17	0.4/38	0.4/116	0.4/79	0.4/10
2	16/377	10/10	16/341	16/21	8/8	16/16	13/13	16/70	16/69	16/177
3	10/170	10/10	10/156	10/32	10/23	10/38	10/34	10/28	10/129	10/107
4	17/67	0/0	17/32	7/7	17/68	17/53	12/12	17/74	17/62	2/2
5	1/1	2/2	5/7	0/0	5/11	5/17	5/223	5/5	5/15	4/4
6	0/0	0/3	0/34	0/39	0/4	0/5	0/72	0/0	0/2	0/9
7	2/5	2/13	2/80	2/28	2/52	2/48	2/182	2/2	0/0	2/5
8	0/4	0/47	0/44	0/17	0/48	0/33	0/35	0/0	0/0	0/213
9	1/2	1/17	1/6	1/279	1/138	1/33	1/9	1/1	0/0	1/489
10	0/0	8/20	1/1	8/50	8/54	8/16	8/118	1/1	2/2	8/102
11	7/7	1/1	7/31	7/57	7/263	7/7	7/25	7/13	0/0	7/8
12	0/554	0/2	0/841	0/63	0/108	0/7	0/10	0/0	0/4	0/1
13	1/123	1/23	1/74	1/57	1/58	1/5	1/89	1/17	1/104	1/0
14	4/336	4/192	0/0	4/110	4/22	4/4	4/17	4/125	4/124	0/0
15	3/349	3/484	1/1	3/69	3/148	3/3	0/0	3/15	3/116	3/753
16	13/260	13/86	2/2	13/219	1/1	7/7	3/3	5/5	13/76	13/1119
17	9/477	9/375	9/157	9/50	2/2	2/2	4/4	9/27	3/3	9/37
18	2/97	2/115	2/263	2/27	2/14	2/2	2/89	2/42	2/10	2/165
19	6/20	6/67	6/163	6/6	6/52	6/6	6/139	6/27	6/13	6/74
20	4/98	4/69	4/125	3/3	4/226	4/129	4/103	2/2	4/5	2/2
21	1/157	1/30	1/84	1/20	1/23	1/70	1/93	1/1	1/142	1/5
22	0/0	3/76	3/131	3/15	3/62	3/205	3/508	3/17	3/128	3/3
23	7/12	1/1	7/27	7/43	2/2	7/30	7/68	7/10	7/40	7/158
24	6/74	6/16	6/190	6/8	6/10	6/6	6/23	6/29	6/17	6/148

Table 4. Volume of down reserve activation provided by the battery per scenario as a portion of the overall activated reserve (rounded to an integer unless close to zero), in MWh.

Hour	s1	s2	s3	s4	s5	s6	s7	s8	s9	s10
1	1/8	1/209	1/6	1/250	1/430	1/37	1/88	1/57	1/10	1/57
2	0/5	0/132	0/4	0/118	0/72	0/43	0/23	0/15	0/10	0/8
3	0/4	0/28	0/5	0/32	0/65	0/20	0/31	0/12	0/6	0/8
4	0/15	0/166	0/5	0/68	0/20	0/27	0/16	0/6	0/12	0/77
5	0/201	0/48	0/9	0/200	0/18	0/49	0/6	0/56	0/81	0/12
6	18/72	5/40	13/48	18/55	18/69	5/40	0/15	18/407	18/247	18/64
7	0/7	0/100	0/4	0/84	0/21	0/66	0/26	0/373	0/214	0/24
8	0/8	0/4	0/14	2/63	2/65	2/112	2/105	2/516	2/477	0/5
9	16/380	0/29	0/27	0/20	0/32	16/173	16/183	16/403	16/181	0/4
10	11/163	11/57	11/71	11/42	0/15	11/444	0/14	11/648	11/237	10/41
11	2/38	2/148	0/12	0/27	0/4	2/455	2/33	2/369	2/333	2/384
12	0/5	6/84	0/4	0/24	0/5	6/108	6/58	6/827	6/65	6/504
13	10/30	10/119	10/28	4/9	10/89	10/54	2/7	10/169	2/7	10/651
14	9/14	11/20	11/246	11/49	11/179	11/387	11/253	11/70	1/6	11/392
15	0/1	0/2	28/60	16/21	28/180	28/198	28/603	4/9	0/5	9/14
16	0/2	0/4	6/74	0/5	6/559	6/99	6/271	6/32	3/8	1/6
17	0/0	0/2	0/36	0/8	0/165	0/394	0/127	0/119	0/134	0/62
18	0/5	0/8	0/6	0/11	0/186	0/292	0/290	0/231	0/284	0/30
19	11/83	11/32	0/14	11/133	11/127	11/546	11/32	11/104	11/113	0/11
20	1/21	0/9	0/5	7/231	0/19	7/121	7/92	7/293	7/85	7/234
21	0/30	0/12	0/5	0/124	0/9	0/16	0/3	0/165	0/6	0/48
22	7/498	0/40	0/16	7/103	0/36	0/45	0/8	7/159	0/18	7/155
23	10/167	10/274	0/27	1/47	10/341	10/89	3/49	10/223	10/118	0/7
24	0/23	2/125	0/15	2/151	2/145	2/141	2/173	2/163	2/155	2/53

To better understand battery storage actions, the prices in different markets are shown in Figure 10. As shown in Table 1, the day-ahead prices are rather low throughout the day, taking the highest values in hours 19–24. The up capacity prices are zero (or slightly positive) throughout the day, which reflects the very low day-ahead market prices. The down capacity prices are much higher, reaching €93/MW in the afternoon hours. The up and down activation prices in Figure 10 are averaged over all scenarios. They are much higher than the day-ahead prices. Despite extremely low up reserve capacity prices, the activation prices are much higher. The peak price €235/MWh is achieved for up reserve activation in hour 16, which is the main reason for the battery storage reserving 13.3 MW of its up capacity and activating it fully in five out of ten scenarios.

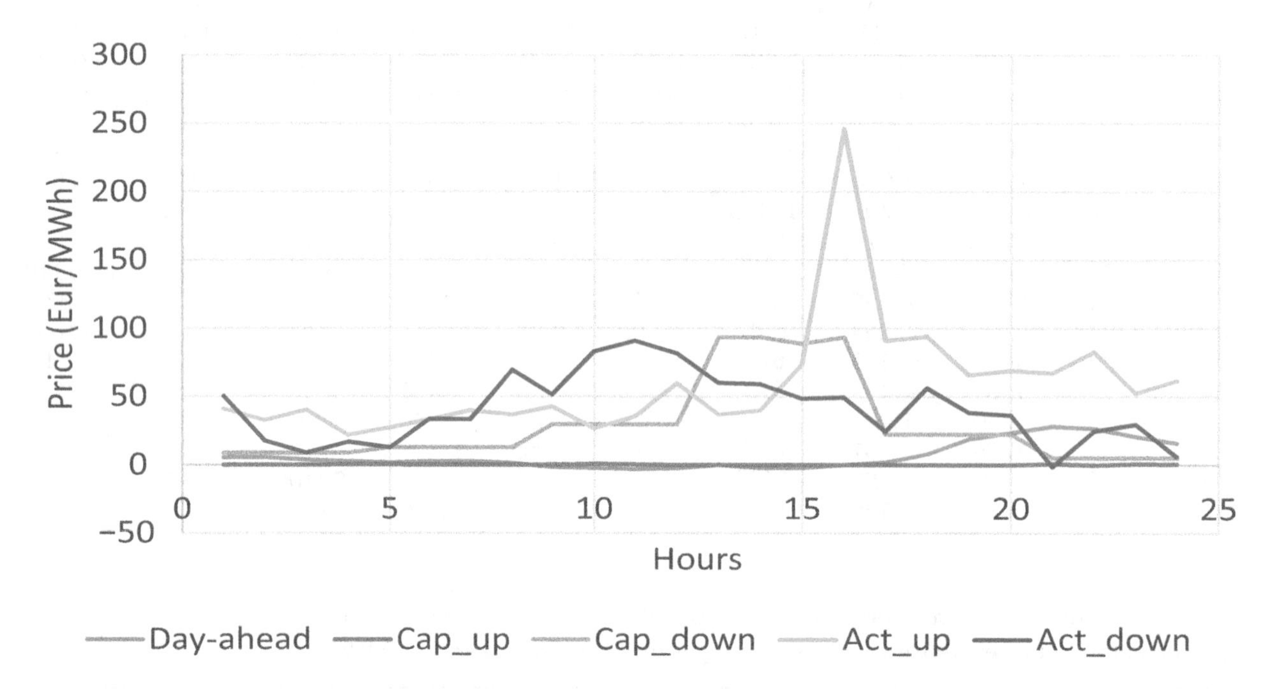

Figure 10. Prices in the day-ahead market (λ_t^{da}), up ($\lambda_t^{cap\uparrow}$) and down ($\lambda_t^{cap\downarrow}$) capacity reservation prices and up ($\lambda_{t,s}^{a\uparrow}$) and down ($\lambda_{t,s}^{a\downarrow}$) activated capacity prices.

4.3. Sensitivity Analysis

This section analyzes the effects of the battery storage bidding prices on its overall profit using the same data as the simulations in the previous section. The sensitivity includes variations in the four bidding parameters related to the reserves market: (i) up capacity reservation price (€ /MW), (ii) down capacity reservation price (€ /MW), (iii) up capacity activation price (€ /MWh), iv) down capacity activation price (€ /MWh). The results presented in Table 5 indicate that, regardless of the bidding prices, the battery storage utilizes the day-ahead market to charge (thus the day-ahead revenue is always negative), while the profit is made in the capacity reservation and activation stage. The only exception is the bidding strategy (10,10,50,−15), which has a high day-ahead positive revenue. This is the result of very frequent down capacity activation (the revenue is €12,416), which, besides that revenue itself, benefits the battery storage by charging it. This energy is discharged in the evening hours with the highest day-ahead prices to bring additional revenue in the day-ahead market.

Up reserve capacity revenue is generally very low, which is a direct consequence of the very low (mostly zero) up capacity reservation prices (see orange curve in Figure 10). However, the up capacity activation prices are high, especially during hour 16, and in most cases this stream of revenue is the highest. Down capacity reservation revenue is usually slightly higher than the activation revenue thanks to the high down capacity reservation prices during the afternoon hours (see the gray curve in Figure 10). The only exception are the last two cases.

Table 5. Effect of the bidding parameters on the battery storage profit (in €); the four numbers in the top cells indicate (i) up capacity reservation price (€ /MW), (ii) down capacity reservation price (€ /MW), (iii) up capacity activation price (€ /MWh), (iv) down capacity activation price (€ /MWh).

	Day-Ahead Revenue	Up Capacity Res. Revenue	Up Capacity Act. Revenue	Down Capacity Res. Revenue	Down Capacity Act. Revenue	Overall Revenue
(0,0,0,0)	−372	21	8507	6724	7291	22,172
(1,1,0,0)	−368	22	8835	6511	7163	22,162
(1,1,25,0)	−354	21	8684	6804	7274	22,429
(1,1,25,15)	−366	21	8316	5568	5630	19,168
(5,5,0,0)	−360	22	8588	6703	7205	22,158
(5,5,25,0)	−327	21	8066	7162	7515	22,438
(5,5,25,15)	−360	22	8096	5713	5698	19,169
(5,5,50,0)	−7	5	5386	5605	6352	17,342
(10,10,50,−15)	1033	7	3585	9172	12,416	26,212

The highest daily profit is achieved for bidding at €10/MW for both up and down capacity reservation, €50/MWh for up reserve activation and −€15/MW for down reserve activation. These bidding prices enable the battery storage to both affect the clearing prices (mostly by increasing them in its favor) and to win the auction in the majority of hours and scenarios. On the other hand, (5,5,50,0) bidding scheme results in the lowest overall profit, mostly because the high up reserve activation price €50/MWh reduced the up capacity activation revenue. However, the down activation bid at €0/MWh is insufficiently low for the battery storage to provide enough down reserve activation revenue to cancel out the negative monetary effects of the high up activation bid. On the other hand, the case with the highest profit (10,10,50,−15) provides sufficiently low down capacity activation bid for the battery storage to be cleared for activation more frequently and results in the highest down reserve activation revenue €12,416. This bidding strategy also results in the highest down capacity reservation revenue.

4.4. Comparison to a Baseline Model

To demonstrate the effectiveness and practical importance of the proposed model, we compare it against a baseline model where the battery storage acts as a price taker in all the markets and disregards its impact on the reserve capacity and activation prices. The baseline model includes only the upper-level problem (1) with capacity reservation and activation prices ($\lambda_t^{\text{cap}\uparrow}$, $\lambda_t^{\text{cap}\downarrow}$, $\lambda_{t,s}^{\text{a}\uparrow}$, $\lambda_{t,s}^{\text{a}\downarrow}$) treated as parameters. The obtained schedule is included in the market-clearing lower-level problem to obtain the actual profitability of the baseline model. The baseline model assessment procedure is described in the following steps:

1. First we solve only the lower-level problem (2) without battery storage bids, i.e., setting $\overline{q}_t^{\uparrow}$ and $\overline{q}_t^{\downarrow}$ to zero. This is needed to obtain the capacity reservation and activation prices $\lambda_t^{\text{cap}\uparrow}$, $\lambda_t^{\text{cap}\downarrow}$, $\lambda_{t,s}^{\text{a}\uparrow}$ and $\lambda_{t,s}^{\text{a}\downarrow}$.
2. Then we solve the upper-level problem (1) using the capacity reservation and activation prices $\lambda_t^{\text{cap}\uparrow}$, $\lambda_t^{\text{cap}\downarrow}$, $\lambda_{t,s}^{\text{a}\uparrow}$ and $\lambda_{t,s}^{\text{a}\downarrow}$ from the previous step. Note that the capacity reservation and activation prices are treated as parameters as opposed to being treated as variables in the proposed formulation. The outcome is the battery storage day-ahead and reserves bids.
3. Finally, we solve the lower-level problem (2) again, but this time with battery storage bids $\overline{q}_t^{\uparrow}$ and $\overline{q}_t^{\downarrow}$ from the previous step. This calculation provides actual reserve capacity and activation prices (note that these may differ from those obtained in step 1) as well as cleared battery storage quantities and profit.

After running Step 2 using the reserve capacity and activation prices from Step 1, the obtained battery storage profit is €57,518, which is more than two and a half times higher than €22,172 obtained using the proposed model. The obtained battery operation schedule for the baseline model is shown in Figure 11. The battery storage very rarely charges in the day-ahead market, the majority in hour 16. The battery charges primarily through the provision of down-regulation capacity.

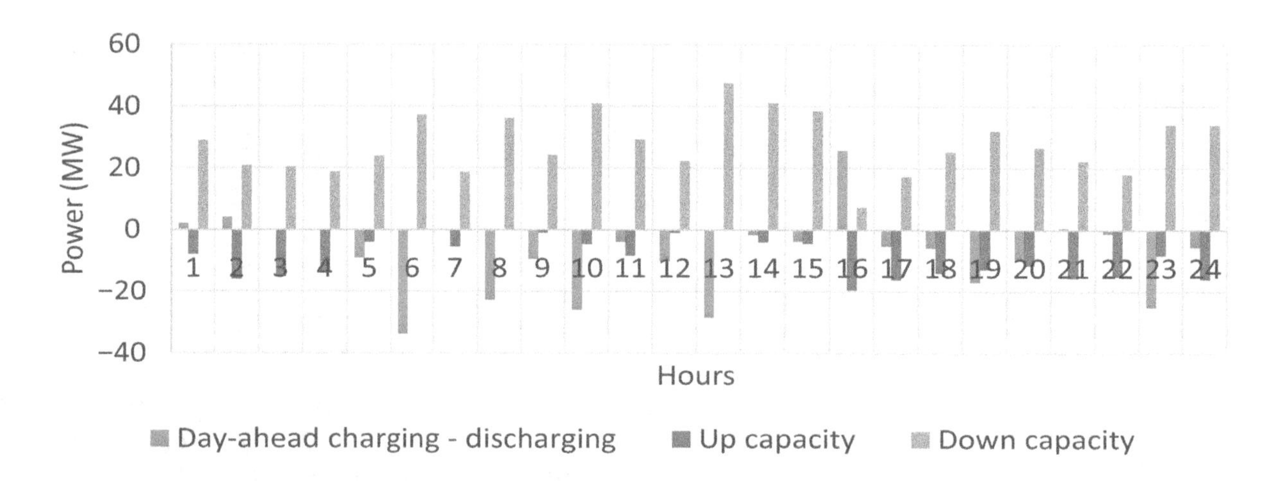

Figure 11. Battery charging day-ahead schedule ($q_t^{ch} - q_t^{dis}$) and up/down cleared reserve capacity q_t^{dis} and q_t^{ch} for the baseline case.

The obtained battery storage bidding schedule is then applied to the lower-level problem to calculate the reserve capacity and activation quantities actually accepted in the market and to deliver the true profit as the actual profit is expected to decrease if the battery storage's bids had an effect on the reserve capacity and activation prices. The obtained actual profit of the battery storage is only € 8856, which is almost three times lower than € 22,172 obtained using the proposed model. Although all battery storage bids were accepted in the market, the obtained baseline battery scheduling process failed to capture the interaction between the battery storage bids and the market-clearing prices. The result is a much lower profit than when using the proposed model, thus proving the effectiveness of the formulation presented in this paper.

5. Conclusions

The paper presented a model for the optimal bidding strategy of battery storage acting in the day-ahead market as a price taker and in the aFRR market as a price maker. The model accurately captures the essence of the electricity market structure in Europe, which is in the process of shifting toward an hourly marginal-price reserve structure. Although the battery storage from the case study is relatively small in size as compared to the overall reserves market volume (50 MW as opposed to over 2.3 GW), the battery storage can significantly affect aFRR reserve market since the activated energy is usually quite low. The bidding prices of the battery storage may have an adverse effect on its profit. Thus, the bidding prices and quantities need to be carefully chosen so the battery storage affects the market prices in a desirable way, but still stays in the money, i.e., gets cleared to provide reserve capacity and, when necessary, becomes activated.

1 May 2020, the day used in the case study, is characterized by a rather low reserve capacity prices. Despite that, the battery storage profit is significant and bidding in the reserves market is much more profitable than bidding only in the energy market. Since most of the days in the year 2020 have higher reserve capacity prices, these results can be considered conservative, i.e., the lower bound on the profits to be achieved in German markets.

The presented model and results should be useful to project developers and battery storage market participants as the battery storage costs are still quite high and accurately seizing all potential revenue streams is essential for the profitability of such investment.

Author Contributions: The authors contributed to the paper in the following capacities: conceptualization, I.A. and H.P.; methodology, I.P.; validation, K.P. and H.P.; model design I.P., I.A., and H.P.; model implementation K.P.; visualization, H.P. All authors have read and agreed to the published version of the manuscript.

Abbreviations

The following abbreviations are used in this manuscript:

AESO	Alberta Electric System Operator
aFRR	Automatic Frequency Restoration Reserve
FCR	Frequency Containment Reserve
KKT	Karush–Kuhn–Tucker
MILP	Mixed-Integer Linear Program
MOL	Merit Order List
SOE	State-Of-Energy
TSO	Transmission System Operator

References

1. Official Journal of the European Union. *Commission Regulation (EU) 2017/2195 of 23 November 2017 Establishing a Guideline on Electricity Balancing*; European Commission, Directorate-General for Energy: Brussels, Belgium, 2017.
2. Koch, C.; Hirth, L. Short-term electricity trading for system balancing: An empirical analysis of the role of intraday trading in balancing Germany's electricity system. *Renew. Sustain. Energy Rev.* **2019**, *113*, 109275. [CrossRef]
3. Lackner, C.; Nguven, T.; Byrne, R.H.; Wiegandt, F. Energy Storage Participation in the German Secondary Regulation Market. In Proceedings of the 2018 IEEE PES Transmission and Distribution Conference and Exposition (T&D), Denver, CO, USA, 16–19 April 2018; pp. 1–9.
4. PICASSO Project TSOs. *Consultation on the Design of the Platform for Automatic Frequency Restoration Reserve (AFRR) of PICASSO Region*; The Platform for the International Coordination of Automated Frequency Restoration and Stable System Operation (PICASSO): Brussels, Belgium, 2017.
5. Figgener, J.; Stenzel, P.; Kairies, K.P.; Linßen, J.; Haberschusz, D.; Wessels, O.; Angenendt, G.; Robinius, M.; Stolten, D.; Sauer, D.U. The development of stationary battery storage systems in Germany—A market review. *J. Energy Storage* **2020**, *29*, 101–153. [CrossRef]
6. Simon, B. The German Energy Storage Market 2016–2021: The Next Energy Transition. In *GTM Research Report*; GTM Research: Boston, MA, USA, 2016.
7. *RTE—Le RéSeau de Transport De L'électricité*; Electricity Report 2018; Réseau de Transport d'Électricité: Paris, France, 2019.
8. Regelleistung.net. Internetplattform zur Vergabe von Regelleistung. 2019. Available online: https://www.regelleistung.net/ext/static/prl (accessed on 5 March 2019).
9. Butler, P.C.; Iannucci, J.; Eyer, J. Innovative business cases for energy storage in a restructured electricity marketplace. In *Sandia National Laboratories Report*; Sandia National Laboratories: Albuquerqe, NM, USA, 2003.
10. Eyer, J.M.; Iannucci, J.; Corey, G.P. Energy storage benefits and market analysis handbook. In *Sandia National Laboratories Report*; Sandia National Laboratories: Albuquerqe, NM, USA, 2004.
11. Sioshansi, R.; Denholm, P.; Jenkin, T.; Weiss, J. Estimating the value of electricity storage in PJM: Arbitrage and some welfare effects. *Energy Econ.* **2009**, *31*, 269–277. [CrossRef]
12. Akhavan-Hejazi, H.; Mohsenian-Rad, H. Optimal Operation of Independent Storage Systems in Energy and Reserve Markets With High Wind Penetration. *IEEE Trans. Smart Grid* **2014**, *5*, 1088–1097. [CrossRef]

13. Kazemi, M.; Zareipour, H.; Amjady, N.; Rosehart, W.D.; Ehsan, M. Operation Scheduling of Battery Storage Systems in Joint Energy and Ancillary Services Markets. *IEEE Trans. Sustain. Energy* **2017**, *8*, 1726–1735. [CrossRef]
14. Fleer, J.; Zurmuhlen, S.; Meyer, J.; Badeda, J.; Stenzel, P.; Hake, J.-F.; Sauer, D. Price development and bidding strategies for battery energy storage systems on the primary control reserve market. *Energy Procedia* **2017**, *135*, 143–157. [CrossRef]
15. Fleer, J. Techno-economic evaluation of battery energy storage systems on the primary control reserve market under consideration of price trends and bidding strategies. *J. Energy Storage* **2018**, *17*, 345–356. [CrossRef]
16. Gomes, I.L.R.; Pousinho, H.M.I.; Melício, R.; Mendes, V.M.F. Stochastic coordination of joint wind and photovoltaic systems with energy storage in day-ahead market. *Energy* **2017**, *124*, 310–320. [CrossRef]
17. Miletić, M.; Pandžić, H.; Yang, D. Operating and Investment Models for Energy Storage Systems. *Energies* **2020**, *13*, 4600. [CrossRef]
18. Alberta Electric System Operator AESO. *Comparison Between Sequential Selection and Co-Optimization Between Energy and Ancillary Service Markets*; Technical Report; AESO: Calgary, AB, Canada, 2018.
19. Pavić, I.; Dvorkin, Y.; Pandžić, H. Energy and Reserve Co-optimization—Reserve Availability, Lost Opportunity and Uplift Compensation Cost. *IET Gener. Trans. Dis.* **2019**, *13*, 229–237. [CrossRef]
20. Ehsani, A. A Proposed Model for Co-Optimization of Energy And Reserve In Competitive Electricity Market. *Appl. Math. Model.* **2009**, *33*, 92–109. [CrossRef]
21. Chen, Y.; Gribik, P.; Gardner, J. Incorporating Post Zonal Reserve Deployment Transmission Constraints Into Energy and Ancillary Service Co-Optimization. *IEEE Trans. Power Syst.* **2014**, *29*, 537–549. [CrossRef]
22. Ela, E.; Milligan, M.; Kirby, B. Operating Reserves and Variable Generation. In *National Renewable Energy Laboratory Report*; National Renewable Energy Laboratory: Golden, CO, USA, 2011.
23. Chen, Y.; Wan, J.; Ganugula, V.; Merring, R.; Wu, J. Evaluating Available Room for Clearing Energy and Reserve Products under Midwest ISO Co-Optimization Based Real Time Market. In Proceedings of the 2010 IEEE PES General Meeting, Providence, RI, USA, 25–29 July 2010.
24. Hassan, M.W.; Rasheed, M.B.; Javaid, N.; Nazar, W.; Akmal, M. Co-Optimization of Energy and Reserve Capacity Considering Renewable Energy Unit with Uncertainty. *Energies* **2018**, *11*, 2833. [CrossRef]
25. Zeh, A.; Müller, M.; Naumann, M.; Hesse, H.C.; Jossen, A.; Witzmann, R. Fundamentals of Using Battery Energy Storage Systems to Provide Primary Control Reserves in Germany. *Batteries* **2016**, *2*, 29. [CrossRef]
26. Goebel, C.; Jacobsen, H. Aggregator-Controlled EV Charging in Pay-as-Bid Reserve Markets with Strict Delivery Constraints. *IEEE Trans. Power Syst.* **2016**, *31*, 4447–4461. [CrossRef]
27. Merten, M.; Rücker, F.; Schoeneberger, I.; Sauer, D.U. Automatic frequency restoration reserve market prediction: Methodology and comparison of various approaches. *Appl. Energy* **2020**, *268*, 114978. [CrossRef]
28. Merten, M.; Olk, C.; Schoeneberger, I.; Sauer, D.U. Bidding strategy for battery storage systems in the secondary control reserve market. *Appl. Energy* **2020**, 268, 114951. [CrossRef]
29. Pandžić, H.; Bobanac, V. An Accurate Charging Model of Battery Energy Storage. *IEEE Trans. Power Syst.* **2019**, *4*, 1416–1426.
30. Conejo, A.J.; Castillo, E.; Minguez, R.; Garcia-Bertrand, R. *Decomposition Techniques in Mathematical Programming*; Springer: Berlin/Heidelberg, Germany, 2006.
31. Pandžić, H.; Dvorkin, Y.; Carrion, M. Investments in merchant energy storage: Trading-off between energy and reserve markets. *Appl. Energy* **2018**, 230, 277–286. [CrossRef]

Real-Time Processor-in-Loop Investigation of a Modified Non-Linear State Observer using Sliding Modes for Speed Sensorless Induction Motor Drive in Electric Vehicles

Mohan Krishna Srinivasan [1,*], Febin Daya John Lionel [2], Umashankar Subramaniam [3], Frede Blaabjerg [4], Rajvikram Madurai Elavarasan [5,*], G. M. Shafiullah [6,*], Irfan Khan [7] and Sanjeevikumar Padmanaban [8]

1. Department of Electrical and Electronics Engineering, Alliance College of Engineering and Design, Alliance University, Bangalore 562 106, India
2. SELECT, Vellore Institute of Technology, Chennai 600127, India; febinresearch@gmail.com
3. Renewable Energy Lab, College of Engineering, Prince Sultan University, Riyadh 11586, Saudi Arabia; usubramaniam@psu.edu.sa
4. CORPE, Department of Energy Technology, Aalborg University, 9000 Aalborg, Denmark; fbl@et.aau.dk
5. Electrical and Automotive parts Manufacturing unit, AA Industries, Chennai 600123, India
6. Discipline of Engineering and Energy, Murdoch University, Murdoch 6150, Australia
7. Marine Engineering Technology Department in a joint appointment with Electrical and Computer Engineering, Texas A&M University, College Station, TX 77843, USA; irfankhan@tamu.edu
8. Department of Energy Technology, Aalborg University, 6700 Esbjerg, Denmark; sanjeevi_12@yahoo.co.in

* Correspondence: smk87.genx@gmail.com (M.K.S.); rajvikram787@gmail.com (R.M.E.); gm.shafiullah@murdoch.edu.au (G.M.S.)

Abstract: Tracking performance and stability play a major role in observer design for speed estimation purpose in motor drives used in vehicles. It is all the more prevalent at lower speed ranges. There was a need to have a tradeoff between these parameters ensuring the speed bandwidth remains as wide as possible. This work demonstrates an improved static and dynamic performance of a sliding mode state observer used for speed sensorless 3 phase induction motor drive employed in electric vehicles (EVs). The estimated torque is treated as a model disturbance and integrated into the state observer while the error is constrained in the sliding hyperplane. Two state observers with different disturbance handling mechanisms have been designed. Depending on, how they reject disturbances, based on their structure, their performance is studied and analyzed with respect to speed bandwidth, tracking and disturbance handling capability. The proposed observer with superior disturbance handling capabilities is able to provide a wider speed range, which is a main issue in EV. Here, a new dimension of model based design strategy is employed namely the Processor-in-Loop. The concept is validated in a real-time model based design test bench powered by RT-lab. The plant and the controller are built in a Simulink environment and made compatible with real-time blocksets and the system is executed in real-time targets OP4500/OP5600 (Opal-RT). Additionally, the Processor-in-Loop hardware verification is performed by using two adapters, which are used to loop-back analog and digital input and outputs. It is done to include a real-world signal routing between the plant and the controller thereby, ensuring a real-time interaction between the plant and the controller. Results validated portray better disturbance handling, steady state and a dynamic tracking profile, higher speed bandwidth and lesser torque pulsations compared to the conventional observer.

Keywords: machine model; adaptive control; model reference; disturbance; stability; real-time; processor-in-loop (PIL); electric vehicles

1. Introduction

Electric vehicles (EVs) have come to occupy considerable space in the transportation sector owing to less harmful emissions, better energy profile, lesser noise and cheaper maintenance and operating costs. However, disadvantages exist in the form of range anxiety, charging infrastructure and battery safety and disposal. An induction motor continues to dominate owing to its robustness, ruggedness, smaller size and plays a major role in the electric transportation domain [1,2]. Additionally, the speed range or bandwidth of the motor plays a major role in an EV. One major aspect, which has often been overlooked, is the space constraints inside the EV. Although the induction motor is compact and eliminates the use of commutator brush assembly (as seen in DC motors), the presence of the speed sensor mounted on the motor shaft adds to the space and additional electronics (sensitive to vehicle vibrations and dynamics) in the EV system. Therefore, it is felt that a sensorless speed estimation system is suitable and also economically and technologically feasible in an EV. Additionally, the use of the speed sensor also adds to the non-linearity of the system by means of the sensor noise, which may affect the gain and dynamic performance of the motor used in the EV [3]. The domain of speed sensorless estimation and control of induction motors has gained popularity for the past decade due to the elimination of the speed sensor owing to cost, reliability and sensitivity constraints [4]. Adaptive speed and parameter observation schemes became more popular owing to their pace of adaptation, ease of use and less computational space [5,6]. The decoupled control strategy also emerged as the most popular one [7,8]. Recently numerous controllers are proposed using wavelet transform and fuzzy tuning (WTFT) [9,10]. Computational intelligence based state estimation also added to the ongoing research on parameter estimation and online adaptation [11,12]. Most of the adaptive schemes follow the concept of model reference adaptive systems (MRASs) [13] as shown in Figure 1. Investigation particularly towards the different configuration schemes based on the state observers is also presented [14]. The extended Kalman filters are widely applied for state estimation, and demanded extensive computational space and a high sampling frequency [15]. Besides, the system dynamics can be linearized for the accuracy. Most of them focused on joint state estimation and adaptation at speeds ranging from low speeds to flux weakening regions [16,17]. Extended Luenberger observer (ELO) had an additional correction term [18] incorporated into the state dynamic equation, which involved the stator current error dynamics and provided more efficient dynamic and robust performance [19,20]. The variable structure concept [21] was also integrated into the ELO to constrain the system state and to reject the effect of the error dynamics, giving rise to sliding mode Luenberger observers (SMLOs) [22,23]. The very essence of observation schemes for the purpose of parameter estimation is brought about in [24]. Some investigations did focus on the estimation of disturbance as a parameter to test the robustness of the observer in offline simulation platforms [25,26]. The amount of non-linearities involved in a closed loop control system was brought about in [27]. Therefore, there was a need to decouple a non-linear system and control it in a linear domain. The emergence of power switching devices and the effect it had on variable frequency control was portrayed in [28]. This also had a negative effect in increasing the number of non-linearities in a drive system.

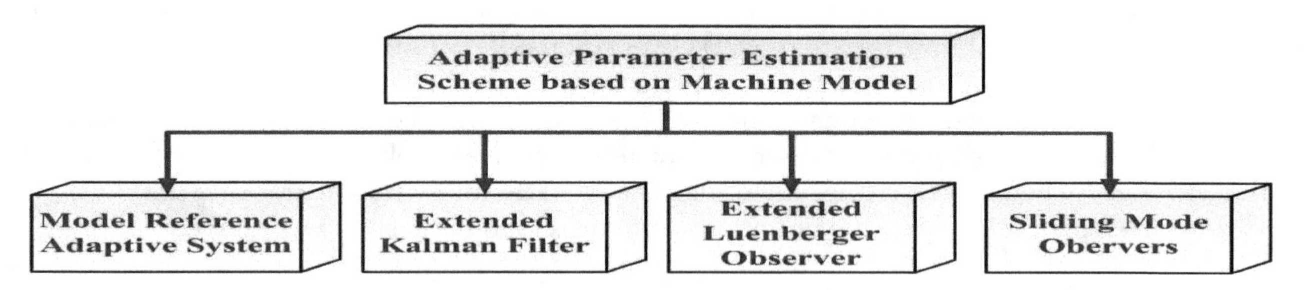

Figure 1. Machine model based adaptive parameter estimation schemes.

Presently, real-time validation platforms have taken over from offline simulation platforms. Several computer languages like Simulink and LabVIEW permit the creation of computer models, which can be connected to real-time embedded systems or electronic control units through digital and analog I/O (input/output) cards. There are many available in the market like xPC and Opal-RT's RT-Lab along with real-time targets [29] that are primarily used for mechatronics, power electronics, electric vehicle drives and power grid protection tests. As part of the model based design strategy for testing computer models, there are several testing levels namely the software in loop (SIL), model in loop (MIL) and hardware in loop (HIL). While SIL is employed for verification of the code with respect to Matlab functions, MIL environment is used for testing the model without any physical hardware components. The PIL strategy is similar to MIL, however, there is a difference in signal routing that is real-time in PIL. Although several observers have been designed by making use of the concepts of sliding modes, artificial intelligence and model reference.

- Very few cater to the handling of measurement and model disturbances and their effects on the steady state and dynamic performance of the drive.
- Existing disturbance observers have short comings in terms of speed bandwidth, parameter estimation, torque profile and stability issues.
- Although many have been validated in an offline simulation platform, some in an experimental test bench, none of them have been tested in a real-time PIL test bench (which is an intermediate between offline simulation and full hardware verification).

The purpose of this work is to design, test and analyze a SMLO estimating the speed and disturbance torque for a three phase speed sensorless induction motor.

- By effective placement of the estimated disturbance torque in the proposed observer state dynamic equation, greater handling of the disturbance is seen as compared to the other disturbance observer whose tracking performance is affected. The speed bandwidth is increased and the torque holding capacity is also good. It is comparatively more stable and the dame has been analyzed through the pole placement technique.
- Furthermore, in a new real-time PIL platform, the plant and the controller are made to interact through digital and analog I/O cards by providing a real-time link. This testing is based on a model based design paradigm and has not been performed in the existing literature.

The paper is organized as follows: Section 1 relates to the motivation, literature survey and the limitations of the existing work and the key contributions of the paper. Section 2 discusses the basic principle of the adaptive system and the structure of the observer using sliding modes, the proposed disturbance rejection mechanisms inside the observers and the stability analysis of the conventional and proposed disturbance observers. Section 3 outlines the mathematical structure of the existing indirect vector control strategy. Section 4 focuses on an elaborate introduction and representation of the real-time test bench based on Opal-RT for the system validation. Section 5 presents the detailed results and analysis of the dynamic and static performance of the observers when subjected to different test cases followed by the pole placement study and performance comparison. Finally, the conclusion section emphasizes the importance of the findings to the existing literature and its significance with respect to vehicle performance and dynamics.

2. Basic Principle of MRAS and Structure of the SMLO

The SMLO with the estimated disturbance torque incorporated into the sliding hyperplane is shown in Figure 2a. It is a multiple input multiple output system (MIMO) where the inputs are the terminal quantities of the motor and the outputs are the estimated speed and the disturbance torque. The state space model of the motor and the observer are used, as it is suited for estimation and control functions. The primary reason of adaptive control is for parameter estimation. It is to match the desired performance (observer model) with that of the process (motor model). This principle can be explained as an optimization problem. Therefore, the essence is to minimize the error for state convergence. The complete system is shown in Figure 2b. Here, 'A' represents the system matrix,

'^' denotes estimated parameters, 'X' represents the state variables comprising of the d and q axes stator and currents rotor fluxes, 'k_{sw}' is the switching gain, it can either be a fixed value or a reduced order matrix. 'J', 'p' and 'B_V' represent the moment of inertia, differential operator and viscous friction coefficient respectively. 'T_e^*' and '$\hat{T}_{dis}$' are electromagnetic and estimated disturbance torque, 'k' is a positive gain.

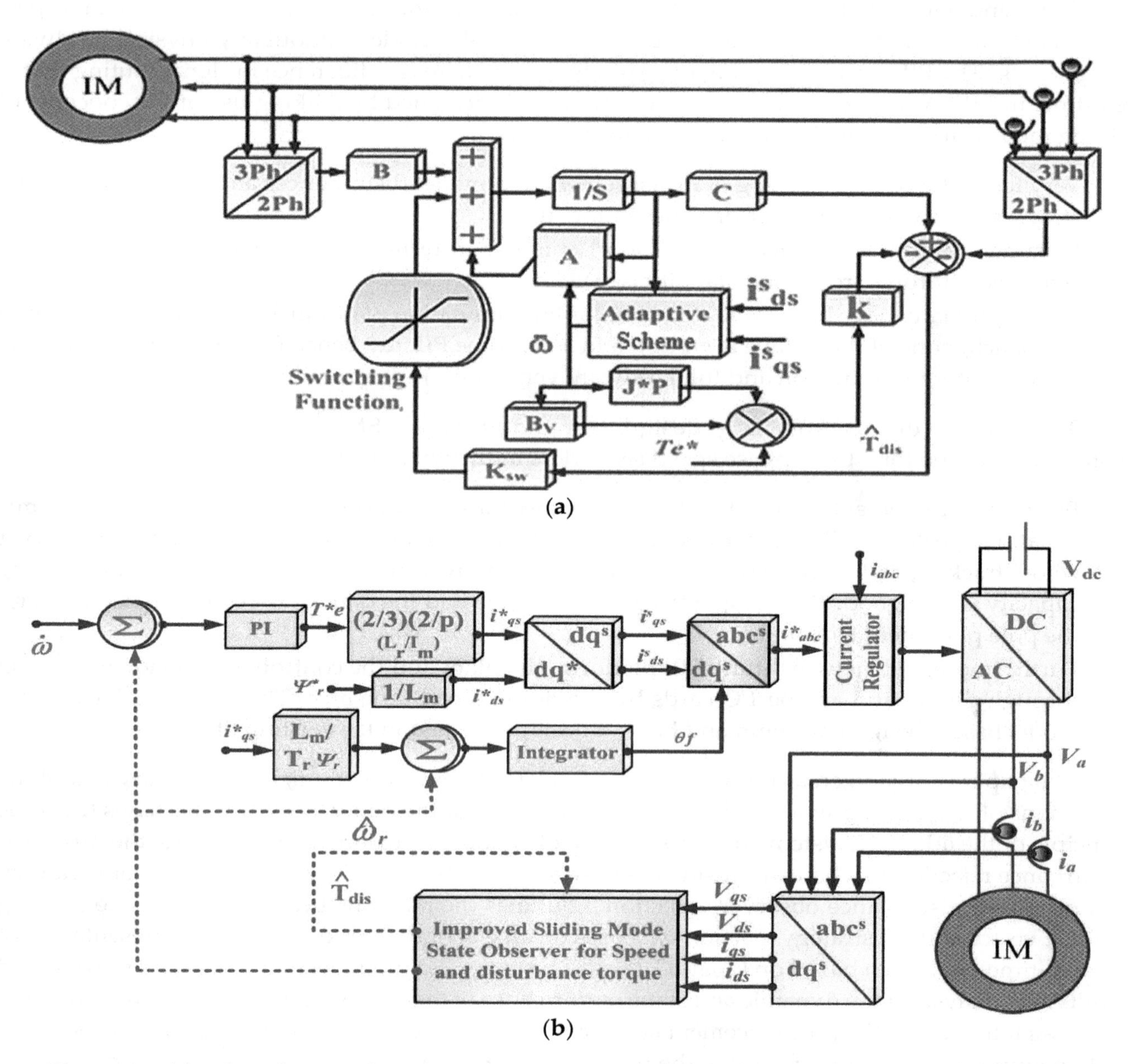

Figure 2. (**a**). Speed sensor-less drive system: sliding mode speed and disturbance observer. (**b**). Schematic of the voltage source inverter fed drive system employing the improved observer.

The selection and the stability conditions of the sliding hyperplane play a major role in observer dynamics. It should be selected such that it satisfies the Lyapunov stability criterion [30]. The sliding hyperplane 'S' and the LFC (Lyapunov function candidate) "V" is a scalar function of S [31].

$$\dot{V}(S) = S(x)\dot{S}(x) \tag{1}$$

The control law is expressed as:

$$u(t) = u_{eq}(t) + u_{sw}(t) \tag{2}$$

The switching vector $u_{sw}(t)$ satisfying stability conditions:

$$u_{sw}(t) = \eta \text{sign}(S(x, t)) \tag{3}$$

where, $\text{sign}(S) = \begin{cases} -1 \text{ for } S < 0 \\ 0 \text{ for } S = 0 \\ +1 \text{ for } S > 0 \end{cases}$, where, η denotes the switching control gain so as to make (1) negative definite. This implies:

$$S(x)\dot{S}(x) < 0 \tag{4}$$

This constrains the disturbance. Chattering is produced due to this non-linear high frequency switching. To remove it, a saturation function with boundary layer of width (Φ) replacing sign (S) with sat (S/Φ):

$$\text{sat}(S/\Phi) = \begin{cases} \text{sign}\left(\frac{S}{\Phi}\right) \text{if } \left|\left(\frac{S}{\Phi}\right)\right| 1 \\ \left(\frac{S}{\Phi}\right) \text{if } \left|\left(\frac{S}{\Phi}\right)\right| < 1 \end{cases} \tag{5}$$

2.1. Motor Model (Reference)

$$\frac{dx}{dt} = [A]x + [B]u \tag{6}$$

$$y = [C]x \tag{7}$$

where,

$$x = \left[i_{ds}^{s}, i_{qs}^{s}, \psi_{dr}^{s}, \psi_{qr}^{s}\right]^{T}, A = \begin{bmatrix} A_{11} & A_{12} \\ A_{21} & A_{22} \end{bmatrix}$$

$$B = \left[\frac{1}{\sigma L_s} I \, 0\right]^{T}, \; C = [I, 0], \; u = \left[v_{ds}^{s} \; v_{qs}^{s}\right]^{T}$$

$$I = \begin{bmatrix} 1 & 0 \\ 0 & 1 \end{bmatrix}, \; J = \begin{bmatrix} 0 & -1 \\ 1 & 0 \end{bmatrix}$$

$$A_{11} = -\left[\frac{R_s}{\sigma L_s} + \frac{1-\sigma}{\sigma T_r}\right] I = a_{r11} I$$

$$A_{12} = \frac{L_m}{\sigma L_s L_r}\left[\frac{1}{T_r} I - \omega_r J\right] = a_{r12} I + a_{i12} J$$

$$A_{21} = \frac{L_m}{T_r} I = a_{r21} I$$

$$A_{22} = \frac{-1}{T_r} I + \omega_r J = a_{r22} I + a_{i22} J$$

2.2. Disturbance Torque Estimation

It is expressed as:

$$\hat{T}_{dis} = T_e^{*} - J\frac{d\hat{\omega}}{dt} - B_V \hat{\omega} \tag{8}$$

2.3. SMLO1

The way the disturbances are handled play a major role in state convergence of an observer system. The disturbance handling method in SMLO1 is similar to many disturbance observers where the estimated disturbance is integrated into the state dynamic equation. In this case, the main difference from the proposed method (SMLO2) is the way it handles the disturbance. Here, it is not constrained in the sliding hyperplane. Therefore, the estimated disturbance is not part of the state convergence of the sliding hyperplane. This is elaborated in the following model:

$$\frac{d\hat{x}}{dt} = [\hat{A}]\hat{x} + [B]u + k_{sw}sat(\hat{i}_s - i_s) + \hat{d} \tag{9}$$

Sliding surface or hyperplane is $s = \hat{i}_s - i_s$ and $\hat{d} = k\,\hat{T}_{dis}$

$$\hat{y} = [C]\hat{x} \tag{10}$$

where $\hat{i}_s$, i_s= estimated and actual stator currents.

$$\hat{A} = \begin{bmatrix} A_{11} & \hat{A}_{12} \\ A_{21} & \hat{A}_{22} \end{bmatrix}$$

$$\hat{A}_{12} = \frac{L_m}{\sigma L_s L_r}\left[\frac{1}{T_r}I - \hat{\omega}_r J\right] = a_{r12}I + \hat{a}_{i12}J$$

$$\hat{A}_{22} = \frac{-1}{T_r}I + \hat{\omega}_r J = a_{r22}I + \hat{a}_{i22}J$$

Using the reduced order matrix:

$$k_{sw} = \begin{bmatrix} k_1 & k_2 \\ -k_2 & k_1 \end{bmatrix}^T \tag{11}$$

The purpose of the switching gain is to make (2) stable through pole placement. The eigenvalues of the observer must be more negative with respect to the motor for faster convergence of the observer and motor states. Consequently,

$$k_1 = (m - 1)a_{r11} \tag{12}$$

$$k_2 = k_p, k_p \geq -1 \tag{13}$$

Therefore, 'm' and 'k_2' are chosen to reflect the placement of the eigenvalues of the observer and the motor. Additionally, the dynamics and damping of the observer are affected by the same thing. 'k_1' is dependent on 'm' and motor parameters.

2.4. SMLO2

Here, the observer state dynamic equation is modified as shown:

$$\frac{d\hat{x}}{dt} = [\hat{A}]\hat{x} + [B]u + k_{sw}sat(\hat{i}_s - i_s - \hat{d}) \tag{14}$$

where, the sliding surface or hyperplane is $s = \hat{i}_s - i_s - \hat{d}$ and $\hat{d} = k\hat{T}_{dis}$

$$\hat{y} = [C]\hat{x} \tag{15}$$

2.5. Adaptive Mechanism with LFC

It is denoted by M:

$$M = e^T e + \frac{(\hat{\omega}_r - \omega_r)^2}{\lambda} \tag{16}$$

'λ', being, a positive constant.

The derivative of the function candidate with respect to t:

$$\frac{dM}{dt} = e^T\left[(A + k_{sw}C)^T + (A + k_{sw}C)\right]e - \frac{2\Delta\omega_r\left(e_{ids}\hat{\varphi}^s_{qr} - e_{iqs}\hat{\varphi}^s_{dr}\right)}{c} + \frac{2\Delta\omega_r}{\lambda}\frac{d\hat{\omega}_r}{dt} \tag{17}$$

where, $e_{ids} = i^s_{ds} - \hat{i}^s_{ds}$ and $e_{iqs} = i^s_{qs} - \hat{i}^s_{qs}$

From (16), the estimated speed expression is obtained.

$$\frac{d\hat{\omega}_r}{dt} = \frac{\lambda}{c}\left(e_{ids}\hat{\varphi}^s_{qr} - e_{iqs}\hat{\varphi}^s_{dr}\right) \tag{18}$$

'c' is arbitrary positive.

2.6. Stability Analysis by the Pole Placement Technique—SMLO1 and SMLO2

For the SMLO1 with conventional disturbance rejection mechanism, we have:

$$\left(A_{11} + k_{sw} + \hat{d}\right) = \begin{bmatrix} a_{r11} + k_1 + \hat{d} & -k_2 + \hat{d} \\ k_2 + \hat{d} & a_{r11} + k_1 + \hat{d} \end{bmatrix} \tag{19}$$

Now, the characteristic equation can be obtained by:

$$SI - \left(A_{11} + k_{sw} + \hat{d}\right) = 0 \tag{20}$$

On solving, we get the characteristic equation of SMLO1:

$$S^2 - 2S\left(a_{r11} + k_1 + \hat{d}\right) + \left(a_{r11} + k_1 + \hat{d}\right)^2 + \left(k_2^2 - \hat{d}^2\right) = 0 \tag{21}$$

From the characteristic equation, the observer poles are obtained:

$$S_1 = \left(a_{r11} + k_1 + \hat{d}\right) + j\left(k_2 - \hat{d}\right) \tag{22}$$

$$S_2 = \left(a_{r11} + k_1 + \hat{d}\right) - j\left(k_2 - \hat{d}\right) \tag{23}$$

For the SMLO2 with improved disturbance rejection mechanism, we have:

$$\left(A_{11} + k_{sw} - k_{sw}\hat{d}\right) = \begin{bmatrix} a_{r11} + k_1 - k_{sw}\hat{d} & -k_2 - k_{sw}\hat{d} \\ k_2 - k_{sw}\hat{d} & a_{r11} + k_1 - k_{sw}\hat{d} \end{bmatrix} \tag{24}$$

The characteristic equation is obtained by:

$$SI - \left(A_{11} + k_{sw} - k_{sw}\hat{d}\right) = 0 \tag{25}$$

Therefore, on solving, the characteristic equation of SMLO2:

$$S^2 - 2S\left(a_{r11} + k_1 - k_{sw}\hat{d}\right) + \left(a_{r11} + k_1 - k_{sw}\hat{d}\right)^2 + \left(k_2^2 - k_{sw}^2\hat{d}^2\right) = 0 \tag{26}$$

The observer poles are obtained as follows:

$$S_1 = \left(a_{r11} + k_1 - k_{sw}\hat{d}\right) + j\left(k_2 - k_{sw}\hat{d}\right) \tag{27}$$

$$S_1 = \left(a_{r11} + k_1 - k_{sw}\hat{d}\right) - j\left(k_2 - k_{sw}\hat{d}\right) \tag{28}$$

3. Indirect Vector Control Strategy—Mathematical Structure

The pulse width modulation (PWM) technique employed here is hysteresis band current control as it has short circuit current protection feature, load independent and good torque response. A discrete PI controller processes and generates the reference torque.

$$e_c = \hat{\omega}_r - \omega^* \tag{29}$$

$$T_e^* = e_c\left[k_p + (k_i/s) * T_s\right] \tag{30}$$

where, e_c, k_p, k_i and T_s denote speed error, proportional and integral gains and sampling time for control algorithm execution.

$$i_{ds}{}^* = \left(\frac{\psi_r}{L_m}\right)\left[1 + \frac{dT_r}{dT_s}\right] \tag{31}$$

$$i_{qs}{}^* = \left(\frac{2}{3}\right)\left(\frac{2}{P}\right)\left(\frac{L_r}{L_m}\right)\left(\frac{T_{ref}}{\psi_r}\right) \tag{32}$$

Stator current components are inversely transformed from synchronously rotating to three phases stationary reference frame making use of the field angle. Therefore, it is obtained from the slip speed, as shown:

$$\theta_f = \theta_{sl} + \theta_r \tag{33}$$

$$i_{as}^* = i_{ds}\sin\theta + i_{qs}\cos\theta \tag{34}$$

$$i_{bs}^* = \left(\frac{1}{2}\right)\left\{-i_{ds}\cos\theta + \sqrt{3}\, i_{ds}\sin\theta\right\} + \left(\frac{1}{2}\right)\left\{i_{qs}\sin\theta + \sqrt{3} i_{qs}\cos\theta\right\} \tag{35}$$

$$i_{cs}^* = -\left(i_{as}^* + i_{bs}^*\right) \tag{36}$$

The generated currents and the actual sensed three phase currents are compared and the current errors are fed to the hysteresis band regulator to generate the switching pulses for the inverter. The hysteresis band value is chosen taking into account the torque and the current pulsations.

4. RT-Lab Based PIL Test Bench

Time critical test and simulation platforms have gained more prominence over offline platforms owing to faster execution, reduced design and development time. They use a fixed step discrete time solver compared to the variable step solvers used in offline. The computer model executed by them is in actual clock time provided the non-linearities and system dynamics are mathematically modeled. It has several features such as hardware-in-Loop testing (HIL), virtual and real control prototyping, data logging, etc. The sensorless drive system built using Simulink blocksets is integrated with RT-Lab blocksets [32,33]. However, the interaction between the plant and the controller is through analog and digital output and input channels and not by Simulink wires. Instead, a real-time link in the form of two loopback adapters along with a 40 pin flat ribbon cable is provided to ensure a real signal interaction. In this real-time link, only signal routing takes place, the signal is not processed. Therefore the estimated speed and actual 3 phase currents are fed via the analog output channels to the controller where it is captured by analog input channels. The switching pulses from the controller is sent via the digital output channels and captured by the digital input channels at the plant side. The output and the input pins and the number of channels are configured accordingly. This is also known as PIL testing [34,35]. The analog loopback is standalone hardware equipment, which does not need a power supply. A +5 V or +12 V source is required for Vsource and Vref for digital feedback. The operation of both of them are shown in the following schematic in Figure 3a,b. Two carriers are used as real-time targets. The OP4500 target has a single processor core activated. The OP5600 target has more processor cores, which makes it possible to have both the plant and controller in two different RT-Lab subsystems. The system model in the PC is connected to the OP4500/OP5600 targets through

TCP/IP. The entire real-time test bench used for PIL testing is shown in Figure 3c,d for OP4500 and in Figure 4 for OP5600 target.

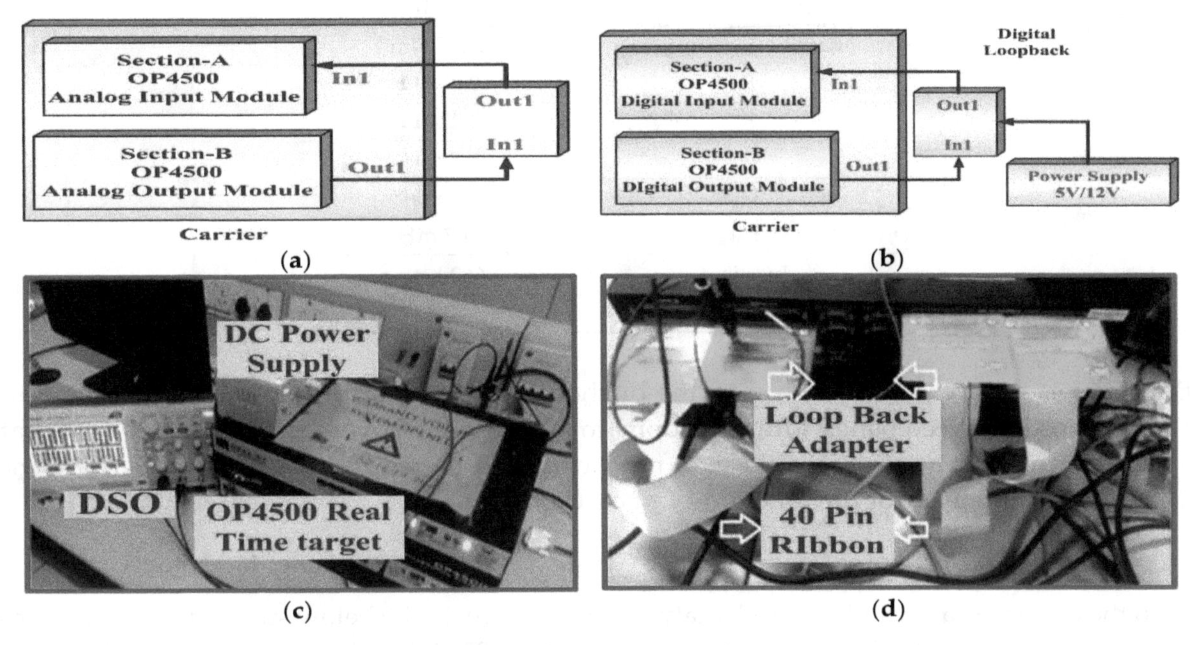

Figure 3. Representation of real-time PIL test bench using OP4500: (**a**) analog loopback; (**b**) digital loopback; (**c**) OP4500 target and power supply for digital loopback and (**d**) rear view of OP4500 with analog and digital loopback.

Figure 4. Representation of the real-time processor-in-loop test bench for a multi core target OP5600.

5. Results (Real-Time Simulation and PIL Based Validation): Analysis and Discussion

The time step used for the discretized model was 50 µs. Both the plant disturbance and the measurement disturbance were introduced in the system. The model was built in Simulink, interfaced with RT-Lab blocksets and the code generated was loaded and executed by the real-time target OP4500. Data logging was done by having OpWritefile blocksets of RT-lab to ensure the real-time data gets populated in mat files from where the real-time results can be extracted. Additionally, the estimated speed (analog output) and the switching pulses (digital output) were extracted from an oscilloscope to emphasize and validate (Hardware verification) the signal routing taking place between the plant and the controller via the Loopback adapter and cables. For, the study, a three-phase, 415 V, 50 Hz, star connected, 4 pole induction motor with the following model parameters considered are given below in Table 1.

Table 1. Model parameters.

Parameters	Ratings
Rated Power	50 HP
Rated Load Torque	237.4 Nm
R_s	0.087 Ω
R_r	0.228 Ω
L_{ls}	0.8 mH
L_{lr}	0.8 mH
L_m	34.7 mH
Inertia, J	1.662 kg m^2
Friction factor	0.1

Both observers were analyzed in terms of their dynamic performance, like tracking ability, disturbance rejection, speed bandwidth, time domain responses like overshoot, etc. Load perturbations and speed command variations could also be considered as model disturbances. The dynamic performance was obtained for the following test cases.

5.1. A Constant Speed Reference of 100 rad/s with a Constant Load Perturbation of 100 Nm

Both the observers are validated and analysed accordingly in the below subsections. For a constant speed command and load, both observers SMLO1 and SMLO2 exhibited similar tracking. The estimated speed, disturbance torque and rotor flux of both the observers were similar at medium speeds. The oscillations in speed tracking and the overshoot were tolerable.

5.1.1. SMLO1

The tracking performance of SMLO1 and the dynamic performance of the same in terms of torque holding capability and flux levels is shown in Figure 5.

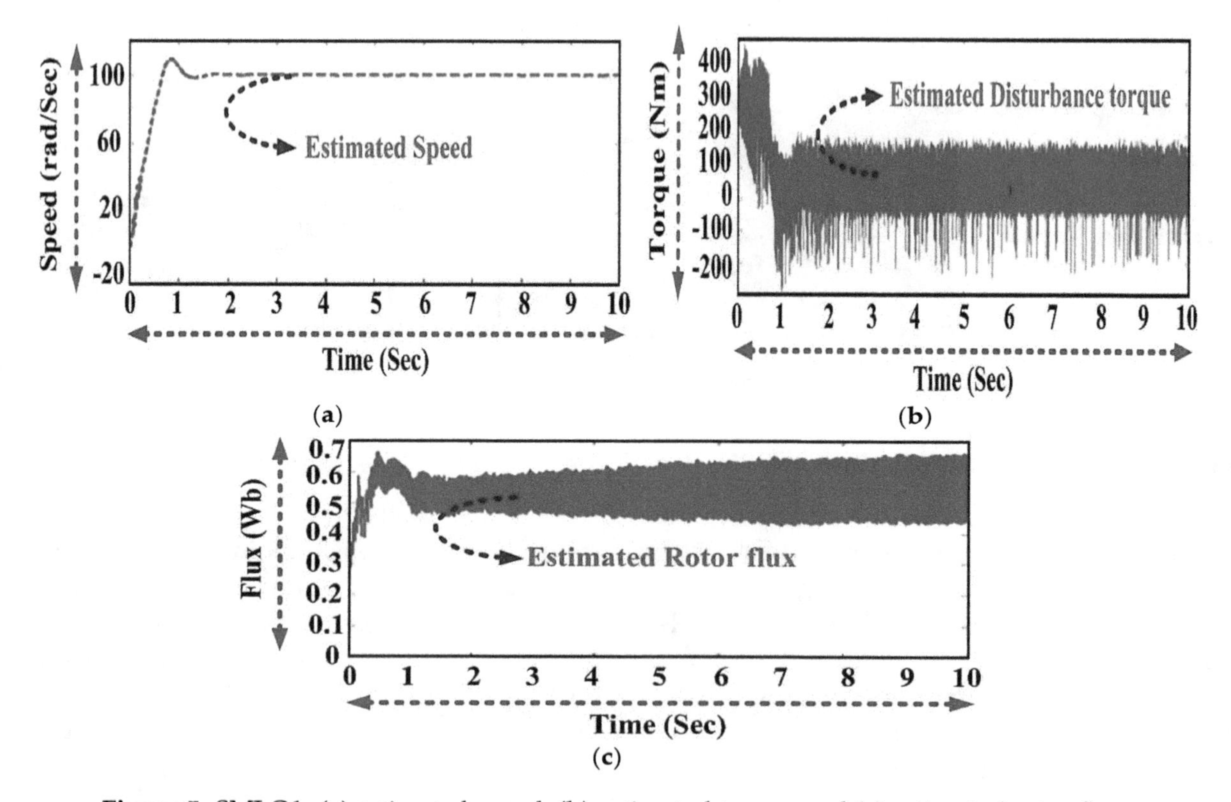

Figure 5. SMLO1: (**a**) estimated speed; (**b**) estimated torque and (**c**) estimated rotor flux.

5.1.2. SMLO2

The tracking performance of SMLO2 and the dynamic performance of the same in terms of torque holding capability and flux levels is shown in Figure 6.

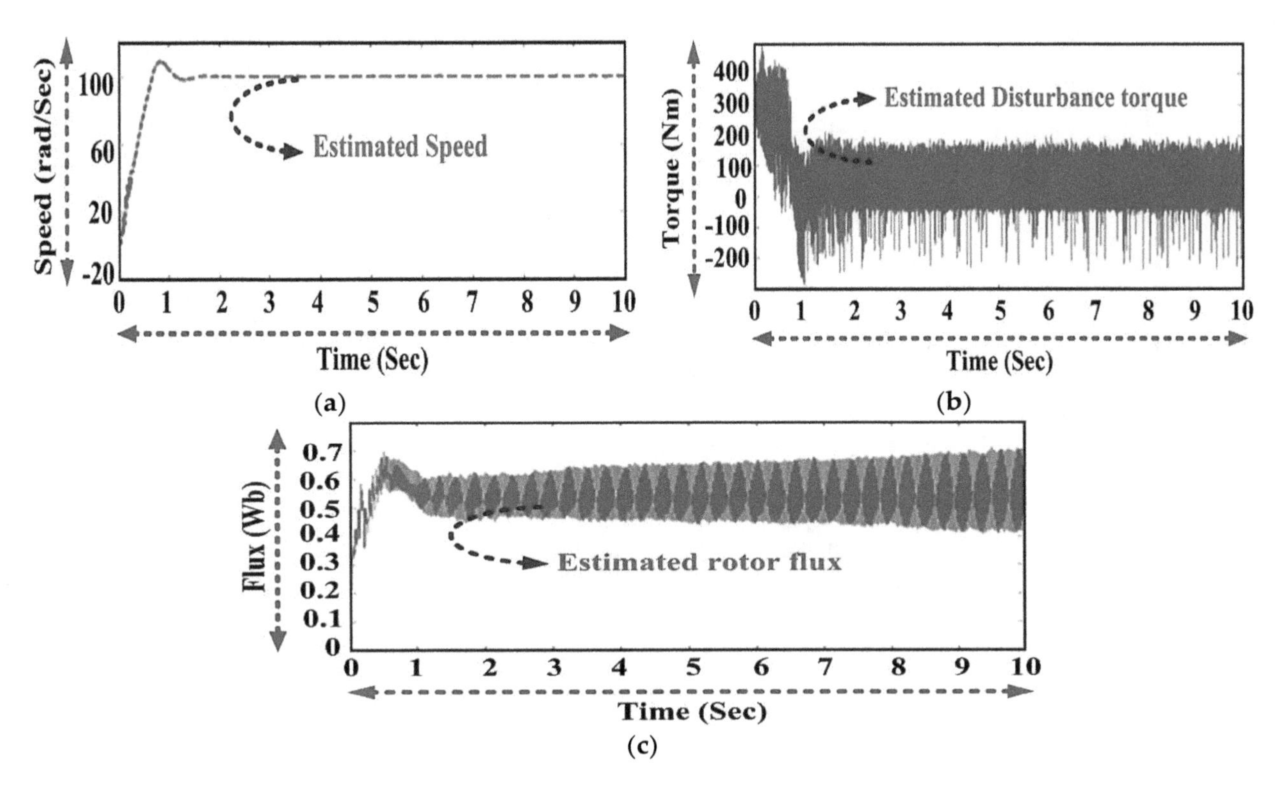

Figure 6. SMLO2: (**a**) estimated speed; (**b**) estimated torque and (**c**) estimated rotor flux.

5.2. A Constant Speed Reference of 100 rad/s with a Step Load Perturbation (Initially at 5 Nm, after a Fixed Time Interval of 15 s, Stepped up to 200 Nm)

Even when the drive was subjected to sudden load step perturbation from light load to rated load, the performance at medium speeds for both the observers were similar. However, there was a slight variation in the estimated flux performance of SMLO2 over SMLO1. The estimated flux of the former stabilized after the load switched to the rated load. This could be attributed to better torque holding capability of the improved disturbance rejection mechanism.

5.2.1. SMLO1

The tracking performance of SMLO1 and the dynamic performance of the same in terms of torque holding capability and flux levels is shown in Figure 7.

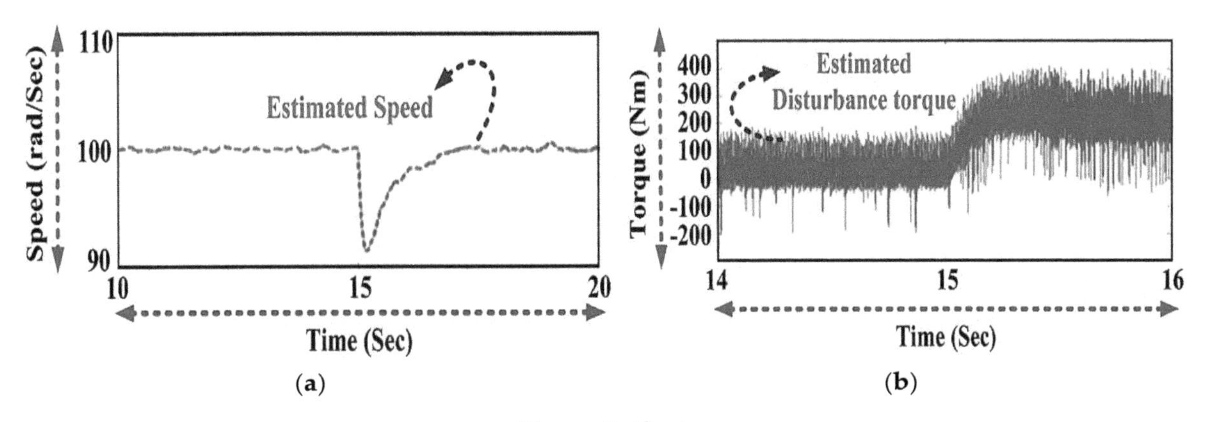

Figure 7. *Cont.*

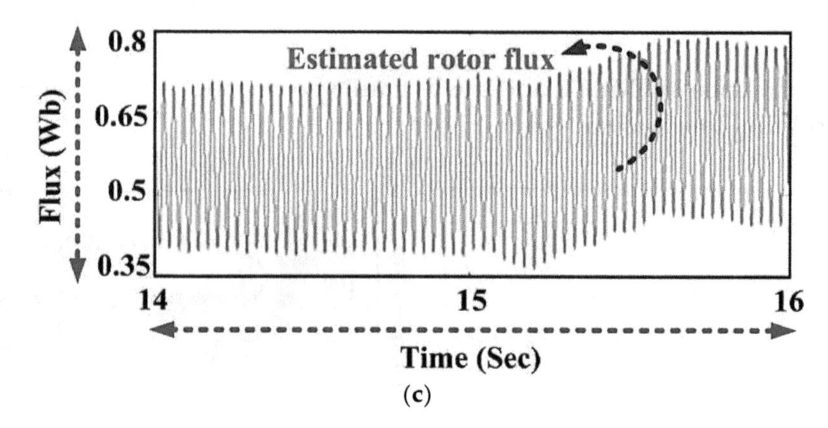

Figure 7. SMLO1: (**a**) estimated speed; (**b**) estimated torque and (**c**) estimated rotor flux.

5.2.2. SMLO2

The tracking performance of SMLO2 and the dynamic performance of the same in terms of torque holding capability and flux levels is shown in Figure 8.

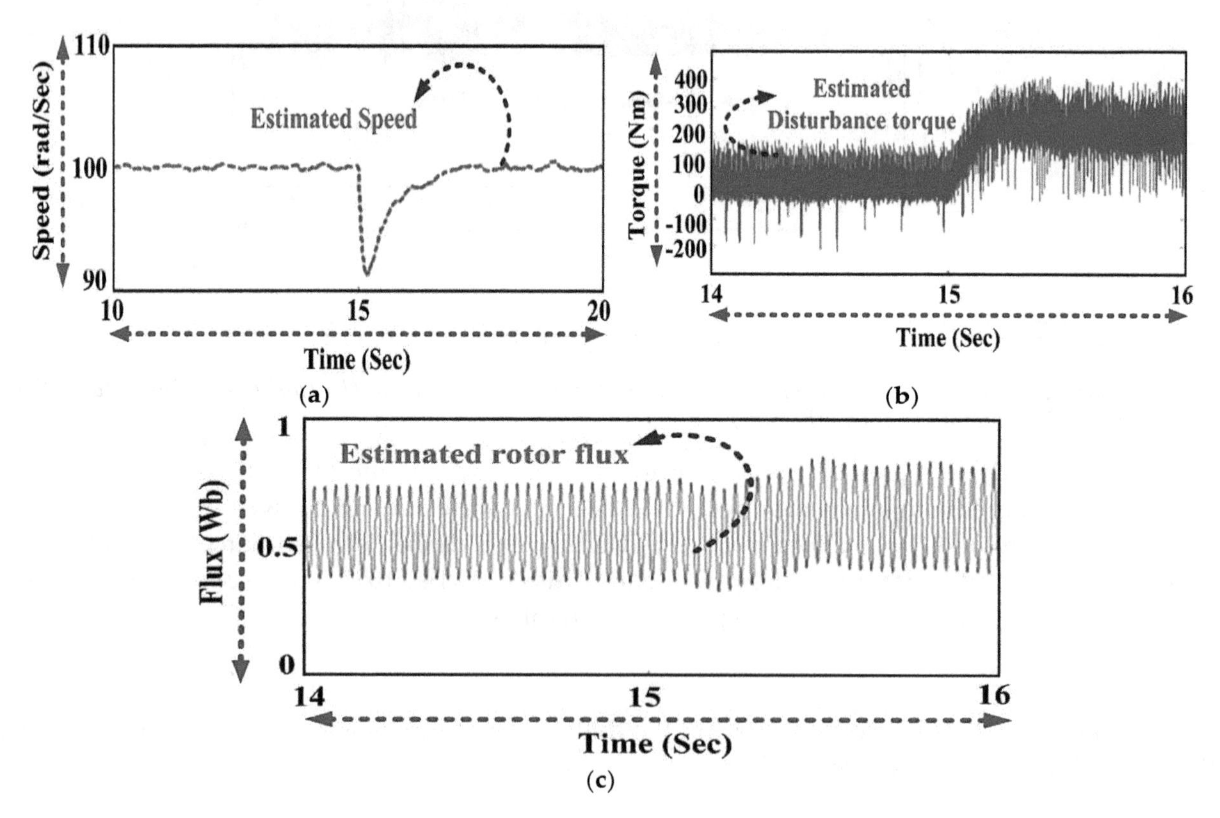

Figure 8. SMLO2: (**a**) estimated speed; (**b**) estimated torque and (**c**) estimated rotor flux.

5.3. A Step Speed Reference with a Constant Load Perturbation of 100 Nm

The difference is observed here when a step speed command is given. Additionally, there were some oscillations present initially when SMLO1 was tracking at 40 rad/s. As compared, the SMLO2 tracked well at speeds as low as 20 rad/s with considerably very few oscillations. The inability to track lower speeds may be due to the state convergence going out of bounds due to magnification of the speed and the stator current errors and also mismatch in critical parameters such as stator resistance, rotor time constant, etc. The flux pulsations were slightly more in SMLO1 as compared to SMLO2 and their profiles were different at low and medium speeds. However, the estimated disturbance torque of both were almost similar due to similar torque error between the motor and the observer errors in SMLO1 and SMLO2, however, the speed bandwidth varied.

5.3.1. SMLO1

The tracking performance of SMLO1 and the dynamic performance of the same in terms of torque holding capability and flux levels is shown in Figure 9.

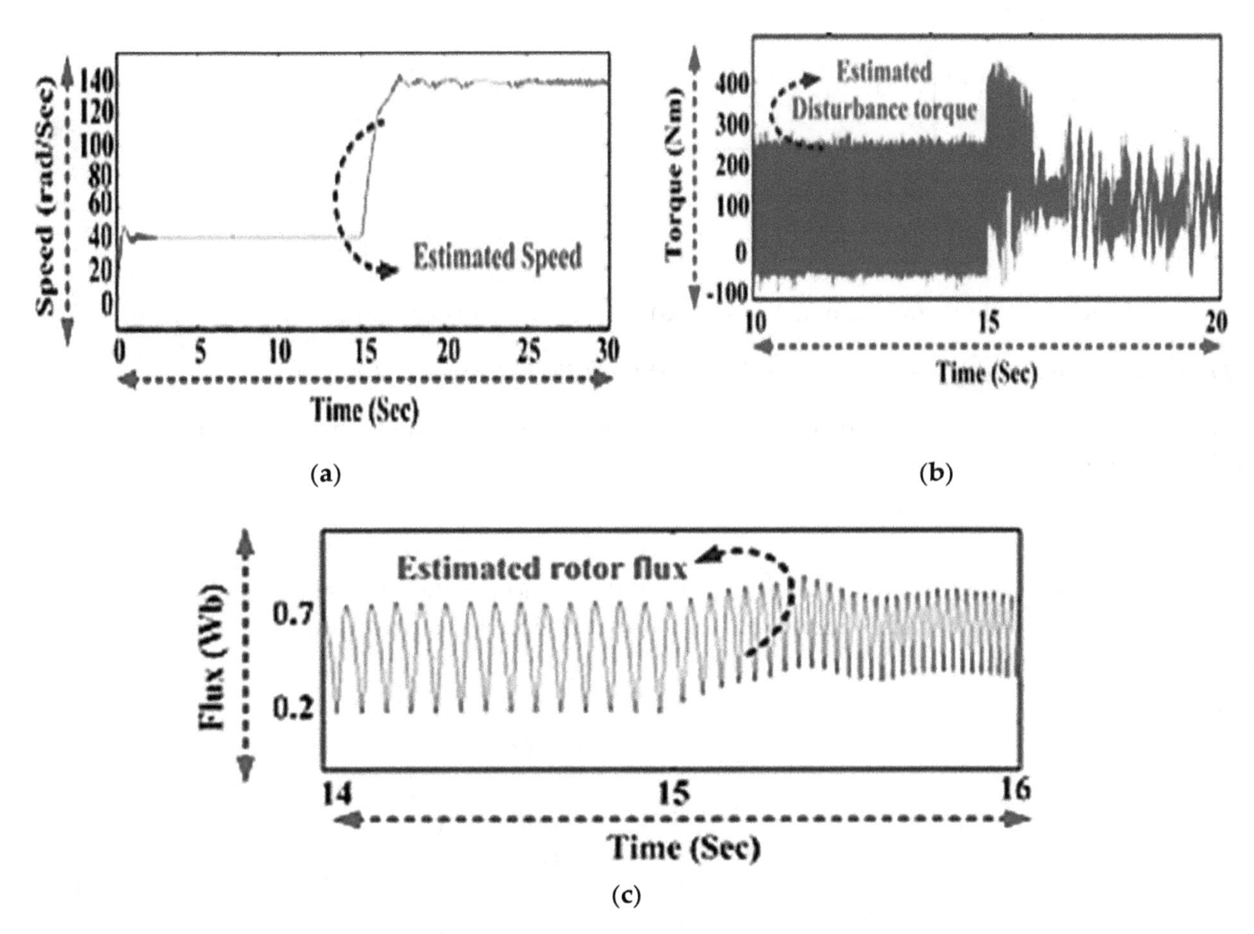

Figure 9. SMLO1: (**a**) estimated speed; (**b**) estimated torque and (**c**) estimated rotor flux.

5.3.2. SMLO2

The tracking performance of SMLO2 and the dynamic performance of the same in terms of torque holding capability and flux levels is shown in Figure 10.

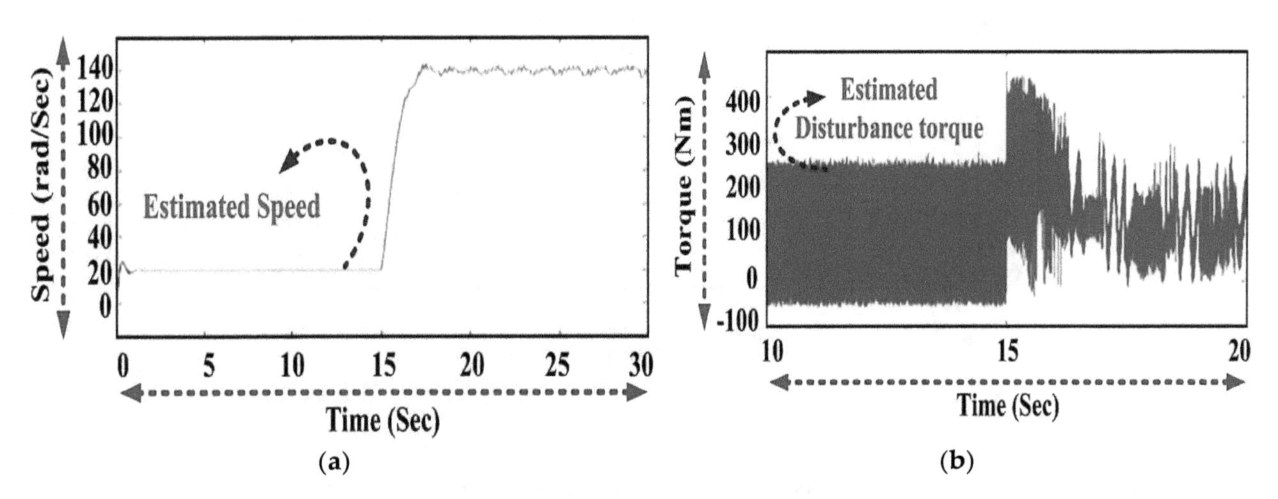

Figure 10. *Cont.*

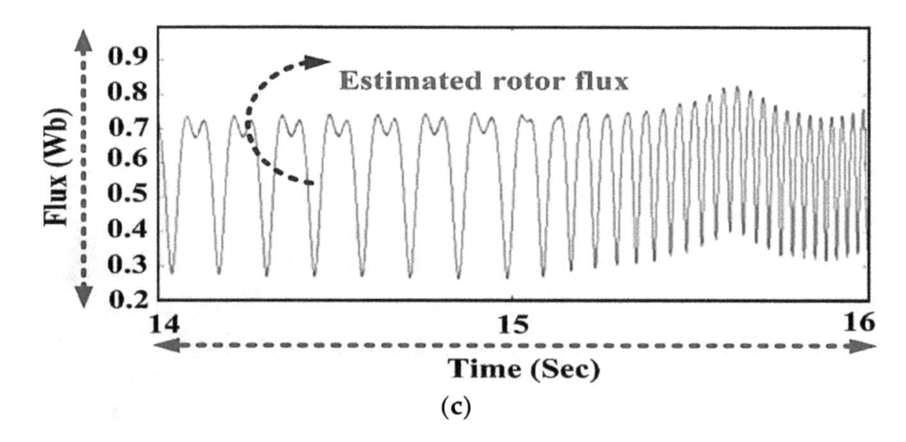

Figure 10. SMLO2: (**a**) estimated speed; (**b**) estimated torque and (**c**) estimated rotor flux.

5.4. Low Speed Command of 30 rad/s with a Constant Load Perturbation of 100 Nm

To verify, the low speed state convergence of both the observers, they were subjected to a low speed command of 30 rad/s at constant load.

5.4.1. SMLO1

It can be clearly seen for SMLO1 in Figure 11 that around a time interval of 1.5 s, all the parameters went out of bounds and became unstable. This only reflected the mismatch in parameter and error dynamics at low speeds.

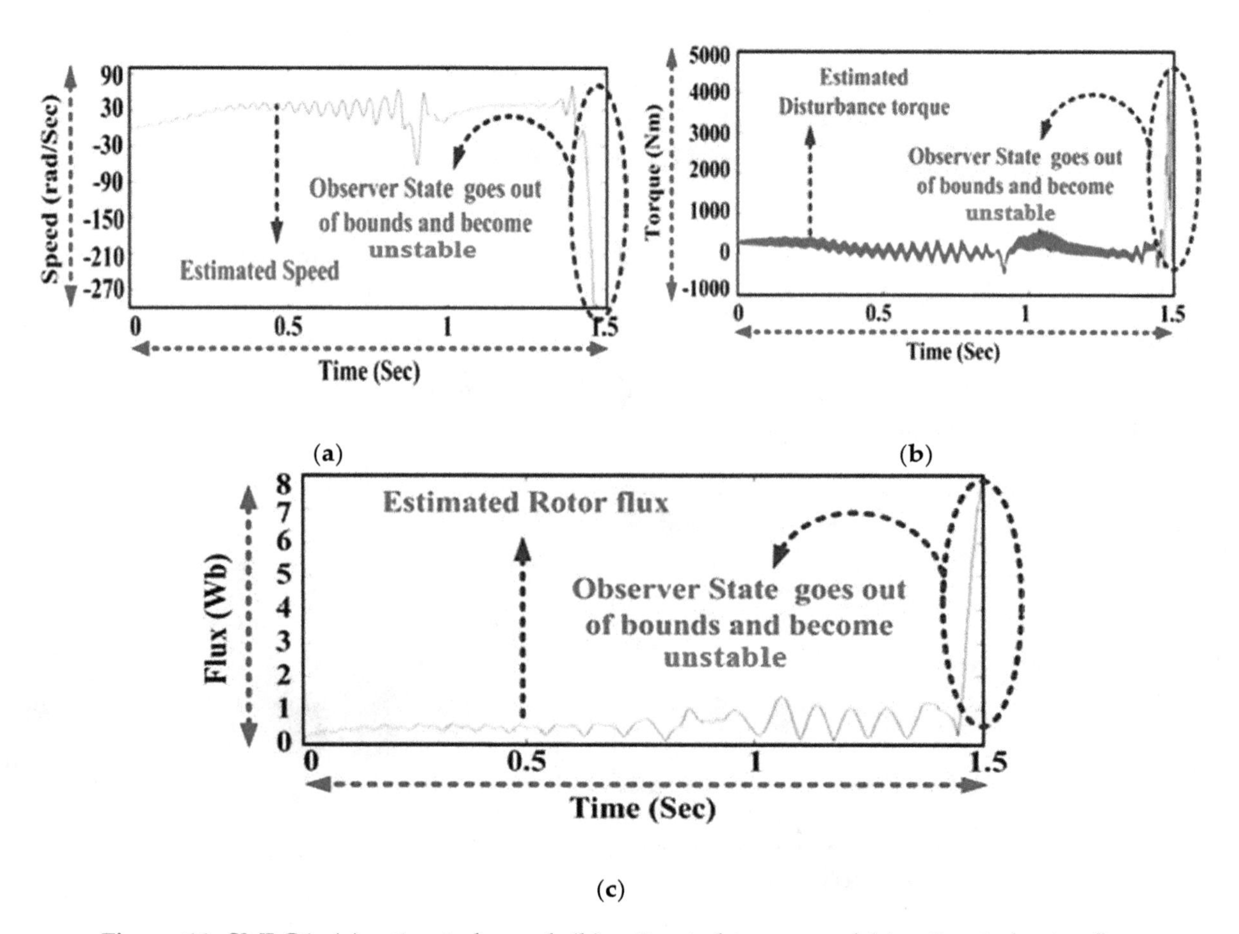

Figure 11. SMLO1: (**a**) estimated speed, (**b**) estimated torque and (**c**) estimated rotor flux.

5.4.2. SMLO2

Analysis Case 1: Real Time Simulation with Processor-in-Loop Validation

Here, for the purpose of adding more weight to the findings, the low speed performance analysis was split into two cases. In case 1, the low speed performance was validated in the real time processor-in-loop platform as was done for all the previous test cases. It can be observed that in analysis case 1, for SMLO2, in spite of the initial high overshoot and undershoot present for a short interval of time in the estimated speed, after 3 s, it settled down and provided a smooth tracking, which is also reflected in the disturbance torque and the estimated flux waveforms shown in Figure 12.

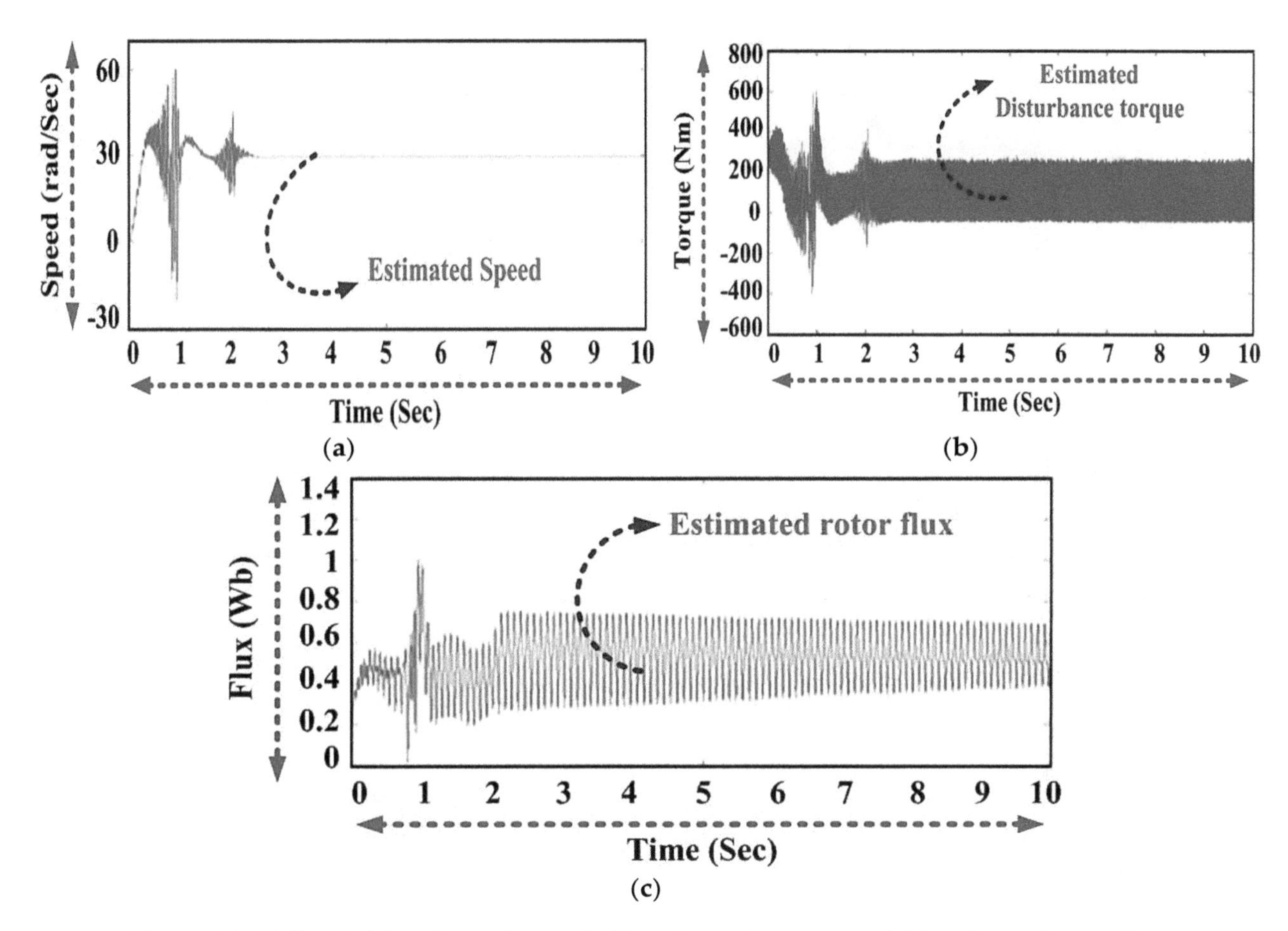

Figure 12. SMLO2: (**a**) estimated speed, (**b**) estimated torque and (**c**) estimated rotor flux.

In analysis case 2, only the low speed performance of the SMLO2 was considered, and an additional fragment for a time period of 4.5–5.5 s was zoomed for the purpose of clarity. Here, the same was tested in just the real time simulation environment (without the processor-in-loop mode). Here, the loop back cables, adapter and the power supply for the same was removed. Although the model was executed in real-time, however, there was no real world signal interaction between the plant and the controller here.

Analysis Case 2: Real Time Simulation without Processor-in-Loop Validation

It can be seen in Figure 13 that the number of overshoots and undershoots were considerably reduced as compared to analysis case 1 and in the additional zoomed fragment of the speed waveform, the tracking performance was very good with a bandwidth ranging from 29.95 to 30.05 rad/s, which only proved the effectiveness of the same in the low speed region.

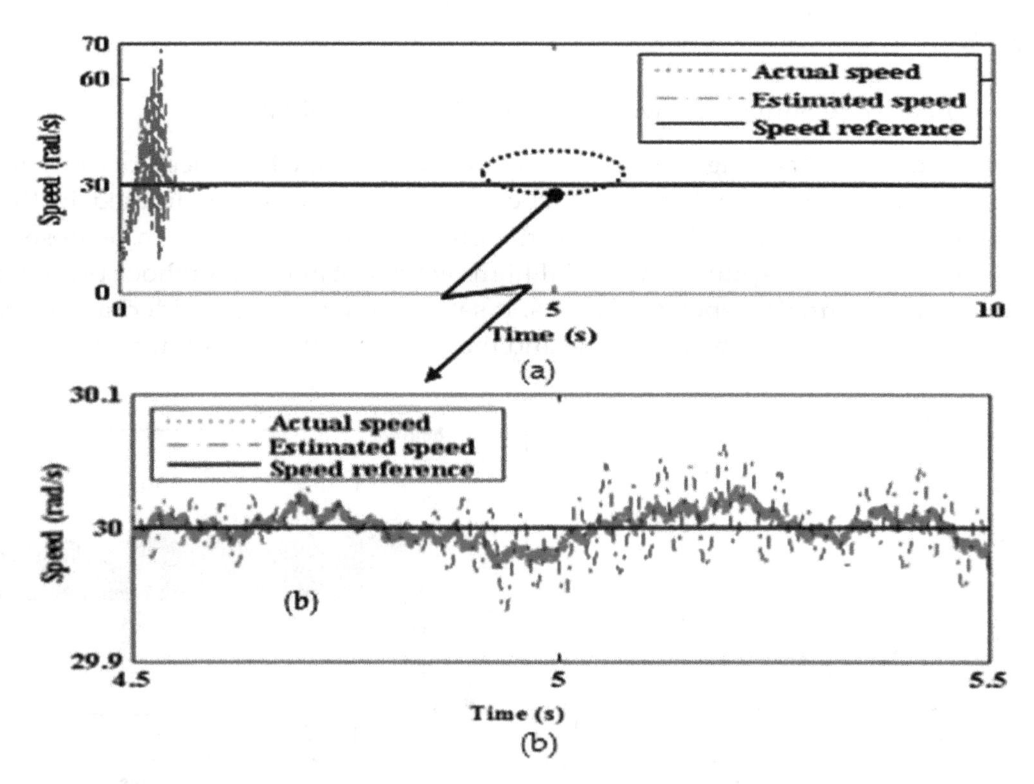

Figure 13. SMLO2: (**a**) estimated speed and (**b**) zoomed version of (**a**).

5.4.3. SMLO1 and SMLO2 Stator Current Error Convergence

The high torque pulsations (estimated disturbance torque) were due to high pulsations in the stator current. The large stator current pulsations and the subsequent stator current error was converged well by the SMLO2 as compared to SMLO1, as shown in Figures 14 and 15.

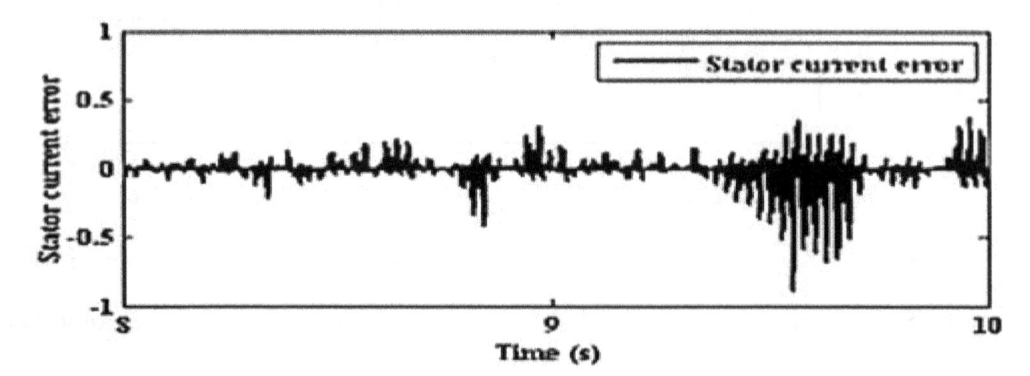

Figure 14. SMLO1 stator current error.

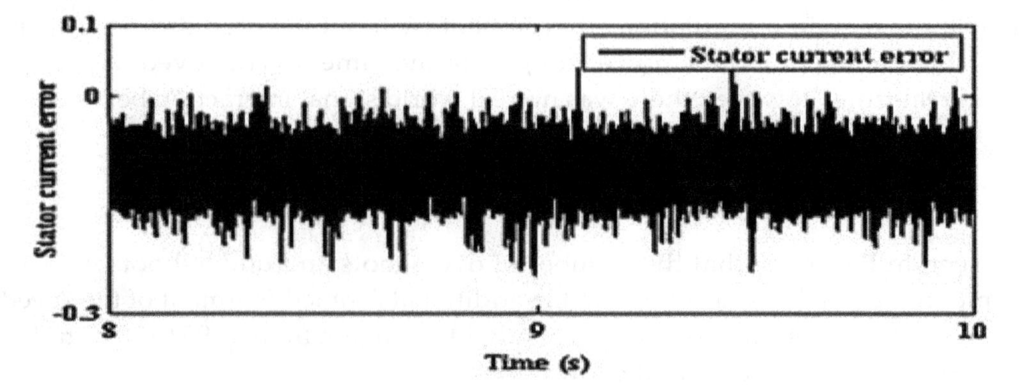

Figure 15. SMLO2 stator current error.

5.5. Estimated Speed Waveforms for a Constant Load Perturbation of 100 Nm as Recorded in Digital Storage Oscilloscope

The estimated speed for different speed commands from the oscilloscope is shown in Figure 16a–d.

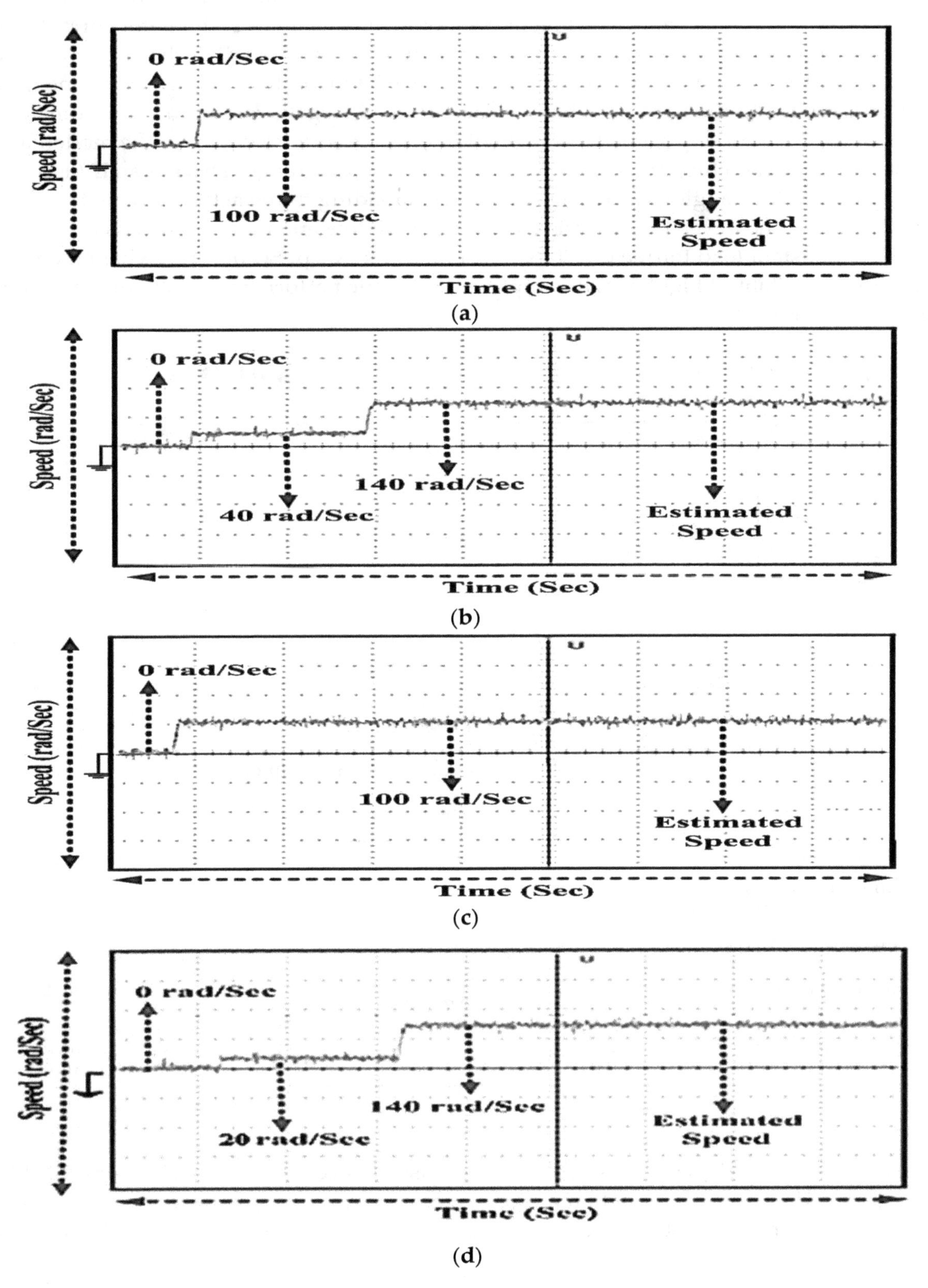

Figure 16. Estimated speed: (**a**) for a constant speed reference of 100 rad/s for SMLO1, (**b**) for a step speed reference of 40–140 rad/s for SMLO1, (**c**) for a constant speed reference of 100 rad/s for SMLO2 and (**d**) for a step speed reference of 20–140 rad/s for SMLO2.

The poles of SMLO2 were shifted to the left of SMLO1 as shown in Figure 17 indicating improved stability performance. The effect of this was more predominant in the low speed regions where the SMLO2 was able to track speeds around 20 rad/s as compared to SMLO1, which is unable to track speeds of the same range. The poles being shifted more to the left of SMLO1 had led to increased speed bandwidth of SMLO2. Owing to high rating of the motor, the dynamics of the stator current impacted the performance of the observer and as a result, the pulsations in the disturbance torque could be attributed to the same. For hysteresis regulation, it is difficult to predict the exact switching frequency since it is not related to the hysteresis band. The OP45OO target provides a maximum switching frequency of (1/(2 × time step)), i.e., 10 kHz. Owing to successful execution of the model, it is to be understood that the switching frequency was within the 10 kHz range. Driving an EV is an endless series of dynamically changing operating states corresponding to the actual driving trajectory and speed of the vehicle, however, the analysis confines itself to low and medium speeds in the motoring mode and for constant load torque of 100 Nm (for all cases except Section 5.2, where there is a step load perturbation). Table 2 highlights the improved dynamic performance of the proposed observer for the different parameters.

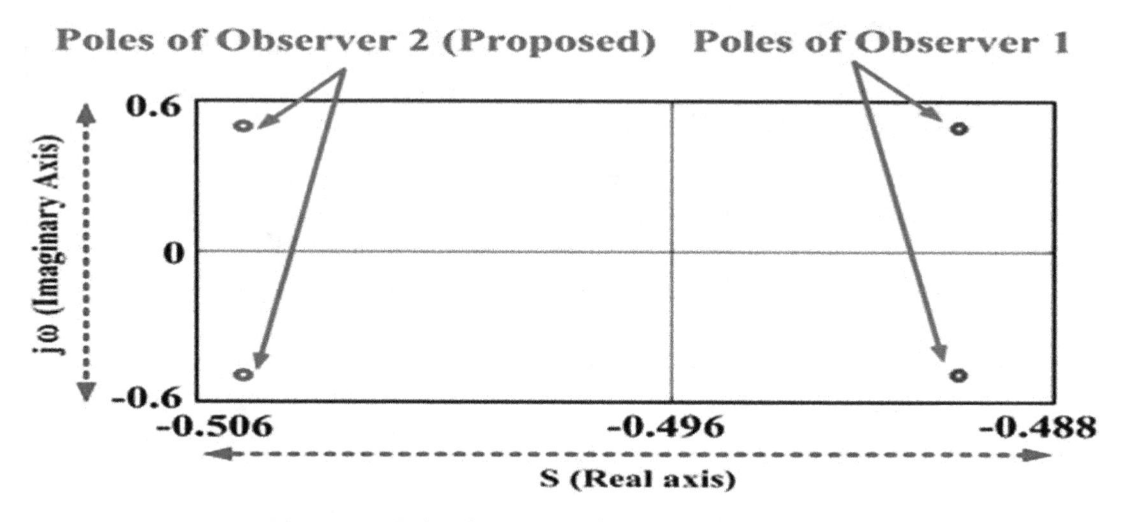

Figure 17. Pole placement of SMLO1 and SMLO2.

Table 2. Performance comparison.

Parameters	SMLO1	SMLO2 (Proposed)
Medium speeds and near synchronous speed (70–150 rps)	Exhibits good tracking performance	Exhibits good tracking performance
Low speeds (<70 rps)	Convergence is affected and does not track well.	Tracks up to 20 rps and considerably better disturbance rejection.
Speed oscillations	Has more oscillations even in the steady state region	Comparatively less oscillations
Observer stability	Less stable than the proposed observer, due to which the speed bandwidth is reduced.	Proposed observer poles are shifted to the left of the SMLO1, which explains the extended speed bandwidth.

6. Conclusions

The two sliding mode observers with different disturbance rejection mechanisms (SMLO1 and SMLO2) for speed sensor-less induction motor drives were designed, tested and analyzed in a new real-time Processor-in-Loop test bench based on a distributed real-time package RT-Lab for application in the EV.

- The purpose of validating the same in a Processor-in-Loop platform is to introduce a real-world signal routing and interaction where the system is placed virtually in the front end.

- The dynamic performance can be treated almost on par with an actual physical system with precise timing requirements.
- SMLO2 displays better performance at low speed regions (less than 70 rps), particularly in terms of tracking, better disturbance handling and stability.
- It also has increased speed bandwidth (20–150 rps) and reduced speed oscillations at lower speeds (20–30 rps).
- The proposed method (SMLO2) would be more ideal for the EV to address the issue of range anxiety. It also delivers considerably better dynamic performance and able to handle disturbances better.

However, it can be further modified considering the real-time trajectory of the EV. This method of hardware verification can also be further extended to having the real world controller interact with the virtual plant or vice versa, where the real plant can interact with the virtual controller placed in the target.

Author Contributions: Conceptualization, M.K.S., F.D.J.L.; Methodology, M.K.S.; Software, M.K.S.; Validation, M.K.S., F.D.J.L.; Writing—original draft preparation, M.K.S.; Visualization, R.M.E., S.P., U.S.; Supervision, R.M.E., U.S., S.P.; Writing—review and editing, M.K.S., U.S., R.M.E., G.M.S., I.K.; Project Administration, F.B. All authors have read and agreed to the published version of the manuscript.

Nomenclature

$i_{ds}{}^{s}$, $i_{qs}{}^{s}$, $i_{dr}{}^{r}$, $i_{qr}{}^{r}$	d- and q-axis stator and rotor currents in the stationary and rotating reference
$v_{ds}{}^{s}$, $v_{qs}{}^{s}$	d- and q-axis stator voltages in stationary reference
T_r, R_s, R_r	Rotor time constant, stator and rotor resistance
σ, L_r, L_m, L_s	Leakage reactance, rotor, magnetizing and stator self inductance
L_{ls}, L_{lr}	Stator and rotor leakage inductances
ω_r, $\hat{\omega}_r$, ω^*, ω_{bsync}	Actual, estimated, reference and base synchronous speed
$\psi_{ds}{}^{s}$, $\psi_{qs}{}^{s}$, $\psi_{dr}{}^{s}$, $\psi_{qr}{}^{s}$	d and q axes stator and rotor flux linkages in stationary reference
$\hat{\varphi}_d$, $\hat{\varphi}_q$	d and q axes estimated rotor flux linkages
θ_f, θ_{sl}, θ_r	Field angle, slip angle and rotor angle
$i_{ds}{}^{*}$, i_{qs}^{*}	d and q axes stator currents in synchronously rotating reference
i_{as}^{*}, i_{bs}^{*}, i_{cs}^{*}	Three phase reference currents

References

1. Tomislav, S.; Siegfried, S.; Wolfgang, G. The Flux-Based Sensorless Field-Oriented Control of Permanent Magnet Synchronous Motors without Integrational Drift. *Actuators* **2018**, *7*, 1–22.
2. Tsuji, M.; Chen, S.; Izumi, K.; Yamada, E. A Sensorless Vector Control System for Induction Motors Using q-Axis Flux with Stator Resistance Identification. *IEEE Trans. Ind. Electron.* **2001**, *48*, 185–194. [CrossRef]
3. Wang, H.; Liu, Y.C.; Ge, X. Sliding-mode observer-based speed-sensorless vector control of linear induction motor with a parallel secondary resistance online identification. *IET. Electr. Power Appl.* **2018**, *12*, 1215–1224. [CrossRef]
4. Hosseyni, A.; Trabelsi, R.; Mimouni, M.F.; Iqbal, A.; Padmanaban, S. Novel Sensorless Sliding Mode Observer of a Five-Phase Permanent Magnet Synchronous Motor Drive in Wide Speed Range. *Lect. Notes Electr. Eng.* **2017**, *436*, 213–220.
5. Hosseyni, A.; Trabelsi, R.; Iqba, A.; Padmanaban, S.; Mimouni, M. An Improved Sensorless Sliding Mode Control/ Adaptive Observer of a Five-Phase Permanent Magnet Synchronous Motor Drive. *Int. J. Adv. Manuf. Tech.* **2017**, *93*, 1029–1039. [CrossRef]
6. Krishna, S.M.; Daya, J.L.F. Effect of Parametric variations and Voltage Unbalance on Adaptive Speed Estimation Schemes for Speed Sensorless Induction Motor Drives. *Int. J. Power Electron. Drives.* **2015**, *6*, 77–85. [CrossRef]

7. Ali Karami, M.; Hamed, T. Estimation of load torque in induction motors via dynamic sliding mode control and new nonlinear state observer. *J. Mech. Sci. Technol.* **2018**, *32*, 2283–2288. [CrossRef]
8. Alexander, I.K.; Alexey, A.E.; Dmitriy, V.S.; Yuriy, N.K. An Approach of the Wavelet-Fuzzy Controller of Vector Control System by the Induction Motor. In Proceedings of the IEEE International Multi-Conference on Industrial Engineering and Modern Technologies (FarEastCon), Vladivostok, Russia, 3–4 October 2018; pp. 1–6.
9. Ho, T.J.; Chang, C.H. Robust Speed Tracking of Induction Motors: An Arduino-Implemented Intelligent Control Approach. *Appl. Sci.* **2018**, *8*, 159. [CrossRef]
10. Zhao, X.; Yu, Q.; Yu, M.; Tang, Z. Research on an equal power allocation electronic differential system for electric vehicle with dual-wheeledmotor front drive based on a wavelet controller. *Adv. Mech. Eng.* **2018**, *10*, 1–24. [CrossRef]
11. Bouhoune, K.; Yazid, K.; Boucherit, M.S.; Chriti, A. Hybrid control of the three phase induction machine using artificial neural networks and fuzzy logic. *Appl. Soft Comput.* **2017**, *55*, 289–301. [CrossRef]
12. Daya, J.L.F.; Padmanaban, S.; Blaabjerg, F.; Wheeler, P.; Ojo, O. Implementation of Wavelet Based Robust Differential Control for Electric Vehicle Application. *IEEE Trans. Power Electr.* **2015**, *30*, 6510–6513. [CrossRef]
13. Chekroun, S.; Zerikat, M.; Mechernene, A.; Benharir, N. Novel Observer Scheme of Fuzzy-MRAS Sensorless Speed Control of Induction Motor Drive. *J. Phys. Conf.* **2017**, *783*, 1–13. [CrossRef]
14. Lamia, Y.; Sebti, B.; Farid, N.; Mihai, C.; Luis Guasch, P. Design of an Adaptive Fuzzy Control System for Dual Star Induction Motor Drives. *Adv. Electr. Comput. Eng.* **2018**, *18*, 37–44.
15. Krisztián, H.; Márton, K. Speed sensorless field oriented control of induction machines using unscented Kalman filter. In Proceedings of the International Conference on Optimization of Electrical and Electronic Equipment (OPTIM) & Intl Aegean Conference on Electrical Machines and Power Electronics (ACEMP), Brasov, Romania, 25–27 May 2017; pp. 523–528.
16. Bo, F.; Zhumu, F.; Leipo, L.; Jiangtao, F. The full-order state observer speed-sensorless vector control based on parameters identification for induction motor. *Meas. Control* **2019**, *52*, 202–211.
17. Ali, H.; Rachid, D.; Othman, H. A New Direct Speed Estimation and Control of the Induction Machine Benchmark: Design and Experimental Validation. *Math. Probl. Eng.* **2018**, *2018*, 1–10.
18. Krishna, S.M.; Daya, J.L.F.; Padmanaban, S.; Mihet-Popa, L. Real-time Analysis of a Modified State Observer for Sensorless Induction Motor Drive used in Electric Vehicle Applications. *Energies* **2017**, *10*, 1077. [CrossRef]
19. Youssef, A.; Mabrouk, J.; Yassine, K.; Boussak, M. A Very-Low-Speed Sensorless Control Induction Motor Drive with Online Rotor Resistance Tuning by Using MRAS Scheme. *Power Electron. Drives* **2019**, *4*, 125–140.
20. Abderrahim, B.; Sandeep, B.; Mustapha, J.; Adil, E.; Mohammed, A. Real Time High Performance of Sliding Mode Controlled Induction Motor Drives. *Procedia Comput. Sci.* **2018**, *132*, 971–982.
21. Kandoussi, Z.; Boulghasoul, Z.; Elbacha, A.; Tajer, A. Real time implementation of a new fuzzy-sliding-mode-observer for sensorless IM drive. *COMPEL* **2017**, *36*, 938–958. [CrossRef]
22. Ilten, E.; Demirtas, M. Fractional order super-twisting sliding mode observer for sensorless control of induction motor. *COMPEL* **2019**, *38*, 878–892. [CrossRef]
23. Tang, J.; Yang, Y.; Blaabjerg, F.; Chen, J.; Diao, L.; Liu, Z. Parameter Identification of Inverter-Fed Induction Motors: A Review. *Energies* **2018**, *11*, 2194. [CrossRef]
24. Krishna, S.M.; Daya, J.L.F. A modified disturbance rejection mechanism in sliding mode state observer for sensorless induction motor drive. *Arab. J. Sci. Eng.* **2016**, *41*, 3571–3586. [CrossRef]
25. Krzeminski, Z. Observer of induction motor speed based on exact disturbance model. In Proceedings of the IEEE 13th International Power Electronics and Motion Control Conference, Poznan, Poland, 1–3 September 2008; pp. 2294–2299.
26. Albu, M.; Horga, V.; Ratoi, M. Disturbance torque observers for the induction motor drives. *J. Electr. Eng.* **2006**, *6*, 1–6.
27. Slotine, J.J.E.; Li, W. *Applied Non Linear Control*; Prentice-Hall: Upper Saddle River, NJ, USA, 1991.
28. Bose, B.K. *Power Electronics and Variable Frequency Drives—Technology and Applications*; Wiley-IEEE Press: River Street Hoboken, NJ, USA, 2013.
29. Mikkili, S.; Panda, A.K.; Prattipati, J. Review of Real-Time Simulator and the Steps Involved for Implementation of a Model from MATLAB/SIMULINK to Real-Time. *J. Inst. Eng. India Ser. B* **2015**, *96*, 179–196. [CrossRef]

30. Mossa, M.A.; Echeikh, H.; Iqbal, A.; Duc Do, T.; Al-Sumaiti, A.S. A Novel Sensorless Control for Multiphase Induction Motor Drives Based on Singularly Perturbed Sliding Mode Observer-Experimental Validation. *Appl. Sci.* **2020**, *10*, 2776. [CrossRef]
31. Krishna, S.M.; Daya, J.L.F. Adaptive Speed Observer with Disturbance Torque Compensation for Sensorless Induction Motor Drives using RT-Lab. *Turk. J. Electr. Eng. Comput. Sci.* **2016**, *24*, 3792–3806. [CrossRef]
32. Krishna, S.M.; Daya, J.L.F. MRAS speed estimator with fuzzy and PI stator resistance adaptation for sesnorless induction motor drives using RT-Lab. *Perspect. Sci.* **2016**, *8*, 121–126. [CrossRef]
33. Ali, H.; Olfa, B. Real-Time Low-Cost Speed Monitoring and Control of Three-Phase Induction Motor via a Voltage/Frequency Control Approach. *Math. Probl. Eng.* **2020**, *2020*, 1–14.
34. Amit Kumar, K.S.; Ilamparithi, T.; Prakash, O.; Belanger, J. Hybrid CPU-Core and FPGA based real-time implementation of a high frequency aircraft power system. In Proceedings of the IEEE Energy Conversion Congress and Exposition, Montreal, QC, Canada, 20–24 September 2015; pp. 5425–5430.
35. Amit Kumar, K.S.; Ilamparithi, T. Real-time studies on an improved modular stacked transmission and distribution system. In Proceedings of the IEEE 16th Workshop on Control and Modeling for Power Electronics (COMPEL), Vancouver, BC, Canada, 12–15 July 2015; pp. 1–8.

10

Optimal Capacity Sizing for the Integration of a Battery and Photovoltaic Microgrid to Supply Auxiliary Services in Substations under a Contingency

Alejandra Tabares [1], Norberto Martinez [1], Lucas Ginez [1], José F. Resende [2], Nierbeth Brito [3] and John Fredy Franco [1,2,*]

[1] Faculty of Electrical Engineering, Campus of Ilha Solteira, São Paulo State University, Ilha Solteira 15385-000, Brazil; tabares.1989@gmail.com (A.T.); abrantemartinez94@gmail.com (N.M.); lucasginezoliveira@gmail.com (L.G.)

[2] School of Energy Engineering, Campus of Rosana, São Paulo State University, Rosana 19274-000, Brazil; jose.resende@unesp.br

[3] INTESA—Integration Power Transmitter S.A.—Power Transmission Utilities, Brasília 70.196-900, Brazil; nierbeth.brito@equatorial-t.com.br

* Correspondence: fredy.franco@unesp.br

Abstract: Auxiliary services are vital for the operation of a substation. If a contingency affects the distribution feeder that provides energy for the auxiliary services, it could lead to the unavailability of the substation's service. Therefore, backup systems such as diesel generators are used. Another alternative is the adoption of a microgrid with batteries and photovoltaic generation to supply substation auxiliary services during a contingency. Nevertheless, high battery costs and the intermittence of photovoltaic generation requires a careful analysis so the microgrid capacity is defined in a compromise between the investment and the unavailability reduction of auxiliary services. This paper proposes a method for the capacity sizing of a microgrid with batteries, photovoltaic generation, and bidirectional inverters to supply auxiliary services in substations under a contingency. A set of alternatives is assessed through exhaustive search and Monte Carlo simulations to cater for uncertainties of contingencies and variation of solar irradiation. An unavailability index is proposed to measure the contribution of the integrated hybrid microgrid to reduce the time that the substation is not in operation. Simulations carried out showed that the proposed method identifies the microgrid capacity with the lowest investment that satisfies a goal for the unavailability of the substation service.

Keywords: auxiliary services; battery; microgrids; photovoltaic generation; substations

---

1. Introduction

Substations are one of the main components of electrical power systems. They serve to modify the voltage level and allow basic maneuvering of power flow within the system. To fulfill their functions, substations require auxiliary services such as monitoring, communications, and maneuvering systems. Other essential loads that must be served in the substation are lighting, heating-cooling, some communication elements, switch operating mechanisms, anti-condensation heaters, and motors. Auxiliary services supply essential trip coils for circuit breakers and associated relays, supervisory control and data acquisition (SCADA), and communication equipment. They are vital for the proper functioning of the substations as allow monitoring, measurement, protection of transformers and buses, supervision of protections and automatic reclosing, remote controls, fault protection of the circuit breaker, monitoring of transformer's overload, voltage control, selective load shedding; they are also involved in alarms and interface systems in the substations' control centers [1,2]. Given their critical importance, the power supply for the auxiliary services at substations must be designed with an appropriate level of redundancy and backup.

Auxiliary services in substations (ASS) can be provided by a low-voltage busbar supplied by a distribution feeder or by a group of diesel generators; the latter has been used as a backup to maintain the energy supply under any condition, especially in the presence of a permanent contingency. Moreover, those services must be economical both in terms of investment and operational costs [3]. However, in the event of a contingency, additional costs are overlooked, as the consequences of not providing auxiliary services, in general, have a high monetary impact. Thus, the objective of alternative backup systems must first ensure their ability to respond in the event of contingencies even if this results in higher operating costs [4].

Some disadvantages of the aforementioned alternatives to provide auxiliary services are the high price of energy when supplied by the medium or low-voltage distribution system, as well as the environmental pollution associated with the operation of diesel generators (which are the vast majority of independent backup solutions) and their high failure probability and maintenance costs. Therefore, the use of alternative sources to supply auxiliary services is justified. Specifically, a suitable alternative is an integration of renewable energy systems (e.g., batteries and photovoltaic systems), which can independently operate the main grid, have a low environmental impact, and present a trending cost reduction in recent years. However, such systems based on renewable energies bring considerable implementation challenges given their non-dispatchable source nature, which should be solved to supply critical loads as are the ASS.

Recently, the microgrid concept has been addressed in the specialized literature to deal with the disadvantages of renewable energy sources (e.g., intermittence and dependence on climatic conditions) so they can be able to participate within multiple applications of power systems [5]. Microgrids are based on hybrid distributed management systems capable of operating in the absence of power supply from the main network and feeding a limited set of loads [6]. This last characteristic has increased the interest in their implementation, allowing for the improvement of the power supply availability, especially for critical loads in the case of permanent contingencies. In addition to their usefulness as back-up systems, a microgrid can be used in a grid-connected mode to take advantage of the generated energy to lower costs required to satisfy the connected loads. Thus, microgrids offer some advantages, e.g., greater penetration of renewable energy resources, reduced energy cost, and reduced greenhouse gas emissions. These advantages satisfy some sustainable development criteria, including economic, environmental, and social aspects [7].

To guarantee that a microgrid provides a reliable operation when the main network is under contingency (islanded operation mode), it is essential that the adoption of storage systems (e.g., a battery) so that critical loads can continue to operate. After recognizing that advantage, a microgrid is an appropriate alternative to provide ASS, either as a primary source or as a backup system after a failure of the distribution feeder. Still, the generation system (e.g., photovoltaic generation), storage (e.g., batteries), and the electronic interface (e.g., inverters) should be properly sized. If the micro-grid is used as the main supplier or as a back-up service of the auxiliary service loads, it should satisfy a certain robustness level to face the occurrence of permanent fault of the distribution feeder [8–10].

Storage systems sized aiming the provision of power to the ASS must ensure that the microgrid works in island mode for as long as necessary during the absence of the main supply source due to shortages. Among storage systems, electrochemical means represent an attractive alternative to other types of storage; the flywheel, which is an electromechanical way of storing energy, has a high installation cost (estimated to be between 1000 and 3900 USD/kWh in 2030) and can have a self-discharge rate of 20% per hour [11]. There is also pumped hydro storage, which has the highest installed power in the world with at least 150 GW in 2016, but needs a favorable geographical location to build the reservoirs and requires a considerable construction area [11].

Electrochemical technologies store energy chemically through various components, and because they are marketed in modules, the desired voltage and current can be configured by making series and parallel connections of several modules until the desired values are reached. The four main types of batteries are lithium-ion, flow, lead-acid, and high-temperature, and each of them can consist of different components; among those types, the vanadium redox (VRFB) and zinc-bromine (ZBFB) flow battery technologies have a depth of discharge (DoD) of 100%, but they have the lowest energy density and power, e.g., 25–70 Wh/L for VRFB. Lead-acid batteries, built in lead-acid flooded (FLA) and valve-regulated lead-acid (VRLA) technologies, have the lowest cost of installation of all and have a better power density than flow batteries, but they have a DoD of 50%, which is a much lower value than the other types. High-temperature batteries, which are built with sodium-sulfur (NaS) and sodium nickel chloride (NaNiCl) technologies are batteries that have a DoD of 100%; NaS has one of the lowest self-discharge rates of 0.05% per day but requires a heating system so that the battery fluid lies in the liquid state. Finally, lithium-ion type batteries have the best specifications because they have the highest energy and power densities and can reach 735 Wh/L for lithium-nickel-manganese-cobalt (NMC) and lithium-manganese-oxide (LMO) technologies. They also have a DoD of 90% of the total energy and a small self-discharge, being less than 0.2% per day. Lithium-ion batteries also feature lithium iron phosphate (LFP), lithium titanate (LTO), and lithium cobalt aluminum (NCA) technologies [11].

An idea behind microgrids is to maximize the integration of distributed energy resources, especially those of renewable nature such as solar energy. Although this requires additional controllability such as the one provided by making use of storage systems, the integration of a renewable energy source in a hybrid system not only improves reliability and efficiency but also reduces the dependence on external supply [12]. Reference [13] performs a comparison of analytical and metaheuristic methods for the sizing of a hybrid system that must operate in stand-alone mode; it is shown that a hybrid system is able to provide energy in a reliable way. In fact, hybrid systems are more reliable than just stand-alone renewable energy systems [7].

In contrast to the usual practice in which the energy for the ASS can be provided by a distribution feeder, from a local microgrid, or even from a dedicated diesel generator, a hybrid microgrid can be adopted aiming the self-assurance of the supply when instabilities or even a complete lack of energy are faced. Hybrid microgrids are combinations of alternative energy sources and energy storage systems to provide energy for a particular purpose. Both resources can be directly connected to the DC bus of the substation, but DC-AC inverters are necessary to power the AC bus; bidirectional inverters are also capable of convert AC to DC, which is convenient in cases where the energy storage system requires to be charged by the AC supply instead of the photovoltaic generation. Another function of the inverter is to keep the DC bus stabilized by controlling the waveform of the injected current. However, for a larger demand than the generation or during intermittency periods, a voltage drop occurs, then the inverter can solve that issue by managing compensation using the external power [14,15].

Different methods have been proposed in the specialized literature for the optimal sizing of microgrids pursuing different types of objectives and constraints. References [16–18] present systematic summaries of the proposed methods. Different metrics to optimize the size of the microgrid have been adopted, most of them related to economic and environmental objectives with restrictions related to dynamic considerations of frequency and voltage stability, and the search to balance the energy management between the generation, load, and storage [17,18]. Many of those proposals consider the stochastic behavior of the generation and the loads within the microgrid using mathematical programming, metaheuristics. and analytical methods, being common the use of the well-established Monte Carlo simulation [16].

There are few works in the state of the art considering objectives based on reliability for the optimal sizing of the microgrids [19,20]. Reference [19] presents a multi-objective metaheuristic based on evolutionary algorithms for the sizing of a hybrid system aiming the minimization of the annualized costs of the system, the loss of power supply probability (LPSP), and the cost of fuel pollution; the algorithm determines the non-dominated solutions sizing photovoltaic panels, wind turbines, batteries, and diesel generators for a DC load profile. Recently, reference [21] proposed a method for the design of a hybrid system composed of photovoltaic panels and biomass generators; a comparative analysis is carried out for different battery technologies based on technical and economic criteria, considering the net present value for a specified LPSP for a set of residential loads. Finally, [20] proposed the minimization of the net present value in the sizing of a microgrid based on photovoltaic panels, wind turbines, and fuel cells considering restrictions of deficit power-hourly interruption probability for residential customers. In conclusion, recent works on the sizing of microgrids consider a specific type of loads, most of them residential, in which the microgrid operates autonomously and is the main source of supply; the issues of continuous power supply are considered through indicators as LPSP.

Based on a mixed linear optimization model, [22] proposes a fault-tolerant supervisory controller for an isolated hybrid ac/dc microgrid seeking robust, efficient, and fault-resilient operation to meet demand with the highest possible utilization of renewable energy even under fault conditions. The reliability of the substation, considering an alternative source of power has been addressed in a few studies. For instance, an analysis of the continuity of the energy supply was done through Markov Chains in a 110/35 kV substation with distributed generation; from the analyzed cases, the more robust was the one in which distributed generation was connected to the low-voltage bus. The use of a compensation device, along with a voltage control system, has been proposed to improve the substation's operation, not only to deal with auxiliary services but also for the energy supplied by the substation [23]; more recently, voltage control was proposed to improve demand response in a smart substation [24].

Few studies have addressed the energy supply of ASS, although alternative means to meet the energy demands of those systems have been discussed and simulated. Specifically, a two-part work makes first a critical analysis of the different types of fuel-cells for energy supply that can be a backup or main source [1]; they can be combined with other technologies such as a photovoltaic generation to support an electrolysis system and also produce hot water for other uses; the second part discusses a case study of the use of fuel-cells in a real substation in Romania, bringing economic information of three possible uses for the fuel-cells [25]. The case with the best economic interest was selected to design a system for the ASS.

This paper proposes a method for the capacity sizing of a microgrid with batteries, photovoltaic generation, and bi-directional inverters to supply the power demanded by ASS under a contingency. A set of alternatives is assessed through exhaustive search and Monte Carlo simulations to cater for uncertainties of contingencies and variation of solar irradiation. The main contribution of the paper is the capacity sizing method along with an unavailability index to measure the contribution of the hybrid microgrid to reduce the time that the substation is not in operation. The highlights of the proposed approach are described as follows:

1. A microgrid based on a hybrid system of photovoltaic energy and batteries is adopted as a backup system for the operation of auxiliary loads in a substation. Generally, substation backup systems use diesel generators without mentioning the possibility of taking advantage of renewable energy sources.
2. The influence of contingency rates and durations to determine the optimal size of the main components of the microgrid is assessed. Unlike other methods, it is unknown a priori the number of hours that the hybrid microgrid should be available to supply the substation loads.
3. An exhaustive search is adopted to identify the optimal size of the main components of the microgrid, such as the photovoltaic panels, the batteries, and the inverter. The election of the technique is justified by the few components that compose the microgrid, which allows focusing on the sensitivity analysis of uncertain parameters of renewable generation and fault duration.

This paper is organized as follows. Section 2 presents the proposed method for the capacity sizing of the microgrid, justifying, and explaining the particularities of each of the elements that are part of the proposed exhaustive search. Section 3 discusses the economic and operation assessment, summarizing, and linking the elements of the proposed sizing method. Section 4 illustrates the application of the method in a case study for auxiliary loads of a substation requiring 12 kW and assuming that the expected fault rates and fault durations of the main substation's feeder are 1 fault/year and 5 h, respectively. Additionally, a sensitivity analysis of these parameters and the costs of the main components is carried out. Conclusions are drawn in Section 5.

2. Capacity Sizing of the Microgrid

Microgrids are configurations that include a set of energy sources and storage, used especially in applications of autonomous generation systems. Their main advantage is the relative simplicity of allocation and autonomy from a single generation resource, which gives them the ability to work independently of weather conditions and time of day. Nevertheless, the effectiveness in the application of hybrid systems depends on the correct sizing of the microgrid resources.

Microgrids are generally designed to increase the integration of renewable sources such as solar and wind energy. Although those sources are cleaner than conventional generators, their unpredictability and climate dependence limit their applications when loads require uninterrupted power. Consequently, and to increase reliable energy delivery, storage systems are used in the configuration of hybrid systems. This paper focuses on the design of a hybrid system consisting of a clean source of solar energy and a storage system based on batteries to be the backup service that provides energy to the ASS loads.

Figure 1 shows the impact of the microgrid in the increasing of the availability time of the auxiliary services when the main supply system is under a contingency state. In Figure 1a, there is a set of auxiliary services supplied only by the distribution network; when the substation's feeder is in a contingency state, the ASS are interrupted. The time gap between the "main grid operation" and "repair" represents the fault detection time. Figure 1b shows the positive influence of the microgrid in a contingency state scenario; in the outage period, the microgrid supplies the loads, therefore increasing the time of availability of the ASS.

Traditionally, the components of hybrid systems are sized based on the assumption that the estimated value of the load and the predefined time to operate in autonomous mode are known [16]. However, a microgrid used as a backup service for ASS has characteristics that make unsuitable the a priori sizing of the microgrid capacity: It is sought that microgrids, as a support service for ASS, have the capacity to autonomously operate for a longer period than any contingency in the main supply system. However, considering that the main supply comes from the distribution system, the contingency durations have a random behavior depending on the feeder of the distribution network. Therefore, the energy that should supply the microgrid is not known in advance.

Although it is desirable that the microgrid is sized to meet any duration of contingency, in the practice investment limitations, physical limitations for installation, and the random behavior of the contingency's duration determine the selection of an appropriate investment. In consequence, the hybrid system must be sized to be able to provide the ASS loads for a desired proportion of the power distribution system contingency scenarios.

In the case of a hybrid system with few components, an exhaustive search is proposed here to assess the cost-effectiveness for each possible combination of photovoltaic systems and batteries. For this purpose, each possible configuration of the microgrid will be evaluated through economic and unavailability indicators. Simulations using the well-known Monte Carlo method are carried out to determine the performance in contingency state for each possible microgrid configuration, taking into account the uncertainties of the occurrence and duration of contingencies, as well as the random behavior of the solar irradiation.

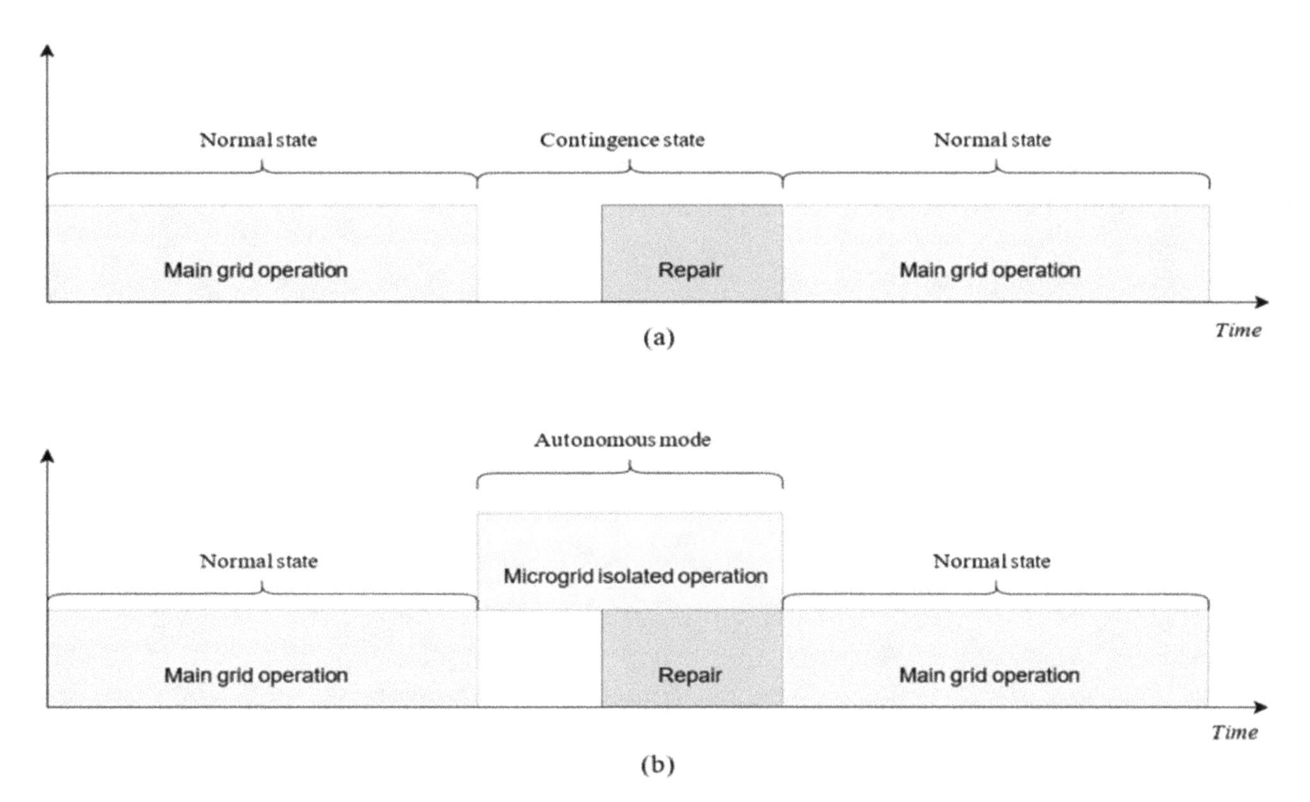

Figure 1. Impact of the microgrid in the increasing of the available time. (**a**) Provision of auxiliary services in the substation without a microgrid. (**b**) Provision of auxiliary services in the substation with a microgrid.

As mentioned above, the possible combinations of microgrids are defined by two main aspects: the number of photovoltaic panels and the size of the battery bank; other components of the hybrid system (e.g., the inverter) take a secondary role in determining the sizing of the backup microgrid. Figure 2 shows how each configuration is created to be assessed by the exhaustive search method: each investment possibility for the photovoltaic panels, identified by letters within blue circles, is combined with each of the investment possibilities for the batteries, identified by numbers within green circles. As a result of all combinations, each microgrid configuration to be evaluated is defined by the letter and number of their main components, within yellow circles. To choose the best microgrid configuration from a set of possible configurations (yellow circles), two indexes are proposed in this work. They allow the compromise analysis between the economic value of the investment and the unavailable time of the ASS due to any contingency of the main system.

2.1. Economic Assessment

The total net cost will be used as an investment index for each possible microgrid configuration represented by the index s. This cost includes the cost of the batteries, the cost of the photovoltaic panel system, and the cost of the inverter. The latter is added, given its large proportion within the overall costs in hybrid systems.

For each case, the investment is calculated according to the number of components of the microgrid. For batteries, the cost is calculated based on their nominal storage capacity ($\overline{E}_s^{bat}$), measured in kWh. The photovoltaic panels cost is calculated according to the units used on each microgrid s (N_s^{pv}), whilst the bidirectional inverter cost is calculated according to its power in kW $\left(P_s^{In}\right)$. Thus, the total investment value I_s for a microgrid configuration s is given by Equation (1). The equipment costs are c^{bat}, c^{pv}, and c^{In} for batteries, photovoltaic panels, and inverter, respectively.

$$I_s = \overline{E}_s^{bat} \cdot c^{bat} + N_s^{pv} \cdot c^{pv} + P_s^{in} \cdot c^{in} \tag{1}$$

In addition to the investment costs in the microgrid, maintenance costs are also considered through the years of the equipment's lifespan (τ). For this purpose, the costs are brought to a present value at an interest rate δ. The annual maintenance cost for each equipment corresponds to a fraction of its investment, i.e., mc^{bat}, mc^{pv}, and mc^{in} for batteries, photovoltaic panels, and inverter. The maintenance cost is described in Equation (2).

$$MC_s = \sum_{i=1}^{\tau} \frac{\overline{E}_{bat} \cdot c^{bat} \cdot mc^{bat} + N_s^{pv} \cdot c^{pv} \cdot mc^{pv} + P_s^{in} \cdot c^{in} \cdot mc^{in}}{(1+\delta)^i} \tag{2}$$

Since the photovoltaic panels produce energy when solar irradiation is available (under fault and also in normal operation), a profit related to the selling of that energy ($Profit^{PV}$) could contribute to reducing the total cost. That profit is calculated by Equation (3) in terms of the mean annual energy generated by a photovoltaic panel ($\overline{E^{PV}}$) and the energy price (π).

$$Profit^{PV} = \sum_{i=1}^{\tau} \frac{N_s^{pv} \cdot \overline{E^{PV}} \cdot \pi}{(1+\delta)^i} \tag{3}$$

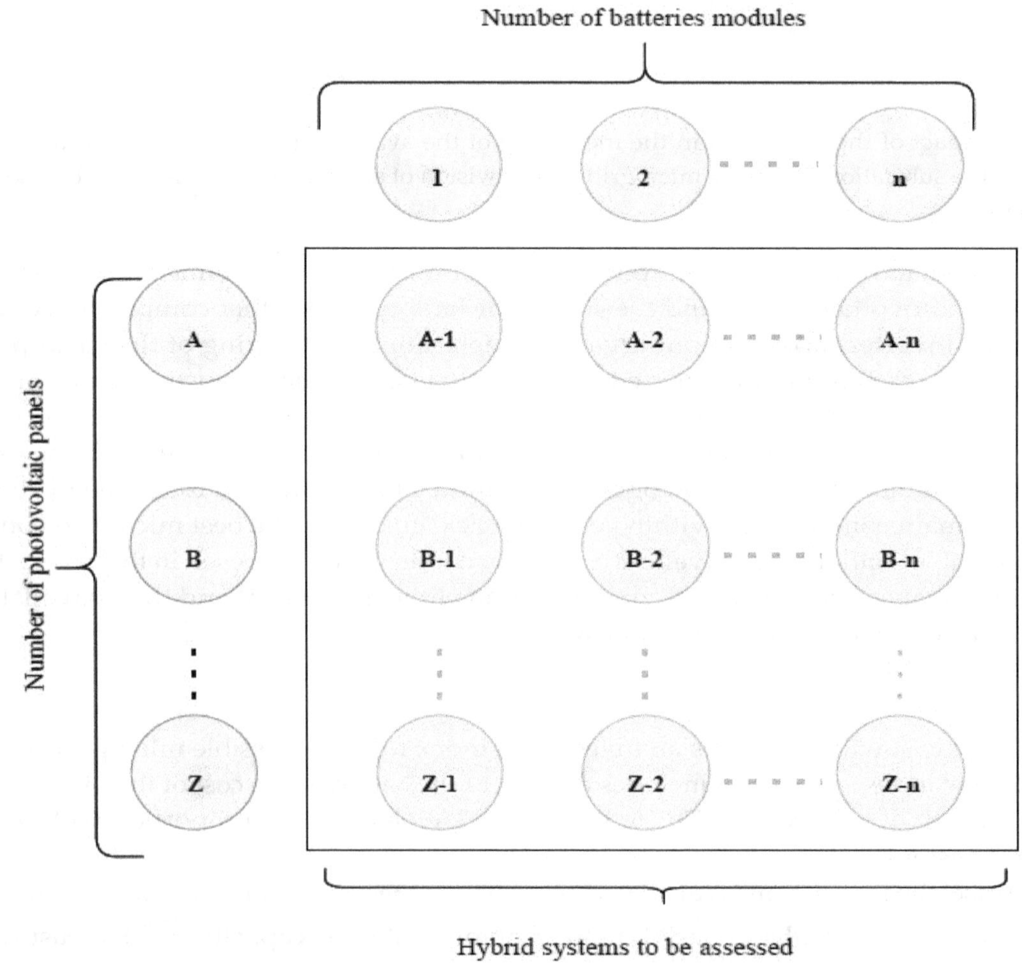

Figure 2. Exhaustive search method for the optimal capacity sizing of the microgrid.

Finally, the economic index is the total cost of the system (TC_s), which corresponds to the sum of the investment and the maintenance costs, as shown in Equation (4).

$$TC_s = I_s + MC_s - Profit^{PV} \tag{4}$$

2.2. Assessment of ASS Unavailability

Generally, the main system for supplying ASS is the distribution system. It is characterized by having a radial topology, where each consumer has a single supply path. This principle is equally scalable when it is the distribution system that supplies the ASS. However, the distribution networks are the main source of interruptions in the power system [26], accounting for around 80% of the interruptions [27]. Moreover, in general, there has been an increase in the monthly half of outages in the United States from 2.5 to 14.5 in the period 2000–2013 [28]. The expectation is that the frequency and severity of the absences will continue to increase [13].

Thus, this paper proposed a method to size the microgrid aiming for the reduction of the impact of distribution system contingencies in the operation of ASS. For this purpose, the contingencies of the main supply system are characterized by their frequencies and durations [27]. These values can be obtained from statistical studies of the distribution system operator or, in the absence of data, can be considered expected values following a normal distribution function.

The unavailability index represents the expected proportion of hours that the substation's auxiliary services will be out of operation for a specific microgrid configuration. For the calculation of this index, it is assumed that (a) the main service feeder of the substation load has a known annual contingency rate λ; and (b) each contingency is characterized by a random duration, which follows a known probability function.

Given the random behavior of the contingency duration and the energy supplied by the photovoltaic panels, it is not possible to guarantee that the available energy of the microgrid in the contingency state is always enough to supply the ASS. Figure 3 shows the most probable cases when the microgrid acts as an autonomous backup for the ASS: (a) the available energy of the microgrid is equal or larger than the ASS energy requirement for the contingency duration of the main supply system, and (b) the microgrid does not have enough energy and therefore the ASS will be unavailable for a time smaller than the duration of the contingency.

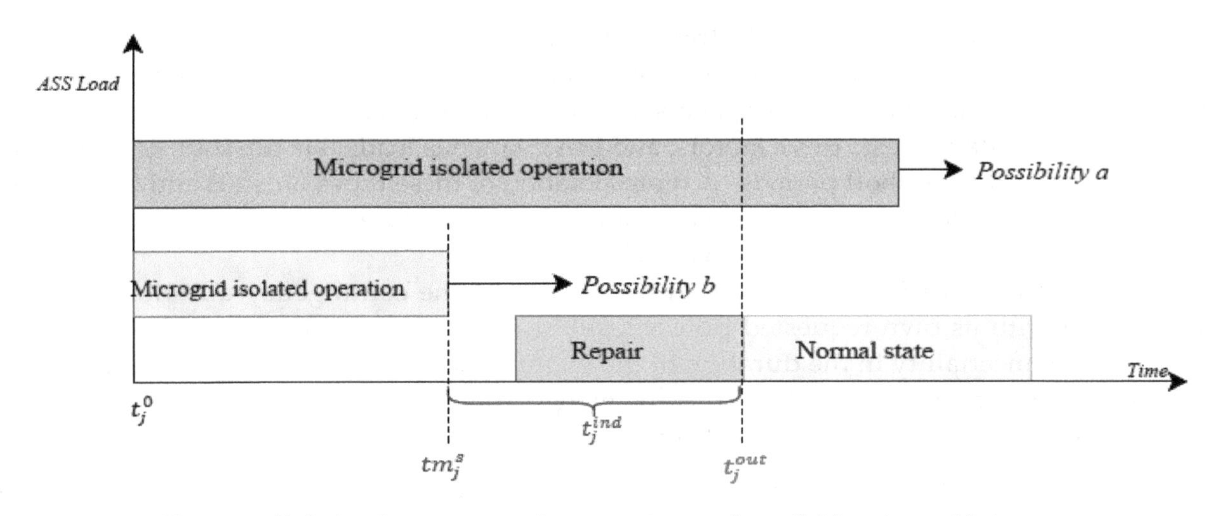

Figure 3. Relation between contingency time and available microgrid time.

The calculation of the unavailability time of ASS t_j^{Ind}, for a contingency j considering these two possible cases, is detailed for a microgrid configuration s. The contingency j has a duration t_j^{out}; the microgrid s has an availability time tm_j^s dependent on the weather conditions at the time of the contingency, and the contingency starts at time t_j^0. Thus, the microgrid has an amount of energy such that:

a. The availability time of the microgrid is equal or longer than the contingency duration $tm_j^s \geq t_j^{out}$. In this case, t_j^{Ind} is zero.

b. The availability time of the microgrid is shorter than the duration of the contingency $tm_j^s < t_j^{out}$. In this case, t_j^{Ind} is equal to the difference between the duration of the contingency and the availability time of the microgrid, i.e., $t_j^{Ind} = t_j^{out} - tm_j^s$.

The calculation of the unavailability index I_s^{ind} assesses the proportion of the unavailable time for the ASS concerning a large number of operating hours of the microgrid. Thus, the numerator calculates the total number of unavailable hours of a microgrid configuration s as the sum of the unavailable time for each contingency t_j^{Ind}. The number of contingencies is defined as the product of the expected feeder contingency rate of the distribution network λ and the number of years to be simulated N^{years}. The denominator of the index defines the number of hours of simulation as the product between a large number of years of operation N^{years} and the parameter α, which is the number of hours in a year. Based on the above, the unavailability index can be expressed by Equation (5). An availability index can be also formulated by Equation (6).

$$I_s^{ind} = \frac{\sum_{j=1}^{N^{years}\cdot\lambda} t_j^{Ind}}{N^{years}\cdot\alpha}\cdot 100\% \tag{5}$$

$$I_s^{disp} = 100 - I_s^{ind} \tag{6}$$

2.3. Energy Analysis for the Autonomous Service of Auxiliary Services

To determine the unavailability time in each interruption it is important to know both the energy requested by the ASS loads and the total energy available from the microgrid, which depends on the energy in the battery system and the energy generated by the photovoltaic panels during the contingency. Both energy components are described in this subsection.

2.3.1. Energy Requested by Auxiliary Services of the Substation

The ASS loads can be divided into three subgroups: permanent loads that are related to the equipment connected continuously as the protection, measurement, and communication devices; temporary loads with high power requirements of short duration and necessary for the reestablishment of service in the substation, e.g., drive motors; and instantaneous loads that are sources of high power requirements in extremely short periods. A representation of these types of loads and their durations is presented on the left of Figure 4 in which permanent loads are represented by green bars, temporary loads by yellow bars, and instantaneous loads by red bars.

Although it is desired to divide the representation of the contingency duration into smaller intervals, each with its own requested power level, it is not practical in terms of planning when considering the uncertainty in the duration of the contingency. Because of that, an equivalent load factor is used to approximate the ASS load requirements. This factor is obtained using the equivalence between energy consumption in Figure 4 as follows:

1. Each demand level for period t, on the left-hand side of Figure 4 can be represented in terms of a load factor f_i related to the nominal power of ASS, i.e., $P_t = f_t \cdot P^{nom}$. Thus, the total energy required by ASS for the contingency illustrated in Figure 4 is equal to $E_{\text{ASS}} = P_1 \cdot t_1 + P_2 \cdot t_2 + P_3 \cdot t_3 + P_4 \cdot t_4 = P^{nom}(f_1 \cdot t_1 + f_2 \cdot t_2 + f_3 \cdot t_3 + f_4 \cdot t_4)$.
2. Equivalently, the power level related to the right-hand side of Figure 4 can be expressed in terms of a global factor F_g related to the nominal power of ASS, i.e., $P_{eq} = F_g \cdot P^{nom}$. Consequently, the total energy required by ASS for the contingency of Figure 4 is equal to $E_{ASS} = P_{eq} \cdot \left(tm_j^s - t_j^0\right). = F_g \cdot P^{nom} \cdot \left(tm_j^s - t_j^0\right)$.
3. Since the energy required by ASS is the same for both representations in Figure 4 the load factor for this example is $F_g = (f_1 \cdot t_1 + f_2 \cdot t_2 + f_3 \cdot t_3 + f_4 \cdot t_4)/\left(tm_j^s - t_j^0\right)$.

Generalizing the calculation above, the general load factor can be written as shown in Equation (7):

$$F_g = \frac{\sum_{t=i}^{T} f_t \cdot t_t}{\left(tm_j^s - t_j^0\right)} \tag{7}$$

Thus, in the proposed method, the nominal power of ASS P^{nom} and the global load factor F_g allow determining the supplied energy by the microgrid in each contingency state using Equation (8).

$$E_{ASS} = F_g \cdot P^{nom} \cdot \left(tm_j^s - t_j^0\right) \tag{8}$$

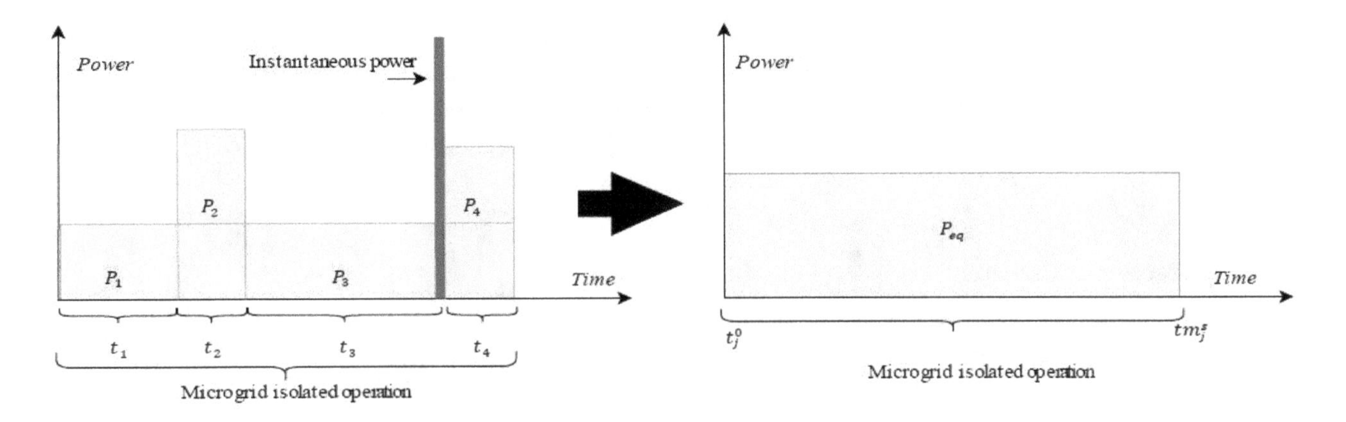

Figure 4. Load types and their contingency duration.

2.3.2. Energy Generated by the Photovoltaic System

The operation of the photovoltaic system determines the power generated by the panels based on solar irradiation values and depending on the temperature of the solar cell. The relation between the solar irradiance and the output power of a solar generating source can be described by the set of Equations (9)–(13) [29].

$$T_c = T_a + \left(\frac{N_{OT} - 20}{0.8}\right) \cdot G_{gh} \tag{9}$$

$$I_c = G_{gh} \cdot [I_{sc} + K_i \cdot (T_c - 25)] \tag{10}$$

$$V_c = V_{oc} + K_v \cdot T_c \tag{11}$$

$$FF = \frac{V_{MPPT} \cdot I_{MPPT}}{V_{oc} \cdot I_{sc}} \tag{12}$$

$$P_{pv}^{oper} = N_s^{pv} \cdot FF \cdot V_c \cdot I_c \tag{13}$$

Equation (9) calculates the temperature in the photovoltaic cell T_c in terms of the ambient temperature T_a, the nominal operating temperature of the cell N_{OT}, and the solar irradiation G_{gh}. Equation (10) calculates the current provided by the photovoltaic cell as a function of its temperature and the temperature coefficient for the current K_i. Similarly, the voltage of the photovoltaic cell is calculated using Equation (11) as a function of T_c and the temperature coefficient for the voltage K_v. The cell efficiency is determined by the fill factor, calculated in Equation (12), in which I_{MPPT} is the current at the maximum power point and V_{MPPT} is the voltage at the maximum power point.

The previous equations allow the calculation of the output power of a set of solar cells P_{pv}^{oper}, corresponding to the product of the output power of each cell and the number of panels of the configuration N_{PV}. Consequently, it is possible to calculate the energy generated by the panels for a desired time interval, i.e., the island operation mode of the microgrid tm_j^s as given by Equation (14).

$$E_j^{pv} = \int_{t_j^0}^{tm_j^s} P_{pv}^{oper} \cdot \Delta t \cdot dt \quad (14)$$

2.3.3. Energy Available from the Storage System

A disadvantage of photovoltaic panels is that their generated energy must be consumed instantly. Moreover, given their dependence on weather conditions, it cannot be guaranteed that they are always continuous, which makes photovoltaic panels irregular and unreliable. Due to the above considerations, electrical energy storage systems are necessary to make better use of the power generated by the photovoltaic system, improving the availability and quality of energy.

In this work, a set of batteries is used as a storage system to jointly act in the microgrid to support the ASS. The use of batteries allows the microgrid to have a controlled power output, capable of reliably providing power to the ASS whenever necessary. For this purpose, it is assumed that the battery is always charged, and its capacity is available to support the ASS. Hence, the available energy of the battery E_{bat} is expressed in Equation (15), in which η^{out} represents the round-trip efficiency of the battery, $\overline{E}_{bat}$ is the nominal capacity of the battery, and DoD represents the depth of discharge.

$$E_{bat} = \eta^{out} \cdot DoD \cdot \overline{E}_{bat} \quad (15)$$

2.3.4. Bi-Directional Inverter

A bidirectional inverter with the ability to operate in grid-connected and island mode is required to operate the microgrid. The inverter allows the operation of the microgrid in autonomous mode to increase the time availability of ASS, the use of the energy of the main grid for charging the batteries, as well as the injection of the power of the solar panels into the main grid. Figure 5 illustrates the bidirectional flows that allow the inverter, either to feed the microgrid into the ASS loads or to subtract or inject power into the main grid from the photovoltaic system.

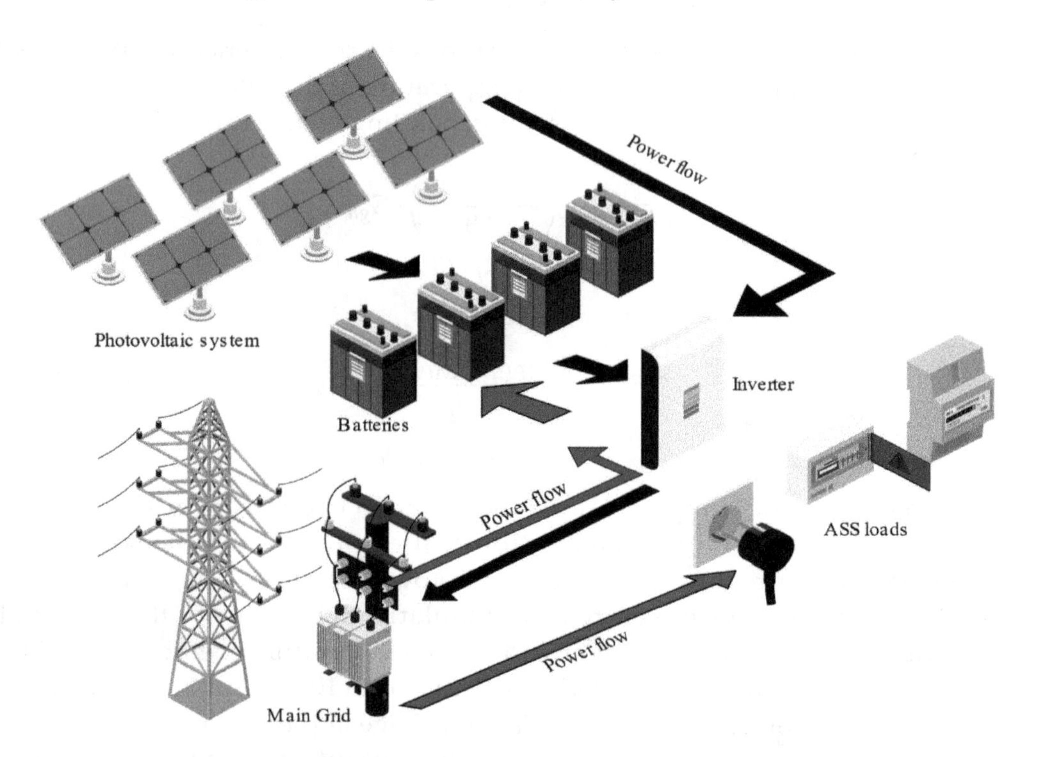

Figure 5. Operation of the bidirectional invert.

Thus, the inverter controls the operation of the microgrid, determining the input and output flows of each of the distributed sources to feed the ASS loads. To define the necessary inverter capacity in the proposed microgrid, two quantities are considered: the installed power of the photovoltaic panels $\overline{P}_s^{PV}$ and the maximum power of the ASS loads. Equation (16) summarizes the criteria to choose the inverter capacity.

$$P_s^{In} = \max\left\{\overline{P}_s^{PV}, P_{eq}\right\} \tag{16}$$

3. Capacity Sizing Method

The solution strategy proposed in this paper for the capacity sizing of the hybrid microgrid is summarized in Figure 6. It calculates economic and unavailability indexes for each of the feasible configurations of the main components of the microgrid. The calculation of the economic index is direct and depends on the dimensions of the two main components of the microgrid: battery banks and photovoltaic panels. On the other hand, the unavailability index is dependent on the capacity of the microgrid in dealing with feeder contingencies, which have a random duration behavior. Thus, the well-established Monte Carlo simulation method [30] is used here to calculate the total unavailability time for each of the possible microgrid configurations under evaluation. The details of the simulation procedure to calculate the unavailability index in Equation (5) are described below:

1. Identify the microgrid configuration to be evaluated from the set of configurations (solar panel/battery banks).
2. Determine the expected value of the substation feeder failure rate λ.
3. Define the number of years to be simulated N^{years} aiming at an appropriate convergence.
4. Calculate the total number of simulations to be performed, i.e., the product between the total number of years to be simulated and the contingency rate ($N^{years} \cdot \lambda$).
5. Generate the duration of the contingency j $\left(t_j^{out}\right)$, as well as the initial time of the contingency $\left(t_j^0\right)$ according to corresponding density probability functions.
6. Calculate the amount of energy requested by the ASS in contingency state j.
7. Calculate the amount of energy available by the batteries in the contingency state j.
8. Calculate the amount of available energy from photovoltaic panels in contingency state j.
9. Determine the difference between the energy required by the substation and the available energy by the microgrid. If the difference is positive, i.e., the hybrid system is unable to supply the ASS load during the entire contingency state j. The microgrid's autonomy time tm_j^s is calculated by equating the available energy with the energy consumed by the load from t_j^{out} to tm_j^s, as follows:

    a. Calculate tm_j^s as the solution of Equation (17).

    $$\eta^{out} \cdot DoD \cdot \overline{E}_{bat} + \int_{t_j^0}^{tm_j^s} P_{pv}^{oper} \cdot \Delta t \cdot dt = F_g \cdot P^{nom} \cdot \left(tm_j^s - t_j^0\right) \cdot \Delta t \tag{17}$$

    b. Take t_j^{ind} as the difference between t_j^{out} and tm_j^s if it is positive, i.e., $t_j^{ind} = \max\left\{t_j^{out} - tm_j^s, 0\right\}$.

10. Accumulate the total unavailable time t_j^{ind} using Equation (5).
11. Repeat steps 5–10 until all contingencies calculated in step 4 are evaluated.
12. Go back to step 1 and choose a new microgrid configuration to evaluate.

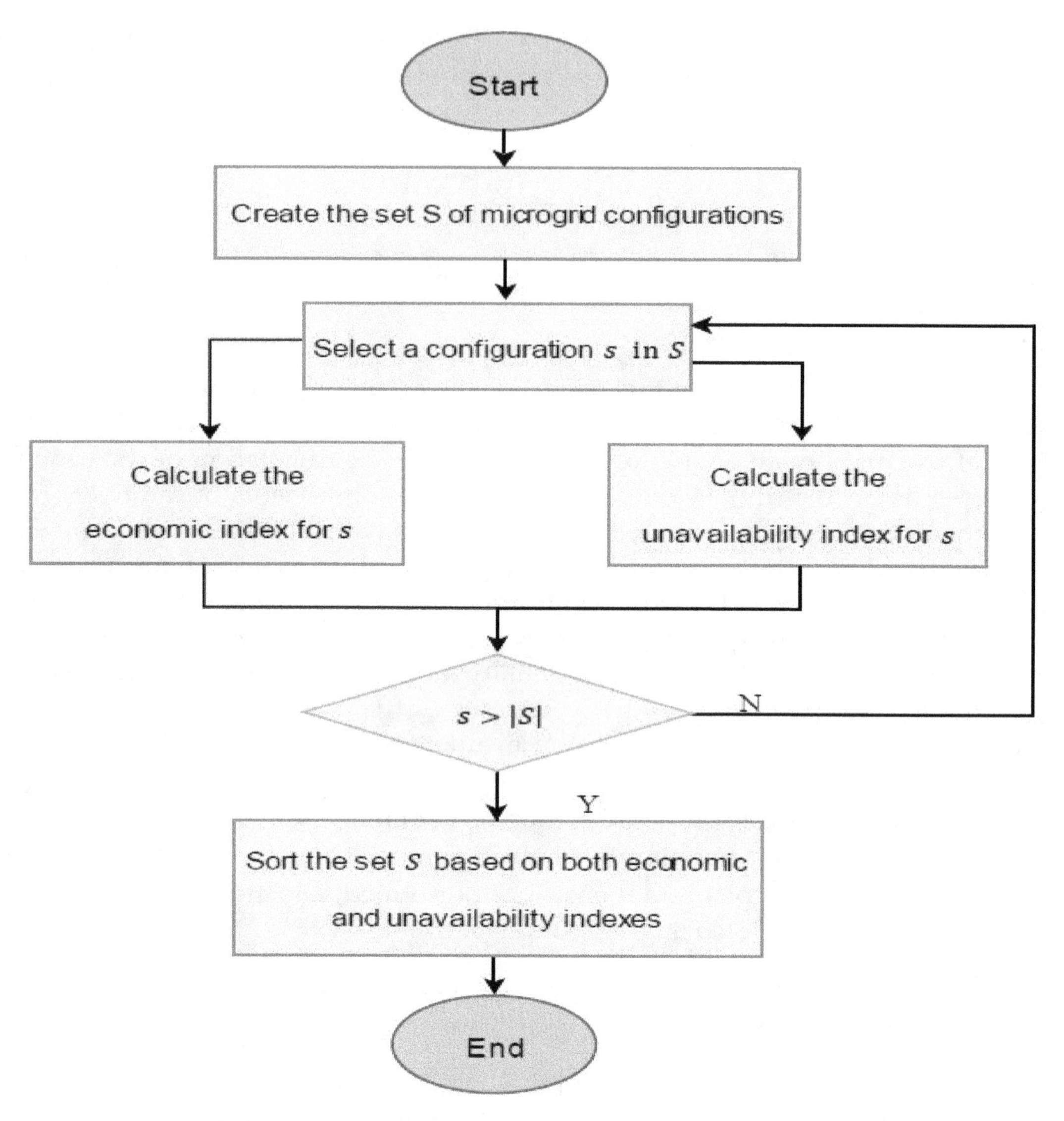

Figure 6. Summary of proposed capacity sizing method.

The aforementioned steps are summarized in Figure 7.

The associated optimization problem can be classified as a stochastic non-convex multi-objective problem, whereby the first objective in Equation (18) minimizes the economic index of a microgrid as a backup system for ASS loads, while the second objective in Equation (19) minimizes the expected value index of the ASS unavailability time due to faults in the main supply system. Both objectives present a conflicting nature since small investments in the microgrid leads to longer unavailability times and larger investments results in shorter unavailability times. Both objectives are subject to the set constraints (7)–(17), summarized in Equation (20), which are related to the operation of the power resources of the microgrid, ASS loads, and the autonomous operation of the microgrid for the contingency states of the ASS main supply.

$$\text{Minimize } TC_s \tag{18}$$

$$\text{Minimize } I_s^{disp} \tag{19}$$

$$\begin{array}{c}\text{subject to: Operation of the power resources of the microgrid} \\ \text{ASS load constraints} \\ \text{Autonomous operation of the microgrid for contingencies}\end{array} \tag{20}$$

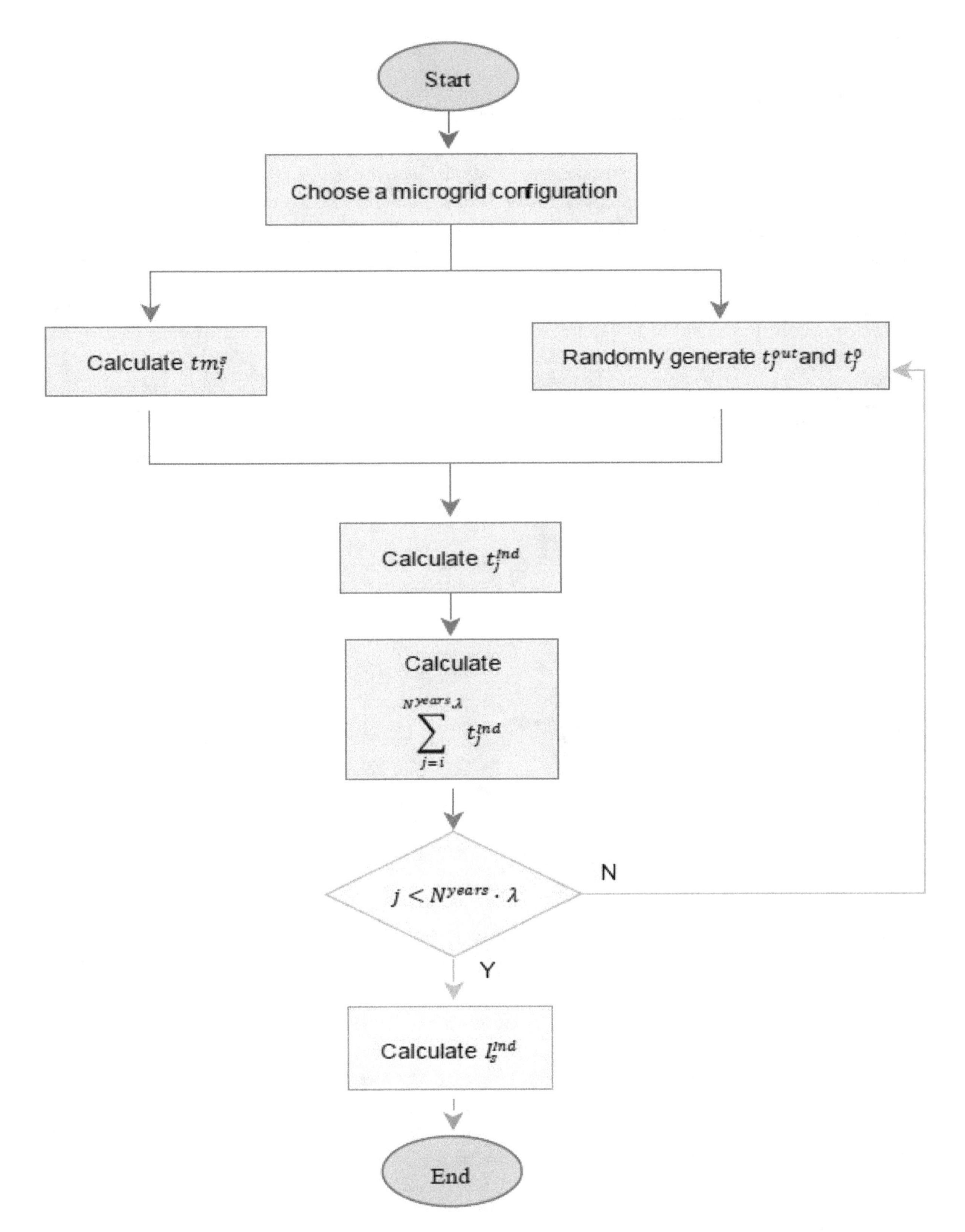

Figure 7. Flowchart for the calculation of the unavailability index.

4. Case Study

The proposed method was applied to define the best combination of batteries and photovoltaic panels for a microgrid by assessing the economic and unavailability indexes presented in Section 2. To consider the randomized behavior of the photovoltaic generation and the duration of the contingencies, Monte Carlo simulations were executed in a computer with an Intel i7-7700K processor using MATLAB [31].

Results of the economic and unavailability indexes for four cases:

- Case I: Goal for the unavailable index.
- Case II: Investment budget for the microgrid configuration.
- Case III: Variation of the fault rate.
- Case IV: Variation of the fault duration.
- Case V: Sensitivity analysis for variations of the battery and photovoltaic panel prices.

The ASS loads are divided into three large groups, each one with nominal power, total power, and load factor, as shown in Table 1. The topology of the sized microgrid is shown in Figure 8.

Table 1. ASS load characteristics.

Load Description	Load Type	Nominal Power (W)	Total Power (W)	Load Factor
Monitoring	Permanent	70	74.90	0.005
Circuit breaker	Temporary	5000	5850.00	0.362
Protections	Permanent	250	235.00	0.015
Measurement	Permanent	120	139.20	0.009
Communication	Permanent	320	291.20	0.018
Illumination & climatization	Permanent	10,000	10,300.00	0.637
Drive motors	Temporary	400	396.00	0.025
Drive coils	Instantaneous	6200	7440.00	0.460

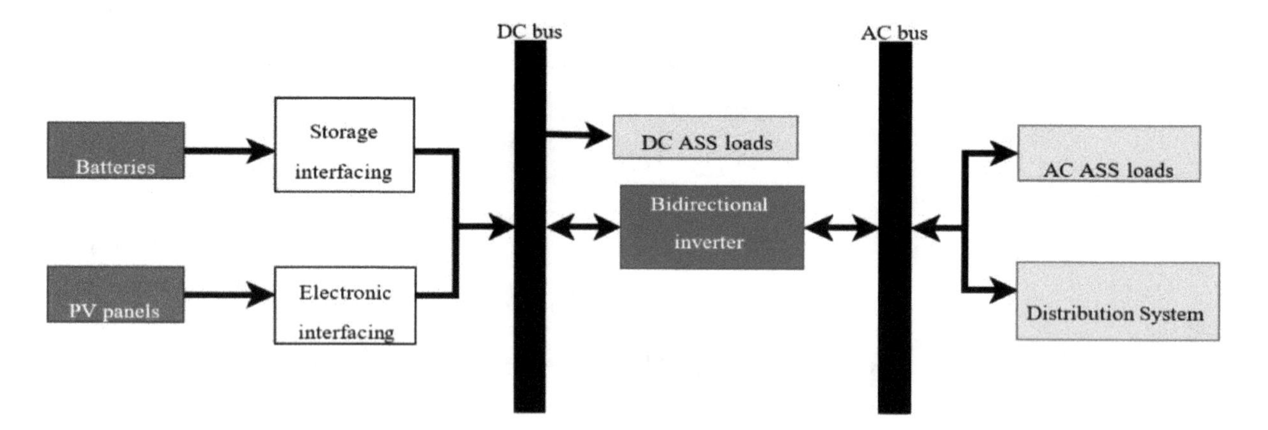

Figure 8. Topology of the microgrid used in the case study.

Based on these data and assuming that the availability time of the microgrid can be divided into four intervals, as illustrated in Figure 4, the global power factor is calculated as shown in Table 2. Load factors for the permanent and temporary loads are in the corresponding interval. It is observed that the temporary loads are in intervals 2 and 4, whereby the power of the circuit breaker is in interval 2 and the drive motors are in interval 4. It is noteworthy that for the global load factor calculation, instantaneous loads are not considered, but are included in the sizing of the inverter. Accordingly, the equivalent power load is equal to 11,918.70 W, which can be rounded to 12 kW.

Table 2. Global power factor bases calculations.

Period	% of Time	Load Factor	Global Load Factor
1	60	0.689	0.738
2	15	0.998	
3	15	0.689	
4	10	0.713	

A state-of-the-art photovoltaic panel manufactured by Panasonic (VBHN330SJ47) was evaluated; it has the highest ratio of power generated per area used [32]. Each panel has 0.33 kWp, a cost per unit equal to $312/panel, and 1% of maintenance cost per year of the total installation cost as considering for the Brazilian commercial sector in 2019 [33]. The relevant data for obtaining the output power for the Panasonic panel model, expressed in (9)–(13), are shown in Table 3, based on the datasheet in [34]. These batteries have a modular capacity of 2 kWh, a cost per kWh equal to 420 $/kWh, a DoD equal to 90%, and a 95% round-trip efficiency. NMC li-ion battery units are considered due to their higher energy density (735 Wh/L) and their low self-discharge (0.1% per day) [11]. Other battery technologies have worst characteristics, as the li-ion NCA, LTO, and LFP (energy density of 620 Wh/L for the LTO option); more importantly, these technologies have a highest cost ($1050/kWh).

Table 3. Information of the photovoltaic panel.

Characteristic	Data
Cost	312 ($/unit)
K_v	−0.174 (V/°C)
K_i	1.82 (mA/°C)
N_{ot}	44 (°C)
V_{oc}	65.8 (V)
I_{sc}	4.89 (A)
V_{mppt}	58.0 (V)
I_{mppt}	5.7 (A)

Moreover, the inverter cost is 105 $/kW, as suggested in [11]. Also, the maintenance cost for the batteries and inverters are assumed to be 1.5% per year of the total installation cost. A lifespan of 20 years is adopted; the maintenance costs of all equipment are calculated using an interest rate of 6%. The energy price is 0.05 $/kWh.

The calculation of the unavailability index was done considering that the contingency durations follow a normal distribution with a mean equal to 5 h and a standard deviation equal to 3 h [35,36]. To guarantee the Monte Carlo convergence, the simulations are carried out for 5000 years. A total of 12,000 photovoltaic irradiation profiles, with a one-minute resolution, were generated using the CREST tool [37], corresponding to 100 monthly profiles; hourly temperature data for one year (2019) was obtained from the Renewables. Ninja online tool in [38]. Those profiles were generated using the geographical information of São Paulo city in Brazil.

With that information, the combination of technologies is analyzed from zero up to 96 battery modules (roughly 16 h of the ASS load) and from zero up to 110 photovoltaic panels (three times the ASS load).

4.1. Case I: Unavailability Index Goal

The optimal size of the solar photovoltaic and batteries is obtained in this case considering a goal of 0.003% for the unavailability index and adopting a fault rate equal to 1 for the feeder of the distribution system. Figure 9 shows the summary results of the unavailability index for each possible configuration of the microgrid. It is clear the inverse relationship between the number of energy resources in the microgrid and the unavailability index, i.e., as more capacity is installed in the microgrid, less time the ASS is unavailable in a contingency state. It is worth mentioning that the number of photovoltaic panels does not generate much influence on the index, contrary to what happens with batteries, whereby the addition of some units produces significant improvements in the index. This fact is explained by the climate dependence on the generation of power of photovoltaic panels and the availability of energy from batteries, which only depends on their state of charge.

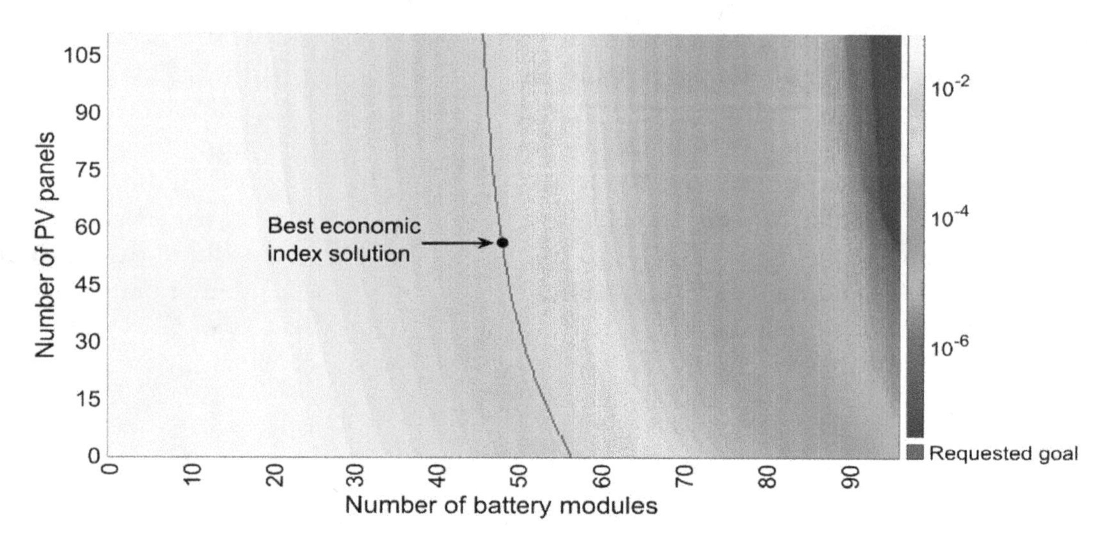

Figure 9. Heatmap for the unavailability index in Case I.

The red line in Figure 9 indicates the solutions that reach the requested unavailability goal. Solutions on the left of that line do not satisfy the goal. From the set of feasible solutions, the one with 48 battery modules and 55 photovoltaic panels, represented by the black circle in Figure 9, has the best economic index ($44.172), i.e., it satisfies the unavailability goal and has the lowest total cost.

The worst unavailability index is the case when no batteries neither photovoltaic panels are present (0.0520%). On the other hand, 96 battery modules and 110 photovoltaic panels result in the lowest unavailability index (0%) but with an economic index of $88,344.

These solutions could be compared to a conventional backup diesel generator, which for a 12 kW/15 kVA power has a cost of about $3000 and consumes approximately 3.2 L/h [39]. Adopting a diesel price in Brazil of $0.55/L and considering that the expected number of fault hours per year is 5, the expected operation and maintenance cost of the diesel generator across the 20-year horizon is just $270.93 [39,40]. Therefore, an equivalent economic index would be just $3270.93. Although that value is just a fraction of the best economic index solution in Figure 9, it is worthy to highlight that the integration of a diesel generator in the microgrid has some disadvantages such as the need for safe storage and handling of 32 L of diesel to keep the service for faults up to ten hours. On the other hand, environmental concerns could inhibit the adoption of a technology that produces green-house emissions (although a small value). Moreover, operation policies could require two or more different backup alternatives, meaning that just a diesel generator would be insufficient to complain that kind of policy. For the particular Brazilian case, the regulation requires that at least two independent sources supply the ASS [41].

4.2. *Case II: Limited Budget*

The optimal size of the photovoltaic panel system and the batteries, in this case, is obtained considering that a limited budget of $40,000 for the economic index. The assumptions for calculating the unavailability are the same as in Case I.

As highlighted in the previous case of study, an increase in the size of the equipment brings a reduction in the index of unavailability. On the other hand, larger dimensions of the equipment require a higher investment, as shown in Figure 10. Thus, on many practical occasions, decision-makers have an investment limit to achieve the lowest values of the unavailability index. Since the budget limitation should be enforced, the solution for this case is the combination that provides the best unavailability index that does not have a total cost above the budget limit. That solution defines the use of the 43 battery modules and 55 photovoltaic panels, leading to an unavailability index of 0.0046%. Note that this solution, shown by the black circle in Figure 10, is worse than the one found in Case I but has an economic index equal to $39,970. It is worthy to highlight that to satisfy the budget limit, fewer battery modules but more photovoltaic panels should be installed.

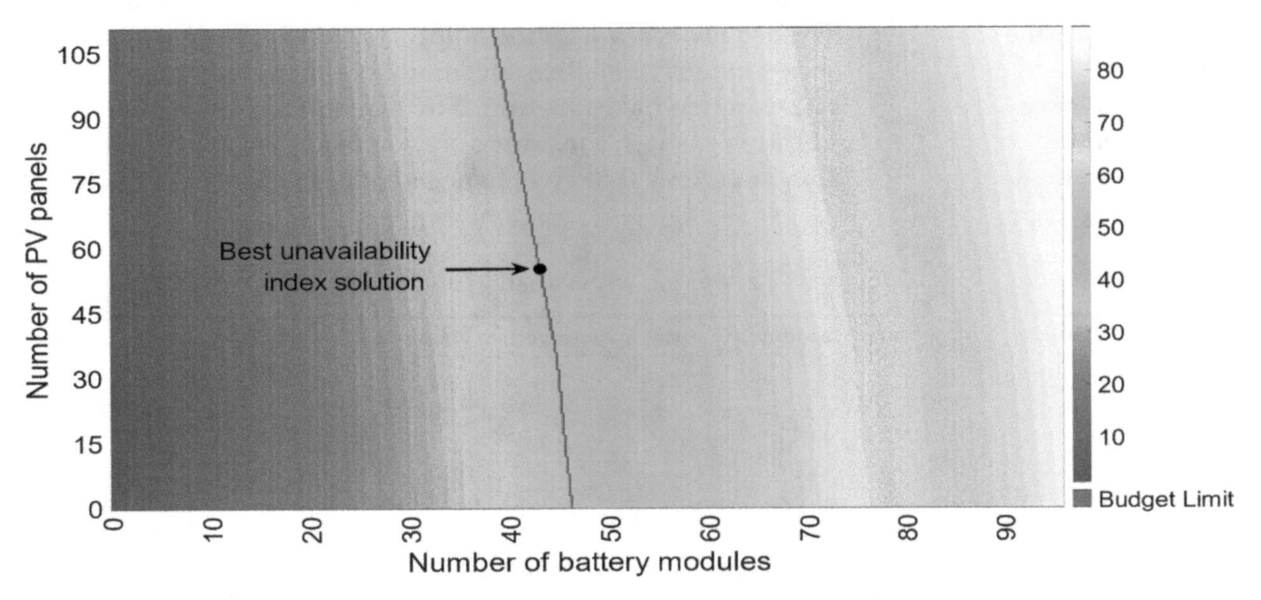

Figure 10. Heatmap for the economic index in Case II.

4.3. Case III: Sensitivity Analysis for the Variation of Fault Rate

One of the most crucial parameters in the unavailability of ASS is the annual fault rate of the distribution system feeder, which in normal operation state supplies the ASS loads. Therefore, different values for the fault rate are analyzed here to find its influence on the microgrid sizing. The same unavailability index goal for Case I is used to defining the best solution for each fault rate.

Table 4 presents the results obtained from the variation of the annual fault rate, in which is possible to verify that to maintain the same level of unavailability, more battery modules are required. It is observed that the economic index increases with the fault rate. It is worthy to highlight that the number of battery modules increases to maintain the required unavailability index; on the other hand, the number of photovoltaic panels varies without a clear trend, being reduced in some cases to save costs without compromising the goal.

Table 4. Sensitivity of the solution with variation of the fault rate.

Fault Rate (Faults/Year)	Battery Modules (Units)	Photovoltaic Panels (Units)	Economic Index ($)
1 *	48	51	44.172
2	55	53	49.915
3	59	52	53.206
4	62	49	55.518
5	64	48	57.129
6	66	41	58.322
7	67	46	59.511
8	68	44	60.212
9	69	43	60.982
10	69	51	61.540

* Annual fault rate for Case I.

4.4. Case IV: Sensitivity Analysis for the Variation of Fault Duration

The duration of interruptions affecting the feeder of the distribution system, which in normal operating state supplies the ASS loads, has a direct relation with the size of the microgrid. Therefore, different values for fault duration are analyzed here in order to find their influence on the microgrid sizing. The same unavailability index objective for Case I is used to define the best solution.

Table 5 presents the results obtained from the variation of the fault duration; the mean fault duration is shown in the first column and the standard deviation is changed proportionally. It is possible to verify that, to maintain the same level of unavailability, more battery modules are needed. It is observed that the economic index also increases with the failure duration. The number of battery modules increases to maintain the necessary unavailability rate; moreover, the number of photovoltaic panels also increases. If the mean fault duration increases from 5 to 10 h, the economic index becomes 95% larger. On the other hand, an improvement in the fault duration from 5 h to 1 h leads to a cost reduction of more than six times. This highlights that an enhancement of feeder reliability results in lower microgrid costs.

Table 5. Sensitivity of the solution with variation of the fault duration.

Mean Fault Duration (h)	Battery Modules (Units)	Photovoltaic Panels (Units)	Economic Index ($)
1	7	0	7.144
2	16	29	15.723
3	27	29	24.967
4	37	45	34.231
5 *	48	51	44.172
6	58	64	53.203
7	68	73	62.234
8	76	100	70.840
9	85	105	78.751
10	94	103	86.175

* Mean fault duration for Case I.

4.5. *Case V: Sensitivity Analysis for the Variation of Battery and Photovoltaic Panel Prices*

Given that the adoption of economies of scale foresees a decrease in the prices of photovoltaic panels and batteries, it is important to analyze the influence of the prices of such equipment in the sizing of the microgrid. For that purpose, price variations from 75% up to 125% of the base photovoltaic panels cost are analyzed (see Table 6). Moreover, price variations from 50% up to 125% of the base battery module cost are analyzed (see Table 7). The same assumptions for the unavailability index goal in Case I are adopted here. All solutions shown have an unavailability index equal to 0.003%, i.e., all satisfy the requested goal.

Table 6. Sensitivity of the solution with variation of photovoltaic panel price.

PV Price (Unit)	Battery Modules (Units)	Photovoltaic Panels (Units)	Economic Index ($)
390.00	51	27	47.171
374.40	50	35	46.689
358.80	50	35	46.143
343.20	49	43	45.517
327.60	49	43	44.846
312.00 *	48	55	44.172
296.40	48	55	43.314
280.80	47	71	42.232
265.20	46	96	40.857
249.60	46	96	39.360
234.00	46	110	37.747

* Solution for Case I.

Higher costs for photovoltaic panels result in a relatively small increase in the economic index but causing the selection of lower panels (27 for a 25% increase). On the other hand, a 25% price reduction leads to using almost all panels studied (110) and has a reduction of almost 15% in the economic index.

Regarding battery modules, there is an influence of the price on the solution, but it is not so strong as seen with the photovoltaic panels. The number of battery modules increases by only two units when the price is reduced by 50%. However, it is remarkable the reduction in the economic index (36%) since the corresponding solution requires less photovoltaic panels. This indicates that, with the ongoing reduction in energy storage prices, technology would be the most cost-effective alternative to supply ASS in substations under a contingency.

Table 7. Sensitivity of the solution with variation of battery price.

Battery Price (kWh)	Battery Modules (Units)	Photovoltaic Panels (Units)	Economic Index ($)
525	48	55	54.252
504	48	55	52.236
483	48	55	50.220
462	48	55	48.204
441	48	55	46.188
420 *	48	55	44.172
399	49	43	42.117
378	49	43	40.059
357	49	43	38.001
336	49	43	35.943
315	49	43	33.885
294	49	43	31.827
273	49	43	29.769
252	50	35	27.705
231	50	35	25.605
210	50	35	23.505

* Solution for Case I.

5. Conclusions

This paper addresses the optimal sizing of a microgrid for the reserve supply of the substation's auxiliary services intending to reduce the time of unavailability of these loads when the main supply is under contingency. Unlike other backup systems, which usually use diesel generators, the backup microgrid is formed by the integration of environmentally friendly technologies such as photovoltaic panels together with battery systems as a unique distributed generation unit named microgrid, which increases the backup system dispatchability.

To deal with the high cost of batteries and the intermittence of photovoltaic generation, a careful analysis determines the capacity of the microgrid identifying the best compromise between the investment and the reduction of the unavailability of auxiliary services. For this purpose, two indexes are proposed to evaluate a set of multiple alternatives using an exhaustive search and Monte Carlo simulations to address the uncertainties of contingencies and variations in solar irradiation.

One of the indexes determines the economic value of the main elements of the microgrid such as the photovoltaic panels, the batteries, and the inverter. Furthermore, an index of unavailability is proposed to measure the contribution of the integrated hybrid microgrid to reduce the time in which the substation is unavailable. The results show the conflicting relationship between both indexes, where a decrease in the unavailability index leads to an increase in the economic index. Hence, the optimal size of the microgrid components is determined by the achievement of the target in the unavailability index with the lowest cost. Besides, the results show the importance of batteries to increase the availability of auxiliary services of the substation above the photovoltaic panels.

Future works may consider the possibility of using the microgrid as the main system for supplying auxiliary services in the substation, considering other components of storage and generation, such as fuel cells, hydrogen storage, and electric vehicles, among others.

Author Contributions: Conceptualization, A.T., N.M. and J.F.F.; Data curation, L.G.; Funding acquisition, J.F.R. and N.B.; Methodology, A.T., N.M., L.G. and J.F.F.; Supervision, J.F.R.; Visualization, N.M.; Writing—original draft, A.T., N.M., L.G. and J.F.F.; Writing—review & editing, A.T., J.F.R. and N.B. All authors have read and agreed to the published version of the manuscript.

Nomenclature

Parameters

α	Number of hours in a year.
δ	Interest rate.
η^{out}	Round-trip efficiency of the battery.
λ	Annual contingency rate.
π	Energy price.
τ	Equipment's lifespan.
c^{bat}	Equipment costs for batteries.
c^{pv}	Equipment costs for photovoltaic panels.
c^{In}	Equipment costs for inverters.
DoD	Depth of discharge.
f_i	Load factor.
F_g	Global load factor.
G_{gh}	Solar irradiation.
I_{MPPT}	Current at the maximum power point.
K_i	Temperature coefficient for the current.
K_v	Temperature coefficient for the voltage.
mc^{bat}	Annual maintenance cost for batteries.
mc^{PV}	Annual maintenance cost for photovoltaic panels.
mc^{in}	Annual maintenance cost for inverter.
N_{OT}	Nominal operating temperature of the cell.
N^{years}	Numbers of years in the Monte Carlo simulation.
P_{eq}	Equivalent power of the ASS loads.
p^{nom}	Nominal power of ASS.
t	Time period.
t_j^0	Contingency start time.
T_a	Ambient temperature.
T_c	Temperature in the photovoltaic cell.
t_j^{out}	Duration of contingency j.
V_{MPPT}	Voltage at the maximum power point.

Variables

I_s	Total investment value for a microgrid configuration.
E_{ASS}	Total energy required by ASS for the contingency.
$\overline{E}_{bat}$	Nominal capacity of the battery.
E_{bat}	Available energy of the battery.
E^{PV}	Annual energy generated by the photovoltaic panels.
E_j^{pv}	Energy generated by the photovoltaic panels for contingency j.
I_s^{ind}	Unavailability index.
MC_s	Maintenance costs.
N_s^{pv}	Number of photovoltaic panels units used on each microgrid.
P_s^{In}	Inverter capacity in the proposed microgrid.
$\overline{P}_s^{PV}$	Installed power of the photovoltaic panels.
P_{pv}^{oper}	Operation power of the photovoltaic panels.
$Profit^{PV}$	Profit related to selling photovoltaic energy.
TC_s	Total cost of the system.
t_j^{Ind}	Unavailability time of ASS for contingency j.
tm_j^s	Microgrid availability time.

References

1. Borlea, I.; Kilyeni, S.; Barbulescu, C.; Cristian, D. Substation ancillary services fuel cell power supply. Part 1. solution overview. In Proceedings of the ICCC-CONTI 2010—IEEE International Joint Conferences on Computational Cybernetics and Technical Informatics, Timisora, Romania, 27–29 May 2010; pp. 585–588.
2. Barbulescu, C.; Kilyeni, S.; Jigoria-Oprea, D.; Chiosa, N. Electric substation ancillary services power consumption analysis. Case study: Timisoara 400/220/110 kV substation. In Proceedings of the ICHQP 2010—14th International Conference on Harmonics and Quality of Power, Bergamo, Italy, 26–29 September 2010; pp. 1–7.

3. Olatomiwa, L.; Mekhilef, S.; Huda, A.S.N.; Sanusi, K. Techno-economic analysis of hybrid PV–diesel–battery and PV–wind–diesel–battery power systems for mobile BTS: The way forward for rural development. *Energy Sci. Eng.* **2015**, *3*, 271–285. [CrossRef]
4. Prostean, O.; Kilyeni, S.; Barbulescu, C.; Vuc, G.; Borlea, I. Unconventional sources for electric substation ancillary services power supply. In Proceedings of the 14th International Conference on Harmonics and Quality of Power—ICHQP, Bergamo, Italy, 26–29 September 2010; pp. 1–6.
5. Bahramara, S.; Mazza, A.; Chicco, G.; Shafie-khah, M.; Catalão, J.P.S. Comprehensive review on the decision-making frameworks referring to the distribution network operation problem in the presence of distributed energy resources and microgrids. *Int. J. Electr. Power Energy Syst.* **2020**, *115*, 105466. [CrossRef]
6. Parhizi, S.; Lotfi, H.; Khodaei, A.; Bahramirad, S. State of the art in research on microgrids: A review. *IEEE Access* **2015**, *3*, 890–925. [CrossRef]
7. Bahramara, S.; Moghaddam, M.P.; Haghifam, M.R. Optimal planning of hybrid renewable energy systems using HOMER: A review. *Renew. Sustain. Energy Rev.* **2016**, *62*, 609–620. [CrossRef]
8. Tong, Y.; Zhang, H.; Jing, L.; Wu, X. Flexible substation and its control for AC and DC hybrid power distribution. In Proceedings of the 13th IEEE Conference on Industrial Electronics and Applications (ICIEA, 2018), Wuhan, China, 31 May–2 June 2018; pp. 423–427.
9. Ahmed, H.M.A.; Eltantawy, A.B.; Salama, M.M.A. A planning approach for the network configuration of AC-DC hybrid distribution systems. *IEEE Trans. Smart Grid* **2018**, *9*, 2203–2213. [CrossRef]
10. Li, Y.; Chen, N.; Zhao, C.; Pu, T.; Wei, Z. Research on evaluation index system of low-carbon benefit in AC/DC hybrid distribution network. In Proceedings of the China International Conference on Electricity Distribution (CICED 2016), Xi'an, China, 10–13 August 2016; pp. 10–13.
11. Ralon, P.; Taylor, M.; Ilas, A.; Diaz-Bone, H.; Kairies, K.-P. *Electricity Storage and Renewables: Costs and Markets to 2030*; International Renewable Energy Agency: Abu Dhabi, UAE, 2017.
12. Neves, D.; Silva, C.A.; Connors, S. Design and implementation of hybrid renewable energy systems on micro-communities: A review on case studies. *Renew. Sustain. Energy Rev.* **2014**, *31*, 935–946. [CrossRef]
13. Cano, A.; Jurado, F.; Sánchez, H.; Fernández, L.M.; Castañeda, M. Optimal sizing of stand-alone hybrid systems based on PV/WT/FC by using several methodologies. *J. Energy Inst.* **2014**, *87*, 330–340. [CrossRef]
14. Schneider, K.P.; Tuffner, F.K.; Elizondo, M.A.; Liu, C.C.; Xu, Y.; Ton, D. Evaluating the Feasibility to Use Microgrids as a Resiliency Resource. *IEEE Trans. Smart Grid* **2017**, *8*, 687–696.
15. Wu, T.F.; Kuo, C.L.; Sun, K.H.; Chang, Y.C. DC-bus voltage regulation and power compensation with bi-directional inverter in DC-microgrid applications. In Proceedings of the IEEE Energy Conversion Congress and Exposition: Energy Conversion Innovation for a Clean Energy Future, ECCE 2011, The Cobo Center1 Washington BlvdDetroit, Detroit, MI, USA, 11–15 November 2011; pp. 4161–4168.
16. Fathima, A.H.; Palanisamy, K. Optimization in microgrids with hybrid energy systems—A review. *Renew. Sustain. Energy Rev.* **2015**, *45*, 431–446. [CrossRef]
17. Alzahrani, A.M.; Zohdy, M.; Yan, B. An overview of optimization approaches for operation of hybrid distributed energy systems with photovoltaic and diesel turbine generator. *Electr. Power Syst. Res.* **2021**, *191*, 106877. [CrossRef]
18. Anoune, K.; Bouya, M.; Astito, A.; Abdellah, A. Ben Sizing methods and optimization techniques for PV-wind based hybrid renewable energy system: A review. *Renew. Sustain. Energy Rev.* **2018**, *93*, 652–673. [CrossRef]
19. Shi, Z.; Wang, R.; Zhang, T. Multi-objective optimal design of hybrid renewable energy systems using preference-inspired coevolutionary approach. *Sol. Energy* **2015**, *118*, 96–106. [CrossRef]
20. Moghaddam, S.; Bigdeli, M.; Moradlou, M.; Siano, P. Designing of stand-alone hybrid PV/wind/battery system using improved crow search algorithm considering reliability index. *Int. J. Energy Environ. Eng.* **2019**, *10*, 429–449. [CrossRef]
21. Eteiba, M.B.; Barakat, S.; Samy, M.M.; Wahba, W.I. Optimization of an off-grid PV/Biomass hybrid system with different battery technologies. *Sustain. Cities Soc.* **2018**, *40*, 713–727. [CrossRef]
22. Hosseinzadeh, M.; Salmasi, F.R. Fault-Tolerant Supervisory Controller for a Hybrid AC/DC Micro-Grid. *IEEE Trans. Smart Grid* **2018**, *9*, 2809–2823. [CrossRef]
23. Paserba, J.J.; Leonard, D.J.; Miller, N.W.; Naumann, S.T.; Lauby, M.G.; Sener, F.P. Coordination of a distribution level continuously controlled compensation device with existing substation equipment for long term VAr management. *IEEE Trans. Power Deliv.* **1994**, *9*, 1034–1040. [CrossRef]

24. Kadurek, P.; Cobben, J.F.G.; Kling, W.L.; Ribeiro, P.F. Aiding power system support by means of voltage control with intelligent distribution substation. *IEEE Trans. Smart Grid* **2014**, *5*, 84–91. [CrossRef]
25. Vuc, G.; Barbulescu, C.; Kilyeni, S.; Solomonesc, F. Substation ancillary services fuel cell power supply. Part 2. case study. In Proceedings of the ICCC-CONTI 2010—IEEE International Joint Conferences on Computational Cybernetics and Technical Informatics, Timisora, Romania, 27–29 May 2010; pp. 589–594.
26. Amanulla, B.; Chakrabarti, S.; Singh, S.N. Reconfiguration of Power Distribution Systems Considering Reliability and Power Loss. *IEEE Trans. Power Deliv.* **2012**, *27*, 918–926. [CrossRef]
27. Billinton, R.; Allan, R.N. *Reliability Evaluation of Power Systems*; Plenum Press: New York, NY, USA, 1996.
28. Narimani, M.R.; Vahed, A.A.; Azizipanah-Abarghooee, R.; Javidsharifi, M. Enhanced gravitational search algorithm for multi-objective distribution feeder reconfiguration considering reliability, loss and operational cost. *IET Gener. Transm. Distrib.* **2014**, *8*, 55–69. [CrossRef]
29. Atwa, Y.M.; El-Saadany, E.F.; Salama, M.M.A.; Seethapathy, R. Optimal renewable resources mix for distribution system energy loss minimization. *IEEE Trans. Power Syst.* **2010**, *25*, 360–370. [CrossRef]
30. Janssen, H. Monte-Carlo based uncertainty analysis: Sampling efficiency and sampling convergence. *Reliab. Eng. Syst. Saf.* **2013**, *109*, 123–132. [CrossRef]
31. The Mathworks. The Language of Technical Computing—MATLAB. Natick, MA, USA, 2019. Available online: https://www.mathworks.com/products/matlab.html?s_tid=hp_products_matlab (accessed on 14 October 2020).
32. Mesquita, D.D.B.; Silva, J.L.d.S.; Moreira, H.S.; Kitayama, M.; Villalva, M.G. A review and analysis of technologies applied in PV modules. In Proceedings of the 2019 IEEE PES Innovative Smart Grid Technologies Conference—Latin America (ISGT Latin America), Gramado City, Brazil, 15–18 September 2019; pp. 1–6.
33. Agency, I.R.E. *Renewable Power Generation Costs in 2018*; International Renewable Energy Agency: Abu Dhabi, UAE, 2018; p. 160.
34. Panasonic, "Photovoltaic Module N330_325_320SJ47-Datasheet". 2019. Available online: https://panasonic.net/lifesolutions/solar/download/pdf/N330_325_320SJ47Datasheet_190226.pdf (accessed on 10 October 2020).
35. Darling, D.; Sara, H. *Average Frequency and Duration of Electric Distribution Outages Vary by States*; U.S. Energy Information Administration: Washington, DC, USA, 2018.
36. Adderly, S. *Reviewing Power Outage Trends, Electric Reliability Indices and Smart Grid Funding*; University of Vermont: Burlington, VT, USA, 2016.
37. Richardson, I.; Thomson, M. *Integrated Domestic Electricity Demand and PV Micro-Generation Model*; Institutional Repository, Loughborough University: Loughborough, UK, 2011.
38. Pfenninger, S.; Staffell, I. Renewables Ninja. Available online: https://www.renewables.ninja/ (accessed on 14 October 2020).
39. Diesel Generator Fuel Consumption Chart in Litres. 2019. Available online: https://www.ablesales.com.au/blog/diesel-generator-fuel-consumption-chart-in-litres.html (accessed on 9 November 2020).
40. Solano-Peralta, M.; Moner-Girona, M.; van Sark, W.G.J.H.M.; Vallvè, X. "Tropicalisation" of Feed-in Tariffs: A custom-made support scheme for hybrid PV/diesel systems in isolated regions. *Renew. Sustain. Energy Rev.* **2009**, *13*, 2279–2294. [CrossRef]
41. Operador Nacional do Sistema Eléctrico Requisitos Mínimos Para Transformadores e Para Subestações e Seus Equipamentos. Available online: http://www.ons.org.br/%2FProcedimentosDeRede%2FMódulo2%2FSubmódulo2.3%2FSubmódulo2.3_Rev_2.0.pdf (accessed on 10 November 2020).

11

Logistics Design for Mobile Battery Energy Storage Systems

Hassan S. Hayajneh and Xuewei Zhang *

College of Engineering, Texas A&M University-Kingsville, Kingsville, TX 78363, USA;
hassan.hayajneh@students.tamuk.edu
* Correspondence: xuewei.zhang@tamuk.edu

Abstract: Currently, there are three major barriers toward a greener energy landscape in the future: (a) Curtailed grid integration of energy from renewable sources like wind and solar; (b) The low investment attractiveness of large-scale battery energy storage systems; and, (c) Constraints from the existing electric infrastructure on the development of charging station networks to meet the increasing electrical transportation demands. A new conceptual design of mobile battery energy storage systems has been proposed in recent studies to reduce the curtailment of renewable energy while limiting the public costs of battery energy storage systems. This work designs a logistics system in which electric semi-trucks ship batteries between the battery energy storage system and electric vehicle charging stations, enabling the planning and operation of power grid independent electric vehicle charging station networks. This solution could be viable in many regions in the United States (e.g., Texas) where there are plenty of renewable resources and little congestion pressure on the road networks. With Corpus Christi, Texas and the neighboring Chapman Ranch wind farm as the test case, this work implement such a design and analyze its performance based on the simulation of its operational processes. Further, we formulate an optimization problem to find design parameters that minimize the total costs. The main design parameters include the number of trucks and batteries. The results in this work, although preliminary, will be instrumental for potential stakeholders to make investment or policy decisions.

Keywords: battery energy storage systems; electric vehicle charging stations; logistics

---

1. Introduction

It is anticipated that Electric Vehicles (EV) penetration will meet an extensive growth in the near future [1] due to their highly promising performance and negligible production of Greenhouse Gas (GHG) emissions [2–4]. However, EV charging still faces issues arising from infrastructure and technology, for instance, the inadequacy of appropriate charging facilities [5] and the long durations of charging activities [6], in contrast to the fast refueling process of conventional Internal Combustion Engine (ICE) vehicles. In addition to these issues, the constraint from the power grid to meet the demands of the Electric Vehicle Charging Stations (EVCSs) has also been a main challenge to the EV industry and it might be responsible for negatively impacting EVs' social acceptance [7,8]. To address this, we introduced and have been exploring the idea of developing grid independent EVCS networks in previous works [7,9–12]. Essentially, mobile Battery Energy Storage Systems (BESSs) that absorb extra (i.e., would be curtailed otherwise) energy from renewable sources (e.g., a wind farm) and are shipped by Electric Semi-Trucks (ESTs) to the EVCSs power the EVCS networks. Designing and managing a logistics system that minimizes the annualized cost for such network became a necessity, which in return promotes the development of EVs industry and penetration. Handling the logistics system of any operational supply and demand environment is in great importance in most fields.

Hence, its significance in the case of EVCSs is not different, since its need stems from the dynamic interaction between the supply side of energy (the energy producer represented by the wind farm in this study) and the demand side (the energy consumer, EVs in this case). The logistics system in this study involves employing ESTs that deliver BESSs between the two sides, as introduced in [7].

Related Works

The planning of diverse types of logistics systems has been extensively studied [13–15]. In an old study that was performed by Powell [16], the author described a heuristic approach that manages a real time dispatching problem for truckload motor carriers (loaded or empty) from one location to the next, the author stated that the most difficult part is to forecast for the future demand because of its uncertainty. From this point, we can sense that the uncertainty of the future demand makes it hard to accurately plan for the operational process and guarantee its convergence, which easily leads to introduce uncertain errors. Recently, numerous studies, including those conducted by Vasilakos et al. [17] and Fanti et al. [18], focus on the Decision Support System (DSS) applications in shipping systems to mitigate uncertain demand problem. The decision support systems are judged to be essential backers of making crucial operational logistics decisions. Giusti et al. [19] introduced a new system for freight logistics planning for the Synchromodal supply chain eco-network, SYNCHRO-NET, which is an optimization toolset that offers tactical level decision support in terms of routes and schedules for synchro-modal freight transportation. The SYNCHRO-NET is projected to achieve notable results regarding the distances travelled, fuel emissions, and costs. In [20], it was reported that the very fast urban growth forms a serious logistics challenge for decision makers, especially when dealing with the social, economic, and environmental issues. It is necessary to plan for it in advance and, as stated in [21], logistic decision-making can precisely define the distribution pattern of shipments within the product's supply chain.

Plant location and its associated services, transportation infrastructure, and demand profiles are important factors shaping the evolution of logistics systems' structure. For the sake of examining the geography of logistics firms, the authors of [22] shed some light over the importance of the location patterns of logistics corporations, and utilized what is called "geo-referenced firm level data" side by side with an updated data system of transportation infrastructure. They concluded that the logistics sector is extremely urbanized, and its firms, as compared to other sectors, are located near highways and other transport infrastructure. The findings of [22] are in agreement with those in [23], where Kumar et al. summarized that maintaining the transportation infrastructure leads to optimistic effects on transportation and logistics clusters. The expansion of the current century's development demands necessitates similar expansions regarding the logistics schemes. Developing a modernized logistics system that reflects the visions for a new future in which the development is sustainable and efficient is an essential requirement.

When dealing with the shipping actions, most of the previous studies consider the scenario of delivering supplies in one direction (i.e., going in one way from the source to the destination as a fully loaded carriers and returning back to the source as an empty load) [24–26]. While other studies tried to enhance the previously described case by finding other applications for the empty loads to gain more profits by reducing the shipping costs, and obtain improved usage, such as the sequence of deliveries scenario, for example [27–30]. In this paper, we deal with a different structure of logistics where we anticipated that the ESTs are equipped to carry only one unit of the mobile BESS; this unit will be fully charged when transferred from the BESS plant to the targeted EVCS while being swapped by the empty one back to the plant. Furthermore, and for the authors knowledge, combining some of the main components that form the conventional logistics structure configuration, as specified in [31], in one component can be a beneficial way for costs reductions. Therefore, we combine the plant and the storage at the same spot and exclude the retailer part, which is almost like the case of Direct to Store Delivery (DSD) [32].

In this study, a model of logistics system for a grid independent EVCSs network is developed. The design can supply a population of ten thousand EVs with their energy charging demands. The optimum design configured employing ten ESTs and 126 mobile BESS units. The principal objective function of the design was to provide every EV with its charging demands at the 27 ECVSs while aiming for minimizing the annual costs. Such design can be implemented by decision makers and planners for future sustainable communities to accomplish greener transportation systems in which the GHG emissions are the lowest levels.

The rest of the paper is organized as follows: In Section 2, we describe the methodology where the optimization problem is introduced along with the objective functions and the constraints. Section 3 provides the main results of the optimization process and their evaluations; the section is further supported by some findings and discussions. The conclusions of the paper and suggestions of further work in this field could be found in Section 4.

2. Methodology

The objective of this study is to design a logistics system model for mobile BESSs powered EVCS network, and then examine its performance when applied into a scenario of supply side, energy produced by renewable energy source (wind energy in this case), and demand side, energy consumed by EVs at the EVCSs. The scheme of EVCSs we study in this paper includes 27 charging stations spread in the city of Corpus Christi, Texas, and their locations are known as obtained from our previous study [9]. These EVCSs are projected to serve a population of about ten thousand EVs. Securing the supply of energy at the EVCSs is maintained by deliveries of mobile BESS units shipped by ESTs. The configuration of the best number of ESTs and mobile BESS units will be optimized while minimizing the annualized costs. The design considers placing a BESS plant, that stores the fully charged BESS, close to the supply side (wind farm). In this study, we assume the availability of boundless energy produced by the wind farm, so, the charging of the mobile BESS units at the BESS plant is conducted without any constraints at any time.

In the simulations, we consider a possible future scenario in which the interaction between the main components (BESS plant, ESTs, mobile BESSs units, EVCSs, and the enormous number of EV arrivals) is described as following. To begin with, this scenario maintains (N_{EV}) as the total number of EVs (10,000 EVs), (N_{EVCS}) as the total number of EVCSs (27 EVCSs), (N_{BESS}) as the number of mobile BESS units (to be optimized) and (N_{EST}) as the number of ESTs (to be optimized). At time step $t = 0$, which represents the beginning of the simulation and by referring to each time step by one *hour*, the capacity of each EV's battery ranges from [60 kWh–120 kWh] and the initial State of Charge (*SoC*) ranges randomly between [0%–100%]. Also, each charging station is considered fully charged by having specific number of mobile BESS units (the capacity of each unit is 5 MWh). For instance, if $EVCS_i$ configures 3 units of mobile BESS, then, it is fully charged at the capacity of 15 MWh initially. The EVCSs are divided into two categories, 15 commercial (COM) EVCSs serving the offices and shopping centers areas, and 12 residential (RES) EVCSs serving the homes and suburban areas. Therefore, each type of EVCSs receives a different trend of EV arrivals during the day [33–35]. As a result, the energy demand profile at each ECVS is changing over time and peaks multiple times during the day. Each EV arrives to an EVCS will be likely to spend specific time while waiting for the charging process, that time is dependent on the type of the EVCS. Figure 1 shows the probabilities for EV arrivals to both commercial and residential EVCSs during the day, while Figure 2 represents the probabilities for the hours spent at each type of the EVCSs at any time.

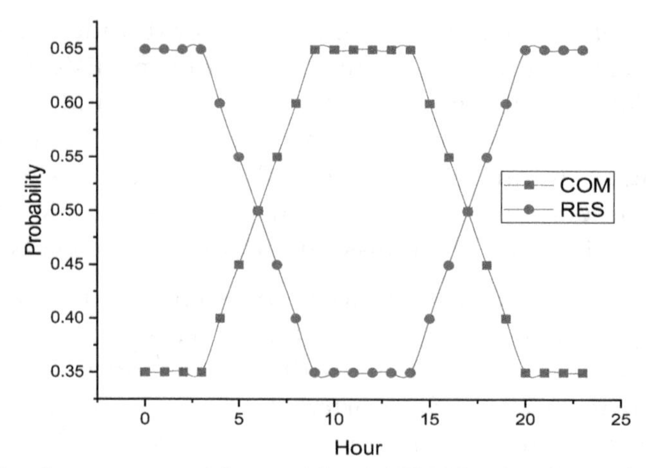

Figure 1. The probability for a commercial or residential EVCS to be chosen by an EV that needs to be charged during the day. The COM EVCS's probability is divided randomly between the 15 stations around their average, while the RES EVCS's probability is divided randomly between the 12 stations around their average.

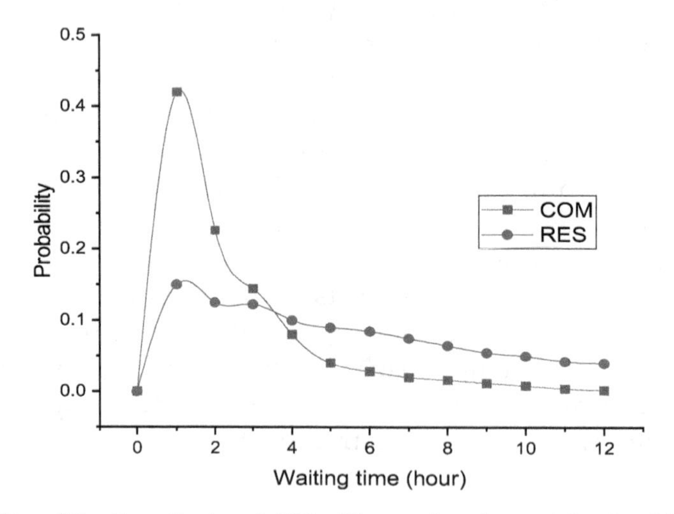

Figure 2. The probability of the time that each EV will spend as the waiting/parking time to get charged at each type of the EVCSs.

To simulate such network of EVCSs that provides service to a population of EVs in a region, the initialization process is as follows: (i) set the number of fully charged mobile BESS units at each EVCS; in this case, the state of charge of each station is at 100% and each EVCS has an initial label of (1) that represents its fully charged status. Later and based on the energy consumption, the label will be switched to (−1) which represents the EVCS's need to be charged (at least one unit of BESS is empty), and a signal to the BESS plant will be sent requesting the mobile BESS unit swapping; (ii) set a number of mobile BESS that is equal to the overall number of BESS units at all EVCSs at the plant as fully charged as well, and ready to secure the shipping process when requested. Each mobile BESS has its own ID and label that represents its status (1; if fully charged, and -1; if needs to be charged). (iii) set the whole fleet of ESTs stationed at the BESS plant and ready to be dispatched to the target EVCS when a request is received. Each EST has a label that represents its availability; (1; if available at the plant, and -1; if not available and on a trip).

After the initialization process is over, the simulation runs for one year (~8760 h or time steps). During each time step, the algorithm updates the followings: (1) the *SoC* of each EV. (2) the *SoC* of each EVCS. (3) the label that represents each EVCS's need to be charged or not. (4) the label that represents each mobile BESS status if full and ready to be shipped or needs charging. (5) the label that represents the availability of each EST. Figure 3 shows the flowchart of the simulation process.

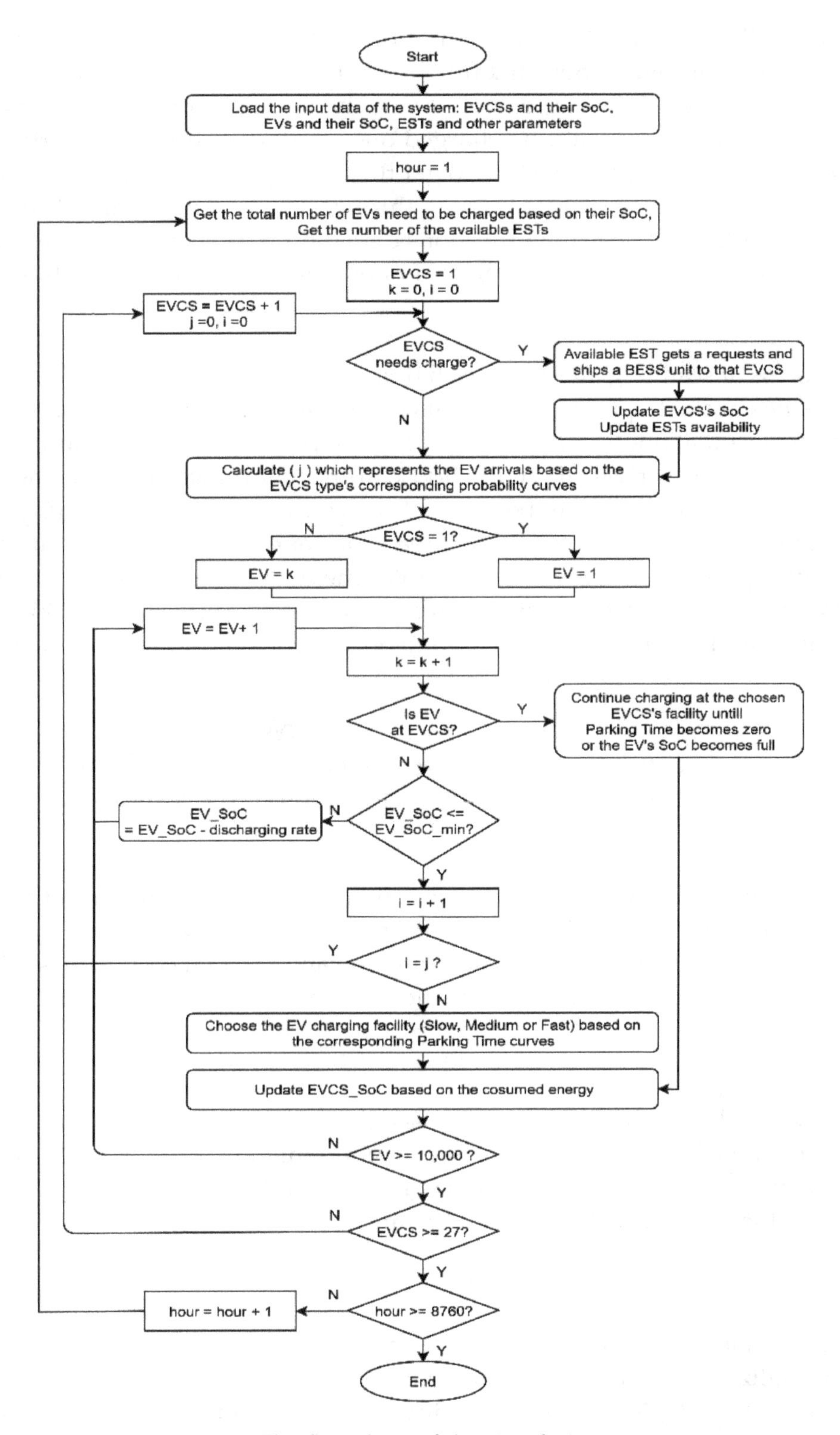

Figure 3. The flowchart of the simulation process.

Loading the input data is the first step of the simulation process; the input data includes but not limited to the 27 EVCSs along with the state of charge status of each station, the population of the EVs along with the state of charge status of each vehicle, and the population of the ESTs along with the availability status of each truck. At each time step of the total simulation time, the number of EVs need to be charged is estimated by comparing each EV's *SoC* to the minimum allowed *SoC* which is ($EV\_SoC_{min} \leq 20\%$) in this case. Then, the total number of EVs that need charging will be spread between the available EVCSs, so each station will receive a portion of the EV arrivals (represented by j in the flowchart in Figure 3). To obtain the number of EVs arrive at each station, the simulation process scans each EVCS through a loop that first of all checks the need of that station to be charged itself (has at least one empty BESS unit) or not, if yes, then a request will be sent to the BESS plant asking to ship and swap that unit by a fully charged one. To find each EVCS's portion of EV arrivals, the whole population of EVs will be checked to get the corresponding j. Two counters (i and k) are initialized at zero value and change while looping through the total number of EVs; i represents the accumulative number of EVs satisfies the conditions demanding its charging process and will reset to zero when it is equal to j, while k represents the number of scanned EVs and will reset to zero at the head of each time step.

During the process of scanning all EVs at each time step, each EV that has a SoC less than or equal to (EV_SoC_{min}) will be checked upon to validate its presence at an EVCS from the previous *hour* and does not need to be placed at another station, or yet to be placed at a station and an occurrence of i is recorded. In that case, the arrived EV will choose an EV charging facility at the EVCS based on the empty capacity of its battery that can be charged which can be estimated by ($Cap_i(1 - SoC_i)$), and the parking time duration (T_i^{dur}). We consider three types of charging facilities at each EVCS: slow charging facility with a power rating of ($P^{SCF} = 10$ kW), medium charging facility with a power rating of ($P^{MCF} = 30$ kW), and fast charging facility with a power rating of ($P^{FCF} = 120$ kW).

Equation (1) as formulated in [33] can be used to obtain the type of the charging facility chosen by that EV at that hour.

$$P_{EV_i} = \begin{cases} P^{SCF}, if\ Cap_i(1 - SoC_i) \le P^{SCF}T_i^{dur} \\ P^{MCF},\ if\ P^{SCF}T_i^{dur} < Cap_i(1 - SoC_i) \le P^{MCF}T_i^{dur} \\ P^{MCF},\ if\ P^{SCF}T_i^{dur} < Cap_i(1 - SoC_i) \le P^{MCF}T_i^{dur} \end{cases} \quad (1)$$

Since the economic feasibility of this design is considered as a very important factor, this study's concern is expanded to optimize for the optimum total number of mobile BESSs units and ESTs required in this design (the control variables) while minimizing the annual total costs as described in Equation (2) and set as the objective function. The optimization process follows those in [10].

$$min\ Cost = C^I + C^{O\&M} \quad (2)$$

where C^I is the annualized investment cost, where its detailed factors are represented in (3), and $C^{O\&M}$ is the annualized operation and maintenance costs as represented in (5). The investment cost includes the costs of the ESTs ($C^{EST} = \mu N_{EST}$) where μ is the price of each EST, and the mobile BESS units ($C^{BESS} = \delta N_{BESS}$) where δ is the price of each mobile BESS unit.

$$C^I = \alpha C^{EST} + \beta C^{BESS} \quad (3)$$

where α and β are two annuity factors associated with the costs of the ESTs and the mobile BESS units respectively to annualize their capital investment costs. They can be obtained by (4) where d is the economical discount rate and y is the corresponding economic lifetime of each element.

$$\alpha, \beta = \frac{d(1+d)^{y_{EST,\ BESS}}}{(1+d)^{y_{EST,\ BESS}} - 1} \quad (4)$$

$$C^{O\&M} = \sigma D_{tot} + \varphi C^{EST} + \omega C^{BESS} + \psi N_{EST} \quad (5)$$

where σ represents the expense of energy consumed by each EST's trip per mile (estimated to be around \$0.165/mile by [36–38]). φ and ω are constants represent the percentages of the investment costs to get the annual maintenance costs for the ESTs and the mobile BESS units respectively (they are chosen to be 5%). ψ characterizes the annual income for each EST operator. The annually total distance (D_{tot}) that ESTs drive while delivering the mobile BESS units is estimated by:

$$D_{tot} = \sum_{i=1}^{N_{EVCS}} l(i) N_{Trips}(i) \quad (6)$$

where $l(i)$ represents the distance between the BESS plant and the ith EVCS which is estimated in Corpus Christi, Texas to be between [10–25 miles], and $N_{Trips}(i)$ is the number of shipments or trips that the ith EVCS requires annually.

3. Results

In this section and by following the model explained above, we present the simulation results. As stated before, each EVCS will receive a unique number EV arrivals during the day. The criteria of choosing the type of the charging facility by each arrived EV is described in Equation (1). It is important to notice that the number of charging facilities at each station is not a point of interest in this study. Therefore, the charging facilities are considered to be available at all times for all EVCSs. In Figure 4a. we present the EV arrivals at one of the RES EVCSs, EVCS #17, to demonstrate the diversity of EV arrivals and their charging facility choices during the day at that station. While in Figure 4b. we show the change over the state of charge of EVCS #17 during the day as the supplying the charging demands of the EVs arrived is taking place. In this case, EVCS #17 maintains two mobile BESS units as its asset. Whenever any of these two units is empty, then the station sends a request to the plant for a BESS unit swap. It can be noticed that EVCS #17 has required to recharge its BESS units for seven times during that day (characterized by the red bars).

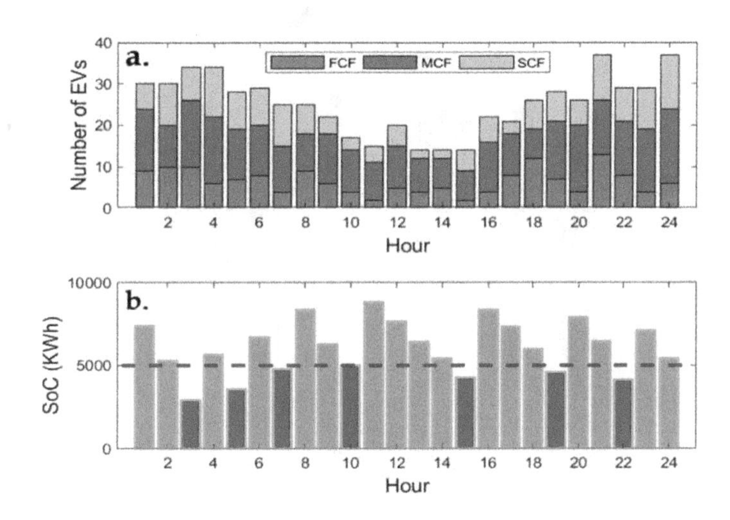

Figure 4. The charging demands and the *SoC* of EVCS #17. (**a**) Illustrates the number of EV arrivals at EVCS #17 in one day, where the total number is represented by arrivals at the three types of the charging facilities. (**b**) Shows the *SoC* of EVCS #17 in the same day, where the availability of energy at the station is represented by the green color and the station's requirement to request a BESS unit swap from the BESS plant is represented by the red color.

To oversee the status of ESTs that serve the logistics system which configures 136 mobile BESS units at the operation mode, a case of employing nine ESTs as the transporters of the BESS units was modeled. Figure 5 presents the availability of these nine ESTs for two days. It is apparent that the ESTs are not available at most of the time because of the high demand at the EVCSs to swap and charge their own batteries.

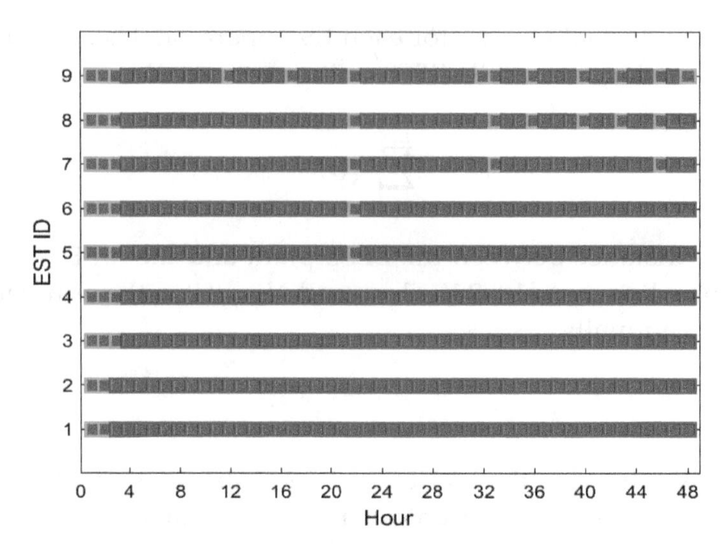

Figure 5. The availability of 9 ESTs serving the logistics system where the red color represents the unavailability of the corresponding EST at the plant (on the road or at an EVCS) and the green color represents their availability.

To see the change over the annualized costs, the effect of both the economic discount rate and the project's lifetime is shown in Figure 6. This case considers deploying 10 ESTs and 140 mobile BESS units. From the economy's point of view, the more discount rate applied, the more annualized costs are required. The opposite is applied to the project's lifetime case, where the longer the lifetime of the project, the less annualized costs.

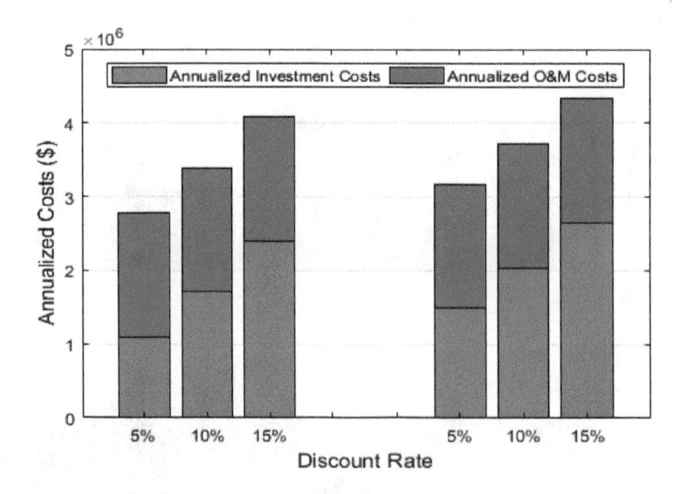

Figure 6. The annualized cost of a logistics system that is composed of 10 ESTs and 140 mobile BESS units. The first group (**left**) considers 25 years as the lifetime of the ESTs and BESS units, while the second group (**right**) considers 15 years.

Obtaining the optimum number of ESTs and mobile BESS units at each station is an optimization problem. The optimization problem is subjected to achieve a scenario of logistics system that supplies the energy demands at all EVCSs, where each EVCS is not allowed to reject any EV by securing its energy demands. Particle Swarm Optimization (PSO) algorithm is used to find the optimal set of the control variables to minimize the fitness function, defined before as the annualized cost in Equations (2)–(6). The simulation model and the optimization process are implemented in MATLAB. The lower bound for the number of ESTs is one, and the upper bound is 27 which is equal to the number of EVCSs. The bounds for the number of mobile BESS units at each station ranges between 1 and 10. We use a typical PSO settings. Figure 7 shows a sample of convergence plot. Usually, the best fitness value is reached within 130 iterations when the average change in the fitness value is almost zero.

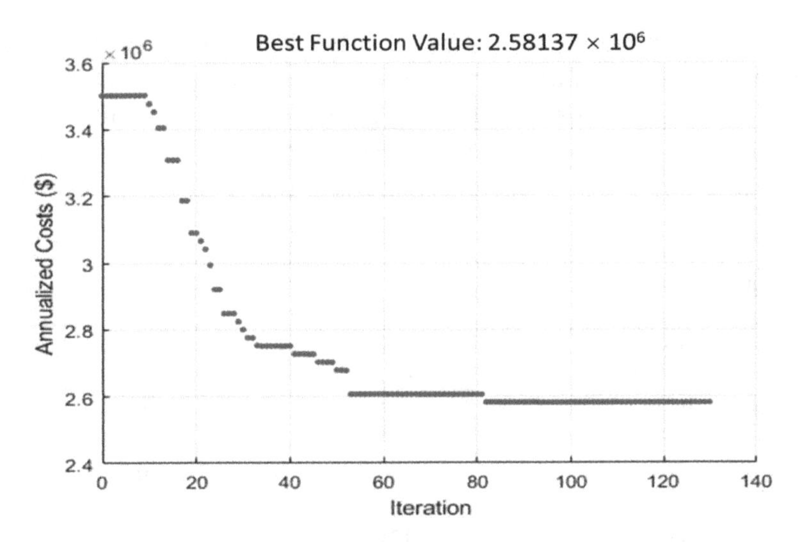

Figure 7. Particle Swarm Optimization (PSO) algorithm's convergence plot.

Now, we present the optimization results of the control variables in three comparable scenarios of different economic discount rates: 5%, 10% and 15% respectively. In each scenario, we obtain the results under two cases of project's lifetime, 15 and 25 years, while fixing the cost of each EST at $150 k and the cost of each mobile BESS unit at $100 k. The optimum control variables are summarized in Table 1 and illustrated in Figure 8.

Table 1. Summary table of the optimization results under varying discount rates and project's lifetime.

Discount Rate	Lifetime of ESTs and Mobile BESS Units (years)	Optimum Number of Mobile BESS Units	Optimum Number of EST	Minimum Annualized Cost ($M)
5%	15	150	9	3.06907
	25	126	10	2.45391
10%	15	142	9	3.49829
	25	136	9	3.07089
15%	15	128	10	3.90900
	25	126	10	3.63470

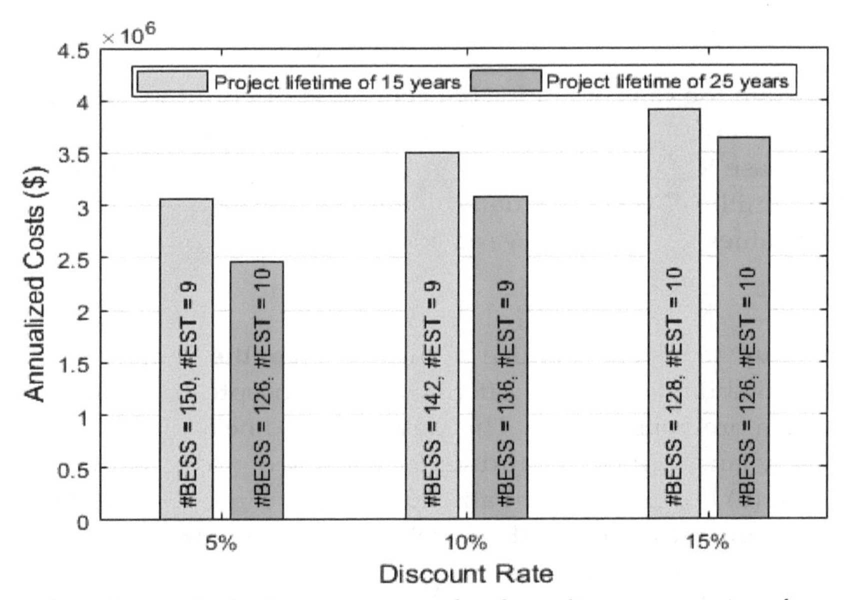

Figure 8. The results of the optimization process under three diverse scenarios of economic discount rates and two cases for the project's lifetime; the anticipated lifetime of both ESTs and mobile BESS units. In this study, the cost of each EST is $150 k while the cost of each mobile BESS unit is $100 k.

For further investigations over the effect of both control variables costs, we present the optimization results in three other comparable scenarios for the cost of an individual EST: $100 k, $150 k and $200 k respectively. In each scenario, we obtain the results under two cases for the cost of an individual mobile BESS unit, $100 k and $150 k years, while fixing the economic discount rate at 5% and the project's lifetime at 25 years. The optimum control variables are summarized in Table 2 and illustrated in Figure 9.

Table 2. Summary table of the optimization results under varying costs of EST and mobile BESS unit.

Cost of One EST ($)	Cost of Mobile BESS Unit ($/Each)	Optimum Number of Mobile BESS Units	Optimum Number of EST	Minimum Annualized Cost ($M)
100 k	100 k	134	9	2.40547
	150 k	124	12	3.28334
150 k	100 k	126	10	2.45391
	150 k	132	9	3.23504
200 k	100 k	134	10	2.61062
	150 k	124	10	3.23957

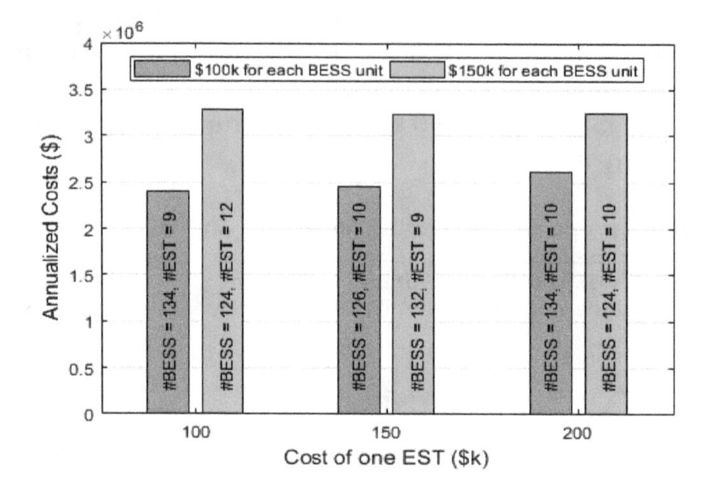

Figure 9. The results of the optimization process under three diverse scenarios for the costs of ESTs and two cases for the cost of mobile BESS units. In this study, the economic discount rate is 5% while the project's lifetime is 25 years.

Finally, we deliberate the case of 5% for the economic discount rate, 25 years for the project's lifetime, $150 k for the cost of each EST and $100 k for the cost of each mobile BESS unit as the most anticipated promising case for the future. By using these values, the optimization results lead to a combination of 2.45391 million dollars annually as the minimum cost, 10 ESTs and 126 mobile BESS units, as the optimum values of the control variables.

4. Conclusions

In summary, this study aims to maximize the utilization of the Renewable Energy Sources (RES) while increasing the profitability of BESS by offering a new conceptual design of mobilizing the BESSs which has emerged from previous studies. This design would be able to tackle three major barriers toward a greener energy landscape in the future: (a) Curtailed grid integration of energy from RES such as wind and solar; (b) Low investment attractiveness of large-scale BESSs; and (c) Constraints from the existing electric infrastructure on the development of charging station networks to meet the increasing electrical transportation demands.

This work develops a logistics system design for mobile BESS implemented within a grid independent EVCSs network. The design achieved supplying a population of ten thousand EVs with their energy charging demands. The optimum attained design configured employing ten ESTs and

126 mobile BESS units while minimizing the annual costs at almost 2.5 million dollars. The principal objective function of the design was to provide every electric vehicle with its charging demands at the 27 ECVSs network. Such design can be implemented by decision makers and planners for future sustainable communities to accomplish greener transportation systems in which the greenhouse gas emissions are at the lowest levels.

The logistics system in which ESTs transport the batteries between the BESS and EVCSs was designed, which facilitates the planning and operation of EVCS networks without constraints from the grid. The design of the logistics system was tested under various scenarios anticipated for the future. These scenarios include executing different values for the economic discount rates, project lifetime, costs of ESTs and costs of mobile BESS units. In this design, a promising case of 5% for the economic discount rate, 25 years for the project's lifetime, $150 k for the cost of each EST and $100 k for the cost of each mobile BESS unit was considered at the optimum setup.

In many regions in the United States (e.g., Texas) where there are plenty of renewable resources and little congestion pressure on the road networks, this solution could be viable. The city of Corpus Christi, Texas and the neighboring Chapman Ranch wind farm were used as the test case, this work implement such a design and analyze its performance based on the simulation of its operational processes. Further, we formulated an optimization problem to find design parameters that minimize the total costs. The main design parameters include the number of trucks and batteries. The results in this work, although preliminary, will be instrumental for potential stakeholders to make investment or policy decisions. Improvements can be made in the future work to (i) consider actual demand profiles of the mobile BESS units at the EVCSs. (ii) include real time travel times to the design model as well as consider a predictable and time-dependent supply profile.

Author Contributions: Conceptualization, H.S.H. and X.Z.; methodology, X.Z.; software, H.S.H.; validation, H.S.H. and X.Z.; manuscript preparation, H.S.H. and X.Z.; project administration, X.Z. All authors have read and agreed to the published version of the manuscript.

Nomenclature and Parameter Values

N_{EV}	The total number of EVs = 10,000 EVs
N_{EVCS}	The total number of EVCSs = 27 EVCSs
N_{BESS}	The total number of mobile BESS units
N_{EST}	The total number of ESTs
Cap_i	The battery's capacity of the *i*th EV
SoC_i	The battery's *SoC* of the *i*th EV
EV_SoC_{min}	The minimum *SoC* of an EV to require charging = 20%
j	The number of EV arrivals at an EVCS
i	A counter for EVs that satisfy the charging requirements and it resets to zero at each EVCS
k	A counter for scanned EVs among N_{EV} and it resets to zero at each hour
T_i^{dur}	The parking time duration of the *i*th EV
P^{SCF}	The power rating of the slow charging facility
P^{MCF}	The power rating of the medium charging facility
P^{FCF}	The power rating of the fast charging facility
P_{EV_i}	The power rating of the charging facility chosen by the *i*th EV
C^I	The annualized investment costs of the logistics system
$C^{O\&M}$	The annualized operations and maintenance costs of the logistics system
C^{EST}	The total costs of the ESTs
μ	The cost of each EST = $150 K
C^{BESS}	The total costs of the mobile BESS units
δ	The cost of each mobile BESS unit = $100 K

α, β	Two economical annuity factors
d	The economical discount rate or the cost of capital = 5%
y_{EST}	The economic life of ESTs = 25 years
y_{BESS}	The economic life of mobile BESS units = 25 years
σ	The cost of energy per mile = $0.165/mile
D_{tot}	The annual total distance driven by ESTs to deliver mobile BESS units
φ	The annual maintenance cost of ESTs = 5% of C^{EST}
ω	The annual maintenance cost of mobile BESS units = 5% of C^{B}
ψ	The average annual salary of an EST operator = $70 K
$l\ (i)$	The distance between the BESS plant and the ith EVCS
$N_{Trips}(i)$	The number of shipments or trips that the ith EVCS requires annually

References

1. Energy Information Administration (EIA). *Annual Energy Outlook 2019 with Projections to 2050 (AEO 2019)*; Energy Information Administration (EIA): Washington, DC, USA, 2019; p. 83. Available online: https://www.eia.gov/outlooks/aeo/pdf/AEO2019.pdf (accessed on 21 December 2019).
2. Li, C.; Negnevitsky, M.; Wang, X.; Yue, W.L.; Zou, X. Multi-criteria analysis of policies for implementing clean energy vehicles in China. *Energy Policy* **2019**, *129*, 826–840. [CrossRef]
3. Qiao, Q.; Zhao, F.; Liu, Z.; He, X.; Hao, H. Life cycle greenhouse gas emissions of Electric Vehicles in China: Combining the vehicle cycle and fuel cycle. *Energy* **2019**, *177*, 222–233. [CrossRef]
4. Li, W.; Long, R.; Chen, H. Consumers' evaluation of national new energy vehicle policy in China: An analysis based on a four paradigm model. *Energy Policy* **2016**, *99*, 33–41. [CrossRef]
5. Hayajneh, H.S.; Zhang, X. Evaluation of Electric Vehicle Charging Station Network Planning via a Co-Evolution Approach. *Energies* **2020**, *13*, 25. [CrossRef]
6. Hou, K.; Xu, X.; Jia, H.; Yu, X.; Jiang, T.; Zhang, K.; Shu, B. A Reliability Assessment Approach for Integrated Transportation and Electrical Power Systems Incorporating Electric Vehicles. *IEEE Trans. Smart Grid* **2018**, *9*, 88–100. [CrossRef]
7. Hayajneh, H.S.; Bashetty, S.; Salim, M.N.B.; Zhang, X. Techno-Economic Analysis of a Battery Energy Storage System with Combined Stationary and Mobile Applications. In Proceedings of the 2018 IEEE Conference on Technologies for Sustainability (SusTech), Long Beach, CA, USA, 11–13 November 2018; pp. 1–6.
8. Barth, M.; Jugert, P.; Fritsche, I. Still underdetected—Social norms and collective efficacy predict the acceptance of electric vehicles in Germany. *Transp. Res. Part F Traffic Psychol. Behav.* **2016**, *37*, 64–77. [CrossRef]
9. Hayajneh, H.S.; Bani Salim, M.N.; Bashetty, S.; Zhang, X. Optimal Planning of Battery-Powered Electric Vehicle Charging Station Networks. In Proceedings of the 2019 IEEE Green Technologies Conference (GreenTech), Lafayette, LA, USA, 3–6 April 2019; pp. 1–4.
10. Hayajneh, H.; Salim, M.B.; Bashetty, S.; Zhang, X. Logistics system planning for battery-powered electric vehicle charging station networks. *J. Phys. Conf. Ser.* **2019**, *1311*, 012025. [CrossRef]
11. Hayajneh, H.S.; Lainfiesta, M.; Zhang, X. Three Birds One Stone: A Solution to Maximize Renewable Generation, Incentivize Battery Deployment, and Promote Green Transportation. In Proceedings of the 2020 IEEE Innovative Smart Grid Technologies Conference North America (ISGT NA 2020), Washington, DC, USA, 17–20 February 2020.
12. Hayajneh, H.S.; Lainfiesta, M.; Zhang, X. A New Form of Battery Energy Storage Systems for Renewable Energy Farms. In Proceedings of the 2020 IEEE PES T & D Conference, Chicago, IL, USA, 20–25 April 2020.
13. Fagerholt, K.; Christiansen, M. A combined ship scheduling and allocation problem. *J. Oper. Res. Soc.* **2000**, *51*, 834–842. [CrossRef]
14. de Oliveira, L.K.; dos Santos, O.R.; de Oliveira, R.L.M.; Nóbrega, R.A.A. Is the Location of Warehouses Changing in the Belo Horizonte Metropolitan Area (Brazil)? A Logistics Sprawl Analysis in a Latin American Context. *Urban Sci.* **2018**, *2*, 43. [CrossRef]
15. Lu, E.H.-C.; Yang, Y.-W. A hybrid route planning approach for logistics with pickup and delivery. *Expert Syst. Appl.* **2019**, *118*, 482–492. [CrossRef]

16. Powell, W.B. An operational planning model for the dynamic vehicle allocation problem with uncertain demands. *Transp. Res. Part B Methodol.* **1987**, *21*, 217–232. [CrossRef]
17. Vasilakos, S.; Iacobellis, G.; Stylios, C.D.; Fanti, M.P. Decision Support Systems based on a UML description approach. In Proceedings of the 2012 6th IEEE International Conference Intelligent Systems, Sofia, Bulgaria, 6–8 September 2012; pp. 041–046.
18. Fanti, M.P.; Iacobellis, G.; Nolich, M.; Rusich, A.; Ukovich, W. A Decision Support System for Cooperative Logistics. *IEEE Trans. Autom. Sci. Eng.* **2017**, *14*, 732–744. [CrossRef]
19. Giusti, R.; Manerba, D.; Perboli, G.; Tadei, R.; Yuan, S. A New Open-source System for Strategic Freight Logistics Planning: the SYNCHRO-NET Optimization Tools. *Transp. Res. Procedia* **2018**, *30*, 245–254. [CrossRef]
20. Rześny-Cieplińska, J.; Szmelter-Jarosz, A. Assessment of the Crowd Logistics Solutions—The Stakeholders' Analysis Approach. *Sustainability* **2019**, *11*, 5361. [CrossRef]
21. Xu, J.; Hancock, K.L. Enterprise-wide freight simulation in an integrated logistics and transportation system. In Proceedings of the 2003 IEEE International Conference on Intelligent Transportation Systems, Shanghai, China, 12–15 October 2003; Volume 1, pp. 534–538.
22. Holl, A.; Mariotti, I. The Geography of Logistics Firm Location: The Role of Accessibility. *Netw. Spat. Econ.* **2018**, *18*, 337–361. [CrossRef]
23. Kumar, I.; Zhalnin, A.; Kim, A.; Beaulieu, L.J. Transportation and logistics cluster competitive advantages in the U.S. regions: A cross-sectional and spatio-temporal analysis. *Res. Transp. Econ.* **2017**, *61*, 25–36. [CrossRef]
24. Zenzerovic, Z.; Beslic, S. Optimization of cargo transport with a view to cost efficient operation of container ship. In Proceedings of the 25th International Conference on Information Technology Interfaces, 2003. ITI 2003, Cavtat, Croatia, 19 June 2003; pp. 531–536.
25. Zenzerović, Z.; Bešlić, S. Contribution to the Optimisation of the Cargo Transportation Problem. *Promet Traffic Transp.* **2003**, *15*, 65–72.
26. Kos, S.; Zenzerović, Z. Modelling the Transport Process in Marine Container Technology. *Promet Traffic Transp.* **2003**, *15*, 13–17.
27. Christiansen, M. Decomposition of a Combined Inventory and Time Constrained Ship Routing Problem. *Transp. Sci.* **1999**, *33*, 3–16. [CrossRef]
28. Krile, S. Application of the minimum cost flow problem in container shipping. In Proceedings of the Elmar-2004. 46th International Symposium on Electronics in Marine, Zadar, Croatia, 16–18 June 2004; pp. 466–471.
29. Kobayashi, K.; Kubo, M. Optimization of oil tanker schedules by decomposition, column generation, and time-space network techniques. *Jpn. J. Indust. Appl. Math.* **2010**, *27*, 161–173. [CrossRef]
30. Kobayashi, H.; Tamaki, H. A method of ship scheduling and inventory management problem for reducing demurrage and freight. In Proceedings of the 2017 IEEE International Conference on Systems, Man, and Cybernetics (SMC), Banff, AB, Canada, 5–8 October 2017; pp. 2790–2795.
31. He, M.; Shen, J.; Wu, X.; Luo, J. Logistics Space: A Literature Review from the Sustainability Perspective. *Sustainability* **2018**, *10*, 2815. [CrossRef]
32. *Direct Store Delivery: Concepts, Applications and Instruments*; Otto, A.; Schoppengerd, F.J.; Shariatmadari, R. (Eds.) Springer: Berlin, Germany, 2009; ISBN 978-3-540-77212-5.
33. Luo, L.; Gu, W.; Zhou, S.; Huang, H.; Gao, S.; Han, J.; Wu, Z.; Dou, X. Optimal planning of electric vehicle charging stations comprising multi-types of charging facilities. *Appl. Energy* **2018**, *226*, 1087–1099. [CrossRef]
34. Zhang, H.; Tang, W.; Hu, Z.; Song, Y.; Xu, Z.; Wang, L. A method for forecasting the spatial and temporal distribution of PEV charging load. In Proceedings of the 2014 IEEE PES General Meeting|Conference Exposition, Washington, DC, USA, 27–31 July 2014; pp. 1–5.
35. Wang, Y.; Guo, Q.; Sun, H.; Li, Z. An investigation into the impacts of the crucial factors on EVs charging load. In Proceedings of the IEEE PES Innovative Smart Grid Technologies, Tianjin, China, 21–24 May 2012; pp. 1–4.
36. Tesla Semi: 500-Mile Range, Cheaper Than Diesel, Quick to Charge—ExtremeTech. Available online: https://www.extremetech.com/extreme/259195-tesla-semi-500-mile-range-cheaper-diesel-quick-charge (accessed on 5 March 2019).
37. "Comparing Energy Costs Per Mile for Electric and Gasoline-Fueled Vehicles," Advanced Vehicle Testing Activity, Idaho National Laboratory. Available online: https://avt.inl.gov/sites/default/files/pdf/fsev/costs.pdf (accessed on 28 December 2019).

Permissions

All chapters in this book were first published in MDPI; hereby published with permission under the Creative Commons Attribution License or equivalent. Every chapter published in this book has been scrutinized by our experts. Their significance has been extensively debated. The topics covered herein carry significant findings which will fuel the growth of the discipline. They may even be implemented as practical applications or may be referred to as a beginning point for another development.

The contributors of this book come from diverse backgrounds, making this book a truly international effort. This book will bring forth new frontiers with its revolutionizing research information and detailed analysis of the nascent developments around the world.

We would like to thank all the contributing authors for lending their expertise to make the book truly unique. They have played a crucial role in the development of this book. Without their invaluable contributions this book wouldn't have been possible. They have made vital efforts to compile up to date information on the varied aspects of this subject to make this book a valuable addition to the collection of many professionals and students.

This book was conceptualized with the vision of imparting up-to-date information and advanced data in this field. To ensure the same, a matchless editorial board was set up. Every individual on the board went through rigorous rounds of assessment to prove their worth. After which they invested a large part of their time researching and compiling the most relevant data for our readers.

The editorial board has been involved in producing this book since its inception. They have spent rigorous hours researching and exploring the diverse topics which have resulted in the successful publishing of this book. They have passed on their knowledge of decades through this book. To expedite this challenging task, the publisher supported the team at every step. A small team of assistant editors was also appointed to further simplify the editing procedure and attain best results for the readers.

Apart from the editorial board, the designing team has also invested a significant amount of their time in understanding the subject and creating the most relevant covers. They scrutinized every image to scout for the most suitable representation of the subject and create an appropriate cover for the book.

The publishing team has been an ardent support to the editorial, designing and production team. Their endless efforts to recruit the best for this project, has resulted in the accomplishment of this book. They are a veteran in the field of academics and their pool of knowledge is as vast as their experience in printing. Their expertise and guidance has proved useful at every step. Their uncompromising quality standards have made this book an exceptional effort. Their encouragement from time to time has been an inspiration for everyone.

The publisher and the editorial board hope that this book will prove to be a valuable piece of knowledge for researchers, students, practitioners and scholars across the globe.

List of Contributors

Kyoungseok Han
School of Mechanical Engineering, Kyungpook National University, Daegu 41566, Korea

Tam W. Nguyen
Department of Aerospace Engineering, University of Michigan, Ann Arbor, MI 48109, USA

Kanghyun Nam
Department of Mechanical Engineering, Yeungnam University, Gyeongsan 38541, Korea

M. S. Hossain Lipu, Aini Hussain and Afida Ayob
Department of Electrical, Electronic and Systems Engineering, Universiti Kebangsaan Malaysia, Bangi 43600, Malaysia

M. A. Hannan
Department of Electrical Power Engineering, College of Engineering, Universiti Tenaga Nasional, Kajang 43000, Malaysia

Mohamad H. M. Saad
Department of Mechanical and Manufacturing Engineering, Universiti Kebangsaan Malaysia, Bangi 43600, Malaysia

Kashem M. Muttaqi
School of Electrical, Computer and Telecommunications Engineering, University of Wollongong, Wollongong, NSW 2522, Australia

Ching-Ming Lai
Department of Electrical Engineering, National Chung Hsing University, Taichung 420, Taiwan

Jiashen Teh
School of Electrical and Electronic Engineering, Engineering Campus, Universiti Sains Malaysia (USM), Nibong Tebal 14300, Penang, Malaysia

Yuan-Chih Lin
Department of Electrical Engineering, National Taiwan University, Taipei 106, Taiwan

Yitao Liu
College of Mechatronics and Control Engineering, Shenzhen University, Shenzhen 518060, China

Yukai Chen, Khaled Sidahmed Sidahmed Alamin, Daniele Jahier Pagliari, Sara Vinco and Massimo Poncino
Department of Control and Computer Engineering (DAUIN), Politecnico di Torino, 10129 Turin, Italy

Enrico Macii
Interuniversity Department of Regional and Urban Studies and Planning (DIST), Politecnico di Torino, 10129 Turin, Italy

Marija Miletić
Faculty of Electrical Engineering and Computing, University of Zagreb, Unska ulica No. 3, 10000 Zagreb, Croatia

Dechang Yang
College of Information and Electrical Engineering, China Agricultural University, No. 17 Qinghuadonglu, Haidian, Beijing 100083, China

Shahid Hussain, Ki-Beom Lee and Young-Chon Kim
Division of Electronic and Information, Department of Computer Science and Engineering, Jeonbuk National University, Jeonju 54896, Korea

Mohamed A. Ahmed
Department of Electronic Engineering, Universidad Técnica Federico Santa María, Valparaíso 2390123, Chile
Department of Communications and Electronics, Higher Institute of Engineering & Technology–King Marriott, Alexandria 23713, Egypt

Miguel Carrión and Ruth Domínguez
Department of Electrical Engineering, University of Castilla – La Mancha, 45071 Toledo, Spain

Rafael Zárate-Miñano
Department of Electrical Engineering, University of Castilla – La Mancha, 13400 Almadén, Spain

Kristina Pandžić
Croatian TSO (Hrvatski Operator Prijenosnog Sustava d.o.o. – HOPS), Zagreb 10000, Croatia

Ivan Pavić and Hrvoje Pandžić
Department of Energy and Power Systems, Faculty of Electrical Engineering and Computing, University of Zagreb, Zagreb 10000, Croatia

Ivan Andročec
Hrvatska Elektroprivreda d.d., Zagreb 10000, Croatia

Mohan Krishna Srinivasan
Department of Electrical and Electronics Engineering, Alliance College of Engineering and Design, Alliance University, Bangalore 562 106, India

Febin Daya John Lionel
SELECT, Vellore Institute of Technology, Chennai 600127, India

Umashankar Subramaniam
Renewable Energy Lab, College of Engineering, Prince Sultan University, Riyadh 11586, Saudi Arabia

Frede Blaabjerg
CORPE, Department of Energy Technology, Aalborg University, 9000 Aalborg, Denmark

Rajvikram Madurai Elavarasan
Electrical and Automotive parts Manufacturing unit, AA Industries, Chennai 600123, India

G. M. Shafiullah
Discipline of Engineering and Energy, Murdoch University, Murdoch 6150, Australia

Irfan Khan
Marine Engineering Technology Department in a joint appointment with Electrical and Computer Engineering, Texas A&M University, College Station, TX 77843, USA

Sanjeevikumar Padmanaban
Department of Energy Technology, Aalborg University, 6700 Esbjerg, Denmark

Alejandra Tabares, Norberto Martinez and Lucas Ginez
Faculty of Electrical Engineering, Campus of Ilha Solteira, São Paulo State University, Ilha Solteira 15385-000, Brazil

José F. Resende
School of Energy Engineering, Campus of Rosana, São Paulo State University, Rosana 19274-000, Brazil

Nierbeth Brito
INTESA – Integration Power Transmitter S.A. – Power Transmission Utilities, Brasília 70.196-900, Brazil

John Fredy Franco
Faculty of Electrical Engineering, Campus of Ilha Solteira, São Paulo State University, Ilha Solteira 15385-000, Brazil
School of Energy Engineering, Campus of Rosana, São Paulo State University, Rosana 19274-000, Brazil

Hassan S. Hayajneh and Xuewei Zhang
College of Engineering, Texas A&M University-Kingsville, Kingsville, TX 78363, USA

Index

Printed in the USA
CPSIA information can be obtained
at www.ICGtesting.com
JSHW051413091023
49903JS00006B/397